Semiology of Graphics

Semiology of Graphics

Diagrams   Networks   Maps

JACQUES BERTIN

TRANSLATED BY
William J. Berg

Esri Press
REDLANDS, CALIFORNIA

Originally published in English by The University of Wisconsin Press
Madison, Wisconsin, USA
uwpress.wisc.edu

Copyright ©1983 by The Board of Regents of the University of Wisconsin System. All rights reserved.

Originally published in French as *Sémiologie graphique* by Jacques Bertin copyright ©1967 Editions Gauthier-Villars, Paris; Editions Mouton & Cie, Paris-La Haye; and École Pratique des Hautes Études, Paris, France

Esri Press, 380 New York Street, Redlands, California 92373-8100
New materials copyright © 2011 Esri
All rights reserved. First edition 2011
15 14 13 12 11   2 3 4 5 6 7 8 9 10

Printed in the United States of America

Library of Congress Cataloging-in-Publication Data
Bertin, Jacques, 1918–
    [Semiologie graphique. English]
    Semiology of graphics : diagrams networks maps / Jacques Bertin ; translated by William J. Berg.
        p. cm.
    Includes index.
    ISBN 978-1-58948-261-6 (hardcover : alk. paper) 1. Graphic methods. 2. Statistics—Graphic methods.
    3. Statistical maps. I. Title.
    QA90.B47513 2010
    001.4'226—dc22   2010031657

The Esri Press edition is published by arrangement with the University of Wisconsin Press. © 2010 by the Board of Regents of the University of Wisconsin System. All rights reserved. Rights inquiries should be directed to the University of Wisconsin Press, Madison Wisconsin, USA, to Rights@uwpress.wisc.edu.

This material is partially based upon work supported by the National Science Foundation under Grants No. SOC76–17768 and SES80–08481 to the Bureau of Social Science Research, Albert Biderman and Howard Wainer, principal investigators.

Any opinions, findings, conclusions, or recommendations expressed in this publication are those of the author(s) and do not necessarily reflect the views of the National Science Foundation.

This work is protected under United States copyright law and the copyright laws of the given countries of origin and applicable international laws, treaties, and/or conventions. No part of this work may be reproduced or transmitted in any form or by any means, electronic or mechanical, including photocopying or recording, or by any information storage or retrieval system, except as expressly permitted in writing.

The information contained in this document is subject to change without notice.

Esri, the Esri Press logo, @esri.com, and www.esri.com are trademarks, registered trademarks, or service marks of Esri in the United States, the European Community, or certain other jurisdictions. Other companies and products mentioned herein are trademarks or registered trademarks of their respective trademark owners.

Ask for Esri Press titles at your local bookstore or order by calling 800-447-9778, or shop online at www.esri.com/esripress. Outside the United States, contact your local Esri distributor or shop online at www.eurospanbookstore.com/Esri.

Esri Press titles are distributed to the trade by the following:

In North America:
Ingram Publisher Services
Toll-free telephone: 800-648-3104
Toll-free fax: 800-838-1149
E-mail: customerservice@ingrampublisherservices.com

In the United Kingdom, Europe, Middle East and Africa, Asia, and Australia:
Eurospan Group
3 Henrietta Street
London WC2E 8LU
United Kingdom
Telephone: 44(0) 1767 604972
Fax: 44(0) 1767 601640
E-mail: eurospan@turpin-distribution.com

July 27, 1918–May 3, 2010

Jacques Bertin was a French cartographer and theorist, and a world-renowned authority on the subject of information visualization. In 1954, he founded the Cartographic Laboratory of the École Pratique des Hautes Études, and was named director of education three years later. Bertin became a professor at the Sorbonne in 1967, then in 1974 he was appointed director of education and director of the Geographical Laboratory of the École des Hautes Études en Sciences Sociales. In the late 1970s, he became head of research at the Centre National de la Recherche Scientifique. In 1993, Bertin received the Mercator-Medaille der Deutschen Gesellschaft für Kartographie. *Semiology of Graphics* represents the first and most far-reaching effort to provide a theoretical foundation for Information Visualization.

Contents

Foreword by Howard Wainer ix
Preface to the 2010 editon of the English translation xi
Preface to the English edition by Jacques Bertin xiii

PART ONE: SEMIOLOGY OF THE GRAPHIC SIGN-SYSTEM

General theory 2
Definition of graphics 2
Annotated table of contents to part one:
   I. Analysis of the information 4
   II. The properties of the graphic system 7
   III. The rules of the graphic system 9

I. Analysis of the information 15
   A. The invariant and the components 16
      Definition 16
      The order of the components 18
      Wording of titles and legends 19
   B. The number of components 28
   C. The length of the components 33
   D. The level of organization of the components 34

II. The properties of the graphic system 41
   A. The scope of the graphic system 42
   B. The plane 44
      (1) Implantation (classes of representation) 44
      (2) The plane is continuous and homogeneous 46
      (3) The level of organization of the plane 48
      (4) Imposition (groups of representation) 50
   C. The retinal variables 60
      1. The level of organization of the retinal variables 64
      2. Characteristics and properties of the retinal variables 71
      Size variation 71
      Value variation 73
      Texture variation 79
      Color variation 85
      Orientation variation 93
      Shape variation 95
      Table of properties of the retinal variables 97

III. The rules of the graphic system 99
   A. The basic graphic problem 100
      A hundred different graphics for the same information 100
      Diagrams 103
      Maps 117

   B. Image theory: Efficiency 139
      (1) Stages in the reading process 140
      (2) Possible questions 141
      (3) Definition of an image 142
      (4) Construction of an image 148
      (5) Limits of an image 154
   C. Three functions of graphic representation 160
      (1) Recording information (inventory drawings) 160
      (2) Communicating information (simplified drawings or "messages") 162
      (3) Processing information (graphics used for processing) 164
   D. General rules of construction 171
      Diagrams 172
      Networks 173
      Maps 173
   E. General rules of legibility (or rules of separation) 175
      (1) Graphic density 176
      (2) Angular legibility 178
      (3) Retinal legibility 180
Summary of the rules of legibility 190

PART TWO: UTILIZATION OF THE GRAPHIC SIGN-SYSTEM

Classification of graphic problems 192

I. Diagrams 193
   A. Diagrams involving two components 195
      1. Nonquantitative problems 196
      2. Quantitative problems 199
   B. Diagrams involving three components 217
      1. Nonquantitative problems 218
      2. A quantitative component 223
      3. Several quantitative components 251
   C. Problems involving more than three components 254
      1. Graphic information-processing 254
      2. The reorderable matrix 256
      3. The image-file 258
      4. The matrix-file 263
      5. An array of curves 263
      6. An ordered table 265
      7. A collection of ordered tables 265
      8. A collection of maps or a matrix permutation? 268

II. **Networks**   269
    Construction and transformation of a network   271
    Application of networks to classifications   275
    Trees   276
    Areas, inclusive relationships   282
    Perspective drawings   283

III. **Maps**   285
    A. External geographic identification   287
        Situational identification: Projections   287
        Dimensional identification: The scale   296
    B. Internal geographic identification   298
        Cartographic accuracy   298
        Cartographic generalization   300
        Base maps   308
    C. Maps involving one component (the geographic component)   318
        GEO (point, line, or area)   318
    D. Maps involving two components   321
        1. Maps GEO $\neq$ 323
        2. Maps GEO O   336
            GEO O: The representation of movement on the plane   342
        3. Maps GEO Q   356
        4. Maps GEO Q area   366
        A regular pattern of graduated circles   369
        Perspective representation   378
        Isarithms   385
    E. Cartographic problems involving more than two components   389
        1. Inventory maps (comprehensive figurations)   391
        2. Processing maps (collection of comprehensive images)   397
        3. Cartographic message (superimposition of simplified images)   408

Appendix: Area-radius table-graph   413
Brief Presentation of Graphics   418
Epilogue: The origins of *Semiology of Graphics*   415
Index   437

Foreword

In his *Atlas* of 1786, William Playfair wrote of the increasing complexity of modern commercial life. He pointed out that when life was simpler and data were less abundant, an understanding of economic structure was both more difficult to formulate and less important for success. But by the end of the eighteenth century, this was no longer true. Statistical offices had been established and had begun to collect a wide variety of data from which political and commercial leaders could base their decisions. Yet the complexity of these data precluded their easy access by any but the most diligent.

Playfair's genius was in surmounting this difficulty through his marvelous invention of statistical graphs and charts. In the explanation of his innovation he tells the viewer: "On inspecting any one of these Charts attentively, a sufficiently distinct impression will be made, to remain unimpaired for a considerable time, and the idea which does remain will be simple and complete, at once including the duration and the amount. Men of great rank, or active business, can only pay attention to general outlines; nor is attention to particulars of use, any further than as they give a general information: And it is hoped, that with the assistance of these Charts, such information will be got, without the fatigue and trouble of studying the particulars of which it is composed."*

Playfair understood the power of his invention to convey information; or, if poorly done, to misinform: "As the purpose of the following Charts is to convey information in a distinct and easy manner, like History, the chief merit that they can have is *truth* and *accuracy*. The mode here adopted for conveying information is accurate in principle, though in execution it may be liable sometimes to error. I have, however, spared no pains to avoid mistake. . . ."

In the course of his great work Playfair illustrates good graphic practice but does not explain why the specific structures of his graphic forms and formats work; nor does he provide any theory to guide future constructors of graphics. A major effort in that regard lay almost two centuries in the future.

Jacques Bertin's *Sémiologie graphique* is the most important work on graphics since the publication of Playfair's *Atlas*. Its arrival could not have been more propitious. The complexity of eighteenth-century Britain and the massiveness of available data are but trifles in comparison to today's complex network of data sources and topics. These data are being transformed into graphic forms at a breathless pace.

This graphic explosion, though caused by the need to present massive amounts of information compactly, is abetted by the computer. Computers can produce instructions for graphic output as easily as they can crunch numbers. Plotters, their partners in this graphic symbiosis, are now available in a dizzying range of capabilities and prices. Moreover, the means of disseminating graphics has advanced almost apace with the means of producing them. Today's printing techniques do not distinguish between word and image—the page is merely a matrix of white and black to be arranged. Thus, as the need for graphics has increased, the means for producing and reproducing them have improved.

All the pieces are here—huge amounts of information, a great need to clearly and accurately portray them, and the physical means for doing so. What is lacking is a deep understanding of how best to do it.

Jacques Bertin here provides a giant step toward such understanding. The *Sémiologie* was first published in French in 1967, with a second edition appearing in 1973. A German translation was published in 1974. This translation is of the second French edition. Since it first appeared the book has become a classic in its own strange and wonderful way. No one has thought at greater length than Bertin about how quantities can be represented on paper. It is without a doubt the most penetrating study ever made of the use of graphics for both analytical and presentational purposes. Nothing that I know approaches it. It is fresh, full of new ideas, and lavishly illustrated.

One of Bertin's insights is his observation that while no one makes the mistake of confusing language with writing, the turn of phrase with calligraphy, few make the analogous distinction in graphics—there is widespread confusion between "the construction of the image and the quality of the stroke." We have too long muddled along in an intuitive or traditional manner. Bertin's book provides a grammar for graphics; at last it is available in English.

This translation was begun in 1977 as part of the National Science Foundation-supported "Graphic Social Reporting Project." This project was done at the Bureau of Social Science Research, Inc., under the direction of Albert D. Biderman and myself. As part of an earlier study at the Bureau in 1971, a survey of the literature on graphic representation in statistics† had disclosed the towering importance of the *Sémiologie* and the lamentable consequences of its inaccessibility to non-Francophones. Translating the *Sémiologie* therefore figured as an essential step in a

---

*William Playfair. *The Commercial and Political Atlas* (London: Corry).

†B. Feinberg and C. A. Franklin, *Social Graphics Bibliography* (Washington, D.C.: Bureau of Social Science Research, 1975).

follow-up proposal to the NSF to further prepare the social sciences for the "graphic revolution" that technology was about to create. We enlisted the aid of Professor William J. Berg, whose expertise in French and background in the theory of perception made him a natural choice for the task.

One of the great rewards of this activity was the opportunity it afforded Berg and myself to meet Jacques Bertin. In the years since this project was started we have met many times in sites as diverse as a Paris bistro, the Toronto airport, a Washington motel, and the lush countryside of the Virginia horse-country. The topic under discussion was always *la graphique*. We often wrestled with certain words, and how to translate them best. The French word *plan* could have been the English "plan," or "plot," or "scheme," or "plane." Context wasn't much help, since we often saw the phrase "the two dimensions of the 'plan.'" All four of the choices fit. The final choice, "plane," resulted after much debate (and after noting that the German translation chose *-ebene*).

On another occasion as we were driving through the rolling hills of Virginia, Berg asked Bertin to point out a *dorsale* should he happen to see one, in hopes that he would then better understand what was the appropriate English word.

To provide a sense of the effort Berg expended in this task I note correspondence from 1979, where Berg described the following problem and its resolution: "One of Bertin's examples mentions in passing the 'foires de Bisenzone… à Plaisance et à Novi de 1600 à 1653.' The problem was to track down and translate the place names, which was complicated by the fact that Bertin's source was probably an Italian work. Is Bisenzone an actual name? The Italian author's rendering of a French word? Bertin's rendering of an Italian word? Etc.? The same holds true for Plaisance, which could be a French city of that name or a French version of the Italian *Piacenza* or the Spanish *Plasencia*." After consulting with Italian and Spanish colleagues, looking for the Italian work Bertin alludes to, and studying linguistic atlases in several languages, Berg tried to track down the solution through the fairs held at that time. Erwin Welsch, a reference librarian at Wisconsin, came up with a probable solution: "There were the 'Fairs of Besançon' held by Genoese merchants (hence the Italian form) from 1535 in that free city in Franche-Comté. Subsequently they moved to Piacenza (Plaisance in the French form) from 1579 to about 1622, but these markets were still called 'Besançon Fairs' because of their origin. The 'Novi' as a city probably refers to Novi di Modeni, which is near Piacenza, but that is not entirely clear."

The recounting of this episode is meant to convey the care that went into this translation, and to allude to the help that many provided (although credit for the mass of the translation should, properly, be Berg's). Among these many are: Albert D. Biderman, who instigated the work; Murray Aborn, of the National Science Foundation, who for a decade provided encouragement for the work centered at the Bureau of Social Science Research on the graphic representation of social data; Arthur H. Robinson, who gave a careful reading and commentary on the work in its final stages; Judy Olson, Jan Mursey, and Edward Tufte, who lent their support and advice at various times during the course of the project; Susan Wise and Norma Chapman, who typed the manuscript from what must have seemed like an endless array of cassettes, and then retyped its revisions.

Among the numerous staff members of the University of Wisconsin Press, the following deserve special credit: Acquisitions Editors Irving Rockwood and Peter Givler for their courage in undertaking such a vast and complex project; Chief Editor Elizabeth Steinberg and her colleagues Ann Ball and Debra Bernardi for carefully overseeing the editing of the text; Gardner Wills, John Delaine, and Marj Winch for patiently preparing the art work; and Bill Day, for pushing the project to completion. And of course, Professor and Madame Bertin for their help and enthusiasm with this project, which they communicated to us with wit and Gallic charm.

Howard Wainer
Princeton
1983

Preface to the 2010 edition of the English translation

Howard Wainer
National Board of Medical Examiners

In the preface to the 1983 translation of the *Semiology*, I emphasized the book's importance as a guide to the future, for it seemed clear that we were on the cusp of an explosion in the use of graphical methods for both the exploration and communication of complex data. This prediction came true. In the intervening twenty-seven years a huge amount of work has been accomplished to help us use graphical methods profitably. This work has been in hardware that can produce graphical displays quickly and inexpensively; software that can translate data files into graphical representations with the click of a mouse; and statistical and perceptual research that helps us know how to use these tools well.

Let me briefly comment only on the latter two aspects. (I choose to refrain from any commentary on hardware because my remarks would almost surely be out of date before they were printed.) First, graphical software: thirty years ago I was enthusiastic and optimistic about the future of graphical use. I thought software would be built with sensible default options, so that when the software was set on maximal stupidity (ours not its), a reasonable graph would result. The software would force you to wring its metaphorical neck if you wanted to produce some horrible, pseudo 3-D, multicolored pie chart. Alas, I couldn't have been more wrong. Instead of making wise, evidence-based, choices, default options seem to have been selected by the same folks who deny landing on the moon, global warming, and evolution. I could not have imagined back then that the revolution in data gathering, analysis, and display taking place in the last decades of the twentieth century would have resulted in the complexity of the modern world being conveyed in bullet-points augmented by PowerPoint and Excel graphics.

The irony of this sorry result is that, since the original publication of the *Semiology,* there have been many wonderful books on graphics—each building on the work of Jacques Bertin as well as John Tukey's practical epistemology. Bill Cleveland's work uses a combination of statistical savvy and experimental evidence to support and expand upon Bertin's foundation. Edward Tufte's brilliant series of books self-exemplifies the advice he gives on how to make evidence beautiful. Leland Wilkinson's *The Grammar of Graphics* provides a codification that is invaluable for both how to think about the structure of graphics and how to automate their construction. And my own work has mixed together statistics, graphics, and history to provide illustrations of what we can accomplish with this marvelous medium. Indeed, the growing popularity and importance of graphics even made it viable for a publisher to reprint William Playfair's eighteenth-century *Atlas,* the work that began it all.

But despite this accumulating graphic wisdom, practice continues to lag. I cannot help but believe that, had Bertin's masterwork not fallen out of print, some of today's graphical ignorance would have been ameliorated. This view was surely shared by the publishing arm of École Pratique des Hautes Études, which republished a fourth edition of the French version in 2005. At about the same time the University of Wisconsin Press was seriously considering the republication of the English translation. Sheila Leary, Director of the University of Wisconsin Press, explained,

"University of Wisconsin Press had been considering bringing *Semiology of Graphics* back into print several years ago and had begun work to bring that about, checking into renewing rights from the French publisher, looking into arranging for a translation of new material from a newer French edition, getting cost estimates, etc. However, several retirements and staff changes sent the project to the back burner for a while."

This delay was unfortunate because the market for a republication seemed more than ready. Today on Amazon, I noticed that one could get a new copy for $550 and the price for a used one began at $399. These remarkable prices merely reflect supply and demand, for those of us who have our own copies are loath to part with them. Happily, Esri decided to do it. They approached the University of Wisconsin Press and, as Leary related,

"As we re-opened the file to get the ball rolling again, we were contacted by Esri expressing their interest in licensing the book for a new English-language edition. When we realized that Esri had a strong publishing program in cartography and cartographic design, the UW Press felt that Esri would be a better home for Bertin's book, where it would receive much more active promotion. …We well know that the *Semiology of Graphics* is an important book with an eager audience, and it seemed to us that Esri would do a fine job of bringing the new edition to that audience."

This publishing generosity is as welcome as it is rare. The reappearance of the *Semiology* did not come too soon, for our need for improving the quality of displays increases apace with the increase in the number of graphs being produced. I am certain that the quality of displays that are prepared to communicate quantitative phenomena will improve in direct proportion to the extent to which this book is read and internalized by those who generate graphics, and more importantly, by those who develop the software that are used to produce them.

The body of the text in this edition is identical to the previous edition: what is new is the postface, which is an update of work done over the decades since the *Semiology* was previously published and a chronology by the author explaining what is new. But what is new is beside the point, for the original work was far ahead of its time. I remember many years ago as Bertin, Bill Berg (the principal translator), and I were discussing the translation, Berg said of one of Bertin's expressions, *"Ça ne se dit pas, en Anglais"* (We don't say that in English); Bertin replied: *"Ne vous inquiétez pas: si ça ne se dit pas, ça se dira"* (Don't worry; if you don't say it now, eventually you will).

Many of us are saying it now, and with the republication of the *Semiology,* many more soon will be.

REFERENCES

Cleveland, W. S. 1994. *The elements of graphing data.* Summit, NJ: Hobart Press.

———. 1994. *Visualizing data.* Summit, NJ: Hobart Press.

Playfair, W. 1801. *The commercial and political atlas and the statistical breviary.* Edited and introduced by Howard Wainer and Ian Spence. New York: Cambridge University Press, 2007.

Tufte, E. R. 1983/2000. *The visual display of quantitative information.* Cheshire, CT: Graphics Press.

Tufte, E. R. 1990. *Envisioning information.* Cheshire, CT: Graphics Press.

Tufte, E. R. 1996. *Visual explanations.* Cheshire, CT: Graphics Press.

Tufte, E. R. 2006. *Beautiful evidence.* Cheshire, CT: Graphics Press.

Tukey, J. W. 1977. *Exploratory data analysis.* Reading, MA: Addison-Wesley.

Wainer, H. 1997. *Visual revelations: Graphical tales of fate and deception from Napoleon Bonaparte to Ross Perot.* New York: Copernicus Books (reprinted in 2000, Hillsdale, NJ: Lawrence Erlbaum Associates).

Wainer, H. 2005. *Graphic discovery: A trout in the milk and other visual adventures.* Princeton, NJ: Princeton University Press.

Wainer, H. 2009. *Picturing the uncertain world: How to understand, communicate and control uncertainty through graphical display.* Princeton, NJ: Princeton University Press.

Wilkinson, L. 2005. *The grammar of graphics.* 2nd ed., New York: Springer-Verlag.

Preface to the English edition

ACKNOWLEDGMENTS

Thanks go first of all to the initiator of this translation, my colleague Howard Wainer, a mainstay of modern graphics, who convinced me that a work which essentially dates from 1965 should be translated today. I would also like to thank Professor Arthur H. Robinson for the encouragement and assistance he brought to this project. I reserve very special thanks for Professor William J. Berg. He ensured the success of an extremely difficult translation, and his determination to be rigorous and precise often led me to clarify my thinking and revise my own text. Finally, let me thank the University of Wisconsin Press for having the courage to publish a work of which half the pages are graphics.

PREFACE

Sémiologie graphique was written in 1965, published in 1967 and revised in 1973; the 1973 edition is translated here. More than ten years have passed. Have the conclusions reached in 1973 been profoundly altered? The publication in 1981 of *Graphics and Graphic Information-Processing* (Berlin and New York: Walter de Gruyter)* shows that the essential points have not changed. However, with the passage of time some analyses have been simplified, certain thoughts are now clearer and better illustrated. But, especially, ideas have been organized differently in light of the evolution of mathematics, computers, and, of course, modern graphics, whose applications are becoming increasingly diverse.

A first reading of the *Sémiologie graphique*

From 1965 to 1973, we were still in the era of complex cartography, of national, regional, and specialized atlases printed in multicolored editions. In numerous disciplines the map came to constitute the basic inventory, the best available "artificial memory." This was also the era of "statistical geography," "statistical history," and we witnessed the first examples of extensive use of computers. In doing our first automated maps, we had to use an IBM that was tens of cubic meters in size. But this computer, however unwieldy, finally made possible the use of complex multivariate analyses, which had begun to be applied in various domains. Finally, this was the era of confrontation between "information theory" and "communication theory," which inspired most graphic research: How should we draw?

---
*Translation by William J. Berg and Paul Scott of Jacques Bertin, *La Graphique et le traitement graphique de l'information* (Paris: Flamarion, 1977).

What should be printed to facilitate "communication," that is, to tell others what we know, without a loss of "information"?

It is in this context that the *Sémiologie graphique* was read. It was seen as a study of the fundamentals of cartography; and reviewers commented especially on the identification of the visual variables. However, the process of reading a graphic and the different properties of the visual variables, which determine the very usefulness of a graphic, did not draw much attention. No more so, it would seem, than the pages devoted to permutation, that is, to the use of the graphic as a tool for research and processing. The statistical and graphical world was still at the stage where the image was printed and fixed, intended as it was for "communication."

Ten years of evolution have brought about an entirely different perspective. Fundamental today are the properties of the visual variables and the processes of graphic classing and permutation. We are entering the era of "operational graphics."

Information processing

Thanks to the computer, information processing has developed prodigiously. We now know that "understanding" means simplifying, reducing a vast amount of "data" to the small number of categories of "information" that we are capable of taking into account in dealing with a given problem. Research in experimental psychology suggests that this number is around three and hardly ever exceeds seven. Information processing involves finding the most acceptable methods for attaining this indispensable simplification. Our forerunners, who did not have the advantage of the computer and were generally unaware of the potential of matrix permutation, proceeded by successive simplifications. The time consumed by such a process severely limited the scale and scope of research possibilities. Now, with the computer, all manner of comparisons seem within rapid reach. The computer provides the necessary simplification, and we believe that that's the end of it! *Vive l'intelligence artificielle!* No further need for thought!

But there is . . . happily! We still need thought because the most observant mathematicians realize that the powerful tools which they have created lead to a lack of reflection on the part of researchers. However, simply "feeding data to the computer" does not in itself constitute scientific work. We have learned that the most important stages are not those that can be automated, but in fact those preceding and following automated processing. Indeed, these stages involve two new questions:

A. What level of simplification is best? In fact, whenever a data table is involved, we encounter different answers according to the type of calculation used. We find ourselves, just like our "ancestors," facing a problem of choice: a choice of pertinent subsets, of weightings, of exclusions; but also a choice of distance calculations and algorithms. No computer will tell us which algorithm is missing.

B. Are the data we feed to the machine pertinent to the problem being considered? In effect, the answers the computer provides fall within the "finite framework" of the data put into it. But is this finite set, extracted from infinity, the best one? The first simplifications stemming from calculation most often lead to criticizing the data and imagining new ones. No computer will tell us which data are missing.

In short, we must call upon external elements—our own knowledge and intuition—in order to imagine data and relationships about which the machine has not yet been instructed. In fact, these two questions send us back to ourselves, that is, back to "natural" intelligence (if in fact intelligence can be clearly defined).

The entire problem is one of augmenting this natural intelligence in the best possible way, of finding the artificial memory that best supports our natural means of perception.

This artificial memory must transcribe a large number of data. It must display groupings of objects and characteristics, as well as exceptions to these groupings. These exceptions are capable of guiding interpretation, of creating nuances, of stimulating new research. Finally, it must be easily modifiable, to accommodate the very studies it has spawned.

The graphic artificial memory
It would appear that the artificial memory that best fulfills all the above conditions is the permutable graphic construction, $x\,y\,z$, that is, the "reorderable matrix."

We no longer speak of a rivalry between graphics and data analysis. Quite the contrary, we recognize the complementary nature of the two languages, particularly at the stage of interpretation. Thus question A, which mathematics seeks to answer by the "calculation of contributions," finds a remarkably efficient answer in graphics.

This discovery of the power and universality of the $x\,y\,z$ construction leads directly to the "matrix theory of graphics," which defines graphics as the sign-system representing any problem conceivable in the form of a double-entry table. This definition excludes "pictography," the sole purpose of which is to define a given set. It applies directly to cartography and, in fact, defines its specific, salient property: a constant $x\,y$ that serves as the basis for comparing different z components. However, this theory also defines the limits of cartography, as we shall see. But these limits prove, if such proof were necessary, that all human logic appears to be based on the visual properties afforded only by the $x\,y\,z$ construction, that is, on the natural and immediate perception of the relationships among three "dimensions." Beyond that lies the fourth dimension, time, accessible only to human memory.

The technological explosion makes questions A and B even more pertinent. The power of ten cubic meters of 1965-vintage computer can now be held in the palm of the hand. Thanks to telephone hookup, the minicomputer presently affords us access to millions of data. Which ones should we use? It offers us thousands of algorithms. Which ones should we choose? It enables us to display the different results of processing. Which ones should we retain and how should we interpret them?

Mathematics and electronics afford us increasingly powerful means of dealing with data. But at the same time they multiply the number of arbitrary choices without changing our natural means of perception in the slightest. It thus comes down to utilizing these natural means in the best way in order to justify the necessary choices.

A new reading of the *Sémiologie graphique*
The passage of time thus leads us to read the *Sémiologie graphique* from a new perspective and to propose the following priorities:

1. The way to "see" a graphic or a map.
    One does not "read" a graphic; one asks three questions of it.
    – What are the x and y components of the data table?
    – What are the groups in x and in y that are constructed by the data in z?
    – What are the exceptions to these groups?

    A graphic should not show only the leaves; it should show the branches as well as the entire tree. The eye can then go from detail to totality and discover at once the general structure and any exceptions to it.

    The questions left without visual answers measure the uselessness of poor constructions. Above all we must learn to pose these three questions. Need we recall that many "experts" remain unaware of them? It would be foolish to count on their opinion if they are ignorant of the true properties of graphics.

2. The $x\,y\,z$ construction of the image.
    This alone permits us to respond to the preceding questions. In any other construction, we see only the leaf, and sometimes the branch; but the tree itself will always remain invisible.

    X and y are the orthogonal dimensions of the table; z is the variation in light energy at each signifying point of the table. This variation can only be obtained by size and value. These, along with the x and y dimensions, constitute the "image variables." The other visual variables—texture, color, orientation, and shape—only serve to vary the quality of the energy, not its quantity. These are the "differential variables."

3. The permutation of the rows and columns of the $x\ y\ z$ matrix.

This is the visual form of information processing. It enables us to discover the groups in x and the groups in y constructed by the data in z, that is, to reduce the vast amount of initial data to an accessible number of categories of information. This manipulation can follow various automated processes. It provides the bases for discussion and decision making. The minicomputer has put this process of interpretation within everyone's reach.

4. The application of matrix theory to cartography.

Matrix theory enables us to read a map and define the two most pertinent questions: What is at a given place? This is the question involving x (objects) of the data table. Where is a given phenomenon? This is the question involving y (characteristics) of the data table. We can thus distinguish two types of maps: maps showing one phenomenon, which allow us to answer both questions; and maps displaying several phenomena, which generally enable us to answer only the first question. By replacing Lasswell's "how" with "why," this analysis controls the study which should precede any cartographic construction.

Applications

Beyond graphic processing, beyond algorithmic processing, which it complements naturally, the "semiology" of graphics—the study of graphics as a sign-system—is being applied in several new directions.

In the visual arts, for example, the semiological approach to graphics provides a rigorous analysis of the visual means used by the artist. It defines the basic properties and laws governing the arts and suggests objective criteria for art criticism.

More substantial yet is its application in pedagogy. The studies of R. Gimeno,* conducted in numerous elementary classes, show that graphics can introduce into all disciplines the bases of logic and the essential processes of analysis and decision making. Graphics can stimulate exceptional motivation, foster better questions, aid in constructing the written text, and . . . reveal the intelligence of so-called "poor students." Learning by graphics is no doubt one of the best answers to the pointed and universal problem of educational renewal and to the question of the role of the computer in the classroom.

Let's return to question B: What data table should we construct? What data should we feed to the computer? For this fundamental question, graphics provides assistance in the form of the "matrix analysis of a problem." This analysis enables us to structure the various ideas which we entertain at this stage in a manner that is coherent and operational but does not limit freedom of thought. This occurs through the drafting of three documents:

– The "apportionment table," which is completely open to all ideas.
– The "homogeneity schema," which recalls the essential constraints of the problem and permits defining an initial coherent table and specifying modes of processing.
– The "pertinency table," which enables us to judge the pertinency of the data retained in relation to the initial questions.

Finally, these tools are completed by the definition of an "aggregation cycle."

The entirety of this analysis figures in *Graphics and Graphic Information-Processing*. It has now been computerized.†

Apart from this last analysis, all the essential points raised in this preface are, in fact, already expressed in the *Sémiologie graphique*, which leads from classic graphics to modern graphics.

Classic graphics involves the FIXED image, which we generally encounter in published form. This is the means of communicating the results of scientific research, sometimes rigorously. But too often useless constructions obscure these results while displaying only the initial elementary data. We must recognize that the reader realizes this. Since time is of the essence, the reader simply ignores the drawings and turns to the text for the simplification necessary for comprehension and communication. The "uninitiated" author, the "habit-ridden" designer, and the "illiterate" layout editor bear a heavy responsibility for this rejection of the potentially incomparable language which is graphics. We must recognize that graphics is learned, not inherited.

Modern graphics involves the TRANSFORMABLE and reorderable image. This is the rigorous research tool that enables the decision maker to discover what should be said and done. The computer, particularly the minicomputer, will find its most complete and powerful means of expression in modern graphics. Moreover, this will take the form of the simplest and thus most communicable image!

It is this image that *Semiology of Graphics* proposes to construct.

Jacques Bertin
Paris
February 1983

*R. Gimeno, *Apprendre à l école par la graphique* (Paris: Retz, 1980).

†Marie M. Rabiller, "Un Outil infographique pour l'organisation des données" [dissertation, Université de Nantes [Unité de recherches en mathématiques], 1982).

Part One

Semiology of the graphic sign-system

General theory

Graphic representation constitutes one of the basic sign-systems conceived by the human mind for the purposes of storing, understanding, and communicating essential information. As a "language" for the eye, graphics benefits from the ubiquitous properties of visual perception. As a monosemic system, it forms the rational part of the world of images.

To analyze graphic representation precisely, it is helpful to distinguish it from musical, verbal, and mathematical notations, all of which are perceived in a linear or temporal sequence. The graphic image also differs from figurative representation, essentially polysemic, and from the animated image, governed by the laws of cinematographic time. Within the boundaries of graphics fall the fields of networks, diagrams, and maps. The domain of graphic imagery ranges from the depiction of atomic structures to the representation of galaxies and extends into the spheres of topography and cartography.

Graphics owes its special significance to its double function as a storage mechanism and a research instrument. A rational and efficient tool when the properties of visual perception are competently utilized, graphics is one of the major "languages" applicable to information processing. Electronic displays, such as the cathode ray tube, open up an unlimited future to graphics.

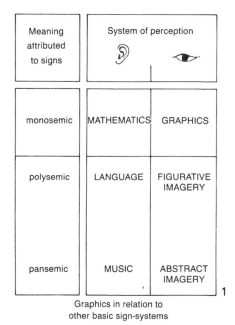

1 Graphics in relation to other basic sign-systems

Definition of graphics

Based on rational imagery, graphics differs from both figurative representation and mathematics. In order to define it rigorously in relation to these and other sign-systems, we shall adopt a semiological approach and begin with two rather obvious statements: (a) the eye and the ear have two distinct systems of perception; (b) the meanings which we attribute to signs can be monosemic, polysemic, or pansemic (figure 1).

A monosemic system

A system is monosemic when the meaning of each sign is known **prior** to observation of the collection of signs. An equation can be comprehended only when the *unique* meaning of each term has been specified. A graphic can be comprehended only when the unique meaning of each sign has been specified (by the legend). Conversely, a system is polysemic when the meaning of the individual signs **follows** and is deduced from consideration of the collection of signs. Signification becomes subjective and thus *debatable*.

Indeed, a figurative image, and for that matter, an ordinary photograph, or an aerial photograph, is always accompanied by a certain amount of ambiguity: "Who is this person?" "What does this black mark or that shape represent?" To these questions, each person will respond individually, since interpretation is linked to the repertoire of analogies and structures characterizing each "receiver." And this repertoire varies from one individual to another, according to personality, surroundings, period, and culture. Faced with the polysemic image, the perceptual process translates into the question: "What does such an element or collection of elements signify?" and perception consists of decoding the image. The reading operation takes place **between the sign and its meaning**. The abstract painting represents an extreme form of polysemy. In its attempt to signify "everything" it no longer signifies anything precise and so becomes "pansemic."

On the other hand, in graphics, with a diagram or map, for example, each element is defined beforehand. The perceptual process is very different and translates into the question: "Given that such a sign signifies such a thing, what are the relationships among all the signs, among all the things represented?" Perception consists of defining the relationships established within the image or among images, or between the image and the real world. The reading operation takes place **among the given meanings**.

This distinction is fundamental because it suggests the true purpose of "graphics" in relation to other forms of visualization.

What does it actually mean to employ a monosemic

system? It is to dedicate a moment for reflection during which one seeks a maximum reduction of confusion; when, for a certain domain and during a certain time, *all the participants* come to agree on certain meanings expressed by certain signs, and *agree to discuss them no further*.

This convention enables us to *discuss the collection of signs* and to link propositions in a sequence which can then become "undebatable," that is, "logical."* This is the object of mathematics, which deals with problems involving a temporal sequence. It is the object of graphics, which operates in areas linked to the tridimensionality of spatial perception. On this point, graphics and mathematics are similar and construct the "rational moment."

A visual system

But graphics and mathematics differ in the perceptual structure which characterizes each of them. It would take at least 20 000 successive instants of perception to compare two data tables of 100 rows by 100 columns. If the data are transcribed graphically, comparison becomes easy; it can even be instantaneous.

As we see in figure 2, auditory perception has only two sensory variables at its disposal: sound and time. All the sign-systems intended for the ear are linear and temporal. (Remember that written transcriptions of music, language, and mathematics are merely formulae for setting down systems which are fundamentally auditory, and that these formulae do not escape from the linear and temporal character of the systems themselves.)

On the other hand, visual perception has at its disposal *three* sensory variables which do not involve time: the variation of marks and the two dimensions of the plane. The sign-systems intended for the eye are, above all, spatial and atemporal. Hence their essential property: in an instant of perception, linear systems communicate only *a single sound or sign*, whereas spatial systems, graphics among them, communicate in the same instant the *relationships among three variables*.

Maximum utilization of this considerable perceptual power within the framework of logical reasoning is the true purpose of graphics, **the monosemic domain of spatial perception.**

Evolution of graphics

The effectiveness of graphics has long been recognized. The most ancient graphic representations which have been discovered are geographic maps engraved on clay, and they probably date from the third millennium before Christ. Graphic images were first conceived, and are still usefully

	System of perception	
	👂	👁
sensory variables	1 variation of sound	1 variation of marks
	1 variation of time	2 dimensions of the plane
total	2 variables	3 variables
instantaneous perception	1 sound	Relationships among 3 variables

2
Perceptual properties
of linear and spatial systems

---

*Monosemy is the fundamental condition of logic, but it also defines its limits. In effect, monosemy can only exist within a finite domain of objects and relationships. Logical reasoning, therefore, can only be a moment of reflection, since there is an infinite number of finite domains, however large they may be. Logic appears, then, as a sequence of rational moments immersed in the infinite continuum of the irrational.

conceived, as reproductions of the visible world, which benefit from only one degree of freedom, that of the scale. In a molecular model, a geometric figure, an assembly diagram, an industrial drawing, a geologic section, or a map, the two dimensions of the coordinate plane are identical to those of the visible space, adjusted for the scale of the drawing.

One had to wait until the fourteenth century to suspect, at Oxford, and until the eighteenth century to confirm, with Charles de Fourcroy (see page 202), that the two dimensions of a sheet of paper could usefully represent *something other than visible space*. This amounts to a transition from a simple representation to a "sign-system" that is complete, independent, and possesses its own laws, thus falling within the scope of SEMIOLOGY.

And now, at the end of the twentieth century, with the pressure of modern information and the advances of data processing, graphics is passing through a new and fundamental stage. The great difference between the graphic representation of yesterday, which was poorly dissociated from the figurative image, and the graphics of tomorrow, is the disappearance of the congenital fixity of the image.

When one can superimpose, juxtapose, transpose, and permute graphic images in ways that lead to groupings and classings, the graphic image passes from the *dead image*, the "illustration," to the *living image*, the widely accessible research instrument it is now becoming. The graphic is no longer only the "representation" of a final simplification, it is a point of departure for the discovery of these simplifications and the means for their justification. The graphic has become, by its manageability, an instrument for information processing. Its study must begin, then, with the analysis of the information to be transcribed.

I. Analysis of the information

Thought can only be expressed within a system of signs. Mimicry is a natural form of coding; verbal language is a code of auditory signs (which must be learned in order to communicate with others); the written language is another code; graphic representation yet another. Memory storage on disks, tapes or in computers necessitates appropriate new codifications.

Graphic representation is the transcription, into the graphic sign-system, of "information" known through the intermediary of any given sign-system.

Graphic representation can thus be approached by semiology, a science which deals with all sign-systems.

Information and representation

Any transcription leads necessarily to a separation of content from form. The "content," those elements of the thought which can remain constant, regardless of the sign-system into which they are translated, must be distinguished from the "container," that is, the means available in a given system and the laws which govern their use. These elements are constant, whatever the thought to be transcribed.

Whether we are studying the means, properties, and

limits of the graphic system, or planning a design, it is first necessary to strictly separate the content (the INFORMATION to be transmitted) from the container (the PROPERTIES of the graphic system).

Generally we will not discuss here the content of the proposed examples. It can be good or bad, "accurate" or "inaccurate." However, what matters to us is the quality and the efficacy of its graphic transcription. Incidentally, it is the singular characteristic of a good graphic transcription that it alone permits us to evaluate fully the quality of the content of the information.

Knowing that each sign-system has its properties, its style, its aesthetic, what constant can be isolated in a thought, throughout its diverse translations? A thought is a relationship among various concepts which have been recognized and isolated at a given moment, from among the multitude of imaginable concepts. Consider the following example: "On July 8, 1964, stock X on the Paris exchange is quoted at 128 francs; on July 9, it is quoted at 135 francs." Whatever form the phrase may take, the content will always involve the pertinent correspondence among certain elements:
(1) the concept "quantity of francs," or a VARIATION in the number of francs;
(2) the concept "time," or a VARIATION of the date;
(3) an element—stock X—of the concept "different stocks quoted on the Paris exchange." This clement is, by definition, INVARIANT, since it constitutes the common ground which relates the francs to the dates.

In graphic representation the translatable content of a thought will be called the INFORMATION. It is constituted essentially by one or several PERTINENT CORRESPONDENCES between a finite set of variational concepts and an invariant.

The information to be transcribed can be furnished in any given sign-system, so long as it is known by the transcriber, that is, the graphic designer. Let us stress once and for all that we will never use the term "information" in the very limited and precise sense which it has in "information-theory," but as a synonym for "data to be transcribed."

A. The invariant and the components (page 16)

Whenever we discuss information to be transcribed, the central notion common to all the pertinent correspondences will be called the INVARIANT. The variational concepts involved will be called the COMPONENTS.

Thus the preceding example can be said to have an invariant—a given stock—which relates two components: a variation in the number of francs and a variation in time. Whatever sign-system is employed, at least two components will be needed to translate the information. In the graphic system two visual components are normally utilized: the two dimensions of the plane.

The wording of the TITLES and LEGENDS is the first application of these notions (page 19).

In order to facilitate explanation, the components of the graphic sign-system will be called VISUAL VARIABLES

(or simply "variables"), and the two variables furnished by the plane itself will be called PLANAR DIMENSIONS.

Information will therefore be formed by pertinent correspondences among given components and its graphic representation by correspondences among given variables.

Visual perception will admit only a small number of variables. Consequently:

B. The number of components (page 28)

The determination of the NUMBER OF COMPONENTS is the first stage in the analysis of the information.
  Components and variables are, by definition, divisible.
  The different identifiable parts of a component or of a variable will be called ELEMENTS or CATEGORIES (or "classes" or "steps"; page 33). We can talk, for example, about categories of "departments" [major administrative subdivisions of France] in a geographic component, about the categories "bovine," "ovine," "caprine," of the component "different domestic animals," about steps of gray in the variable "value," about annual classes in the component "time," or about elements of the component "different persons."

The complexity of a figure is linked to the number of categories in each component:

C. The length of the components (page 33)

We will use the term LENGTH to describe the number of elements or categories which we are able to identify in a given component or variable. This is the second stage in the analysis of the information.
  Thus the binary component "sex" can be said to have a length of two; the geographic component "departments of France" a length of ninety. In a quantitative component we must not confuse the "length," the number of useful steps, with the RANGE of the series, the ratio between the largest and smallest numbers in the statistical series.

D. The level of organization of the components (page 34)

What is properly called graphics depicts only the relationships established among components or elements.
  These relationships define three LEVELS OF ORGANIZATION, and each component, each visual variable, is located on one of these levels:
  THE QUALITATIVE LEVEL (or nominal level; page 36) includes all the concepts of simple differentiation (trades, products, religions, colors . . .). It always involves two perceptual approaches: This is similar to that, and I can combine them into a single group (association). This is different from that and belongs to another group (differentiation).
  THE ORDERED LEVEL (page 37) involves all the concepts that permit a ranking of the elements in a universally acknowledged manner (e.g., a temporal order; an order of sensory valuations: cold–warm–hot, black–gray–white, small–medium–large; an order of moral valuations: good—mediocre—bad . . .). This level includes all the concepts which allow one to say: This is more than that and less than the other.
  THE QUANTITATIVE LEVEL (interval-ratio level;

page 38) is attained when one makes use of a countable unit (this is quarter, triple, or four times that).

These levels are overlapping: What is quantitative is likewise ordered and qualitative. What is ordered is also qualitative. What is qualitative is neither quantitative nor ordered, but is arbitrarily reorderable.

THE LEVELS OF ORGANIZATION form the domain of universal meanings, of fundamental analogies, in which graphic representation can stake a claim. This is the third stage in the analysis of the information.

All forms of signification other than the above *relationships* are in fact exterior to graphics and merely serve to link the graphic system to the world of exterior concepts. They must rely either on an explanation coded in another system (legends) or on a FIGURATIVE ANALOGY of shape or color (symbols), which is based on acquired habits or learned conventions and can never claim to be universal.

Each visual variable has its particular properties of length and level. It is important that each component be transcribed by a variable having at least a corresponding length and level.

Graphics is concerned with the representation of these three levels of organization. However, the relationships of similarity and order, based on metrics, are those which constitute the foundation for information processing and analysis

II. The properties of the graphic system

A. The scope of the graphic system (page 42)

What variables does the graphic sign-system have at its disposal? The eye is the intermediary in a great number of perceptions. Not all of these concern the system we are studying; the intervention of real movement, for example, although perceptible by vision, would make us pass from the graphic system (atemporal) into film, whose laws are very different. We will only consider that which can be represented by readily available graphic means, on a flat sheet of white paper of standard size and under normal lighting.

Within these limits, we will consider that the graphic system has at its disposal EIGHT VARIABLES. A visible mark expressing a pertinent correspondence can vary in relation to the TWO DIMENSIONS OF THE PLANE. It can further vary in SIZE, VALUE, TEXTURE, COLOR, ORIENTATION, and SHAPE. Within the plane, this mark can represent a point (a position without area), a line (a linear position without area), or an area.

B. The plane (page 44)

The type of signification—point, line, or area—assigned to a visible mark on the plane will be called "IMPLANTATION"* or class of representation.

---

*This term will no doubt appear strange to English-speaking cartographers and statisticians. Bertin uses it as a contrast to "imposition" (the type of graphic—diagram, map, etc.) and in order to avoid speaking of point, line, and area "symbols" (translator's note).

A given French department can be represented either by a point or a line, as in a diagram, or by an area, as in a map. These "implantations" are the three moments of the sensory continuum applied to the plane. They constitute the three elementary figures of geometry.

THE LEVEL OF ORGANIZATION OF THE PLANE is maximum (page 48). Its two dimensions furnish the only variables which can correctly represent any component of the information, whatever its level of organization.

The utilization of the two dimensions of the plane will be called "IMPOSITION" or group of representation (page 50).

This utilization depends upon the nature of the pertinent correspondences expressed on the plane and enables us to divide graphic representation into four groups. In effect, the correspondences on the plane can be established:

– among all the elements of one component and all the elements of another component. *The construction is a* DIAGRAM. Example: Variation in the quotations for stock X on the Paris exchange. Any date (component: time) can correspond a priori to any price (component: quantity of francs), and there are no grounds for predicting a correspondence between two dates or between two prices.

– among all the elements of the same component. *The construction is a* NETWORK. Example: The relationships of conversations among individuals situated around a table. Any individual (of the component "different individuals") is capable of communicating with any other individual (of the same component).

– among all the elements of the same geographic component, inscribed on the plane according to the observed geographic distribution. *The network traces out a* GEOGRAPHIC MAP.

– between a single element and the reader (road signs, various codes based on shape, industrial color codes, etc.). The correspondence is exterior to the graphic image. *This is a problem involving* SYMBOLISM, which relies upon figurative analogies.

In diagrams and networks (page 52), the free disposition of the dimensions of the plane leads us to distinguish arrangements dispersed over the entire plane from those which structure it in some manner (rectilinear, circular, orthogonal, polar) and to define TYPES OF CONSTRUCTION which can be characterized by SCHEMAS OF CONSTRUCTION.

C. The retinal variables (page 60)

We will term "ELEVATION" the utilization of the six variables other than those of the plane, that is, the RETINAL variables. A qualitative variation between two cities can be represented on a map by a variation in size, value, texture, color, orientation, shape, or by a combination of several of these variables.

Retinal variables must be utilized whenever a third component appears in the information. But none of these variables has the capability of the plane to represent any component of the information, whatever its level of organization. We must, therefore, determine the LEVEL OF ORGANIZATION (page 64) for each variable, its PROPERTIES OF LENGTH (page 71) and its applicability.

III. The rules of the graphic system

A. The basic graphic problem (page 100)

The great diversity of graphic constructions, within a group or even from one group to another, is due to the designer's apparent freedom to represent a given component by using any one of the eight visual variables or a combination of several of them.

Faced with such a choice, the graphic designer can, for example, represent a geographic component by a single dimension of the plane, thereby constructing a diagram; or by both dimensions of the plane, constructing a map. A variation in color or value could also be used. In fact, to construct 100 DIFFERENT FIGURES from the same information requires less imagination than patience. However, certain choices become compelling due to their greater "efficiency."

B. Image theory (page 139)

EFFICIENCY is defined by the following proposition: If, in order to obtain a correct and complete answer to a given question, all other things being equal, one construction requires a shorter period of perception than another construction, we can say that it is more efficient for this question.

This is Zipf's notion of "mental cost," applied to visual perception (see page 139). In most cases the difference in perception time between an efficient construction and an inefficient one is considerable.

The RULES OF CONSTRUCTION enable us to choose the variables which will construct the most efficient representation.

Efficiency is linked to the degree of facility characterizing each stage in the reading of a graphic. The body of remarks leading to the rules of construction form the IMAGE THEORY. Five aspects of this theory are discussed here:

1. Stages in the reading process (page 140)
To read a drawing is to proceed more or less rapidly in three successive operations:

EXTERNAL IDENTIFICATION: What components are involved? It is necessary to clearly define and situate the concepts proposed for examination.

INTERNAL IDENTIFICATION: By what variables are the components expressed? For example, quantities by the vertical dimension of the plane, time by the horizontal dimension; or, alternatively, quantities by the length of the radius, time by the length of the arc subtended by the angle about the point of origin.

These operations link the graphic system to other systems through the use of written notations (titles and legends) or figurative analogies (shape and color). The two stages of identification are indispensable and must precede any study of the information itself:

PERCEPTION OF PERTINENT CORRESPONDENCES: "On a given date what is the price of stock X?" This perception is always the result of a QUESTION, conscious or not. What are the questions which one can ask in approaching the information?

2. Possible questions—levels of reading (page 141)
In the preceding example two types of questions are possible:
– On a given date what is the price of stock X?
– For a given price, on what date(s) was it attained?

There are as many TYPES OF QUESTIONS as components in the information, but within each type there are numerous possible questions:

(a) Questions introduced by a single element of a component, for example: "On a given date . . ." and resulting in a single correspondence. This is the ELEMENTARY LEVEL OF READING. Here, questions tend to lead outside of the graphic system.

(b) Questions introduced by a group of elements in the component, for example: "In the first three days what was the trend of the price?" Answer: "The price rose." Such questions are quite numerous, since we can form highly diverse groups. This is the INTERMEDIATE LEVEL OF READING. Here, questions tend to reduce the length of the components.

(c) A question introduced by the whole component: "During the entire period, what was the trend of the price?" Answer: "General upward movement." This is the OVERALL LEVEL or "global" reading. Such a question tends to reduce all the information to a single ordered relationship among the components. We can say that:

– **there are as many TYPES OF QUESTIONS as components in the information;**
– **for each type there are THREE LEVELS OF READING: the elementary level, the intermediate level, and the overall level;**
– **any question can be defined by its type and level.**

These reading levels are comparable to the integrative levels of the mind. Their analysis permits knowing in advance the *totality of the questions* which any given information can generate; as a result it permits studying the probability of their occurrence and, if appropriate, of taking them into account in the graphic construction.

3. Definition of an image (page 142)

Answering a given question involves: (a) an input identification: "On a given date . . . ?"; (b) perception of a correspondence between the components; (c) an output identification: the answer "so many francs." This implies that the eye can isolate the input date from all the other dates and, DURING AN INSTANT OF PERCEPTION, see only such correspondences as are determined by the input identification, but see all of these. During this instant, the eye must disregard all other correspondences. This is visual SELECTION. We can state that in certain graphic constructions, the eye is capable of including all the correspondences determined by an input identification within a single "glance," within a single instant of perception. The correspondences can be seen in a single visual form.

The meaningful visual form perceptible in the minimum instant of vision will be called the IMAGE.

In this sense, IMAGE corresponds to "form" in "form theory," to "pattern" and to "Gestalt." A synonym would be "outline." Other constructions do not permit the inclusion of all the correspondences within a single instant of perception; the entire set of correspondences could only be constructed in the memory of the reader, by the addition of the various images perceived in succession. It is therefore obvious that:

The most efficient constructions are those in which any question, whatever its type or level, can be answered in a single instant of perception, in A SINGLE IMAGE.

The image, the temporal unit of meaningful visual perception, must not be confused with the FIGURE, which is the apparent and illusory unit defined by the sheet of paper, by a linear frame or by a geographic border.

4. Construction of an image (page 148)

The image is built upon three homogeneous and ordered variables: the two dimensions of the plane and a retinal variable.

The RULES OF CONSTRUCTION (see page 171) thus lead the designer to utilize the two dimensions of the plane in a homogeneous, rectilinear, and orthogonal manner and also to employ an ordered retinal variable, such as size, value, or texture.*

Consequently:

---

*Bertin draws a rigorous distinction between texture (*grain* in French) and pattern (*texture* in French):

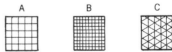

A and B differ in their texture, but there is no difference in pattern. The elementary shapes are the same. The notion of pattern explains the difference between A and C. The elementary shapes are different. A difference in "pattern" is essentially a difference in shape (translator's note)

All information with three components or less can be represented as a single image (page 148).

It is necessary and sufficient that the rules of construction be respected. In this case, whatever the type or level of the question, the answer will be seen in a single instant of perception. We can say that the graphic representation is AN IMAGE.

In any construction not respecting these rules, certain types and levels of questions will necessitate the perception of several images in succession; there will be a high mental cost. The formulation of the answer will be very difficult and often impossible. We will call these graphic constructions FIGURATIONS. They are obviously less efficient than constructions involving a single image.

5. Limits of an image (page 154)

An image will not accommodate more than three meaningful variables. As a result, all information with more than three components cannot be constructed as an image. This means that for certain questions identification will necessitate several instants of perception, several images and:

In information with more than three components, it is necessary to CHOOSE PREFERRED QUESTIONS and construct the graphic so that they can be answered in a single instant of perception. At the same time one reserves input identifications that necessitate several instants of perception for questions which are less useful or less likely to be posed.

Visual efficiency is inversely proportional to the number of images necessary for the perception of the data; it is this rule which, in the final analysis, governs the choice of preferred questions and leads to identifying the three functions of graphic representation:

C. Three functions of graphic representation (page 160)

(1) RECORDING INFORMATION: creating a storage mechanism which avoids the effort of memorization. The graphic utilized for this purpose must be comprehensive and may be nonmemorizable in its totality.

(2) COMMUNICATING INFORMATION (page 162): creating a memorizable image which will inscribe the information in the viewer's mind. The graphic used here must be memorizable and may be noncomprehensive. The image should be a simple one.

(3) PROCESSING INFORMATION (page 164): furnishing the drawings which permit a SIMPLIFICATION and its justification. The graphic should be memorizable (for comparisons) and comprehensive (for choices).

Information with three components or less, constructed as a single image, can fulfill all three functions of graphic representation. But information involving more than three components must be constructed differently, according to the intended function, that is, according to the nature of the preferred questions.

D. Rules of construction (page 171)

The RULES OF CONSTRUCTION, represented by STANDARD SCHEMAS, define the most efficient construction for a given case.

E. Rules of legibility (or rules of separation) (page 175)

The rules of construction govern the choice of visual variables. Once chosen, however, the variables can still be utilized well or poorly. Efficiency also depends on the sensory differentiation which we can obtain from each variable or combination of variables; this differentiation will increase or reduce the capacity for "separation" within the variables. For example, sensory differentiation is greater between blue and red than between blue and green, between black and white than between black and gray. . . .

The observations which permit us to best accomplish sensory differentiation will be called RULES OF LEGIBILITY.

Linked to the faculties of human perception, these rules apply to each variable as well as each combination of variables and are related to their LENGTH. But length in turn varies according to the level of organization involved; selective perception, for example, calls for the greatest amount of differentiation.

I. Analysis of the information

A rigorous definition of the components of the information, specifying their number, length, and level, must precede any graphic construction.

A. The invariant and the components
B. The number of components
C. The length of the components
D. The level of organization of the components

A. The invariant and the components

DEFINITION

Information is a series of correspondences observed within a finite set of variational concepts of "components." All the correspondences must relate to an invariable common ground, which we will term the "invariant." A precise analysis of these terms is the only means:
– to understand complex information
– to determine the best graphic representation of it
– to word its title and legend.
The following examples will enable us to specify the nature of the two concepts: invariant and components.

EXAMPLES

Example 1: The trend of stock X on the Paris exchange.

The INVARIANT is the complete and invariable notion common to all the data. In figure 1, it is the "quotation in new francs for stock X, cash payment, closing price on the Paris exchange."

Indeed, we cannot mix together within this information, cash and time payments, old and new francs, stocks X and Y, the London and Paris exchanges.

The COMPONENTS are the variational concepts. In this case the variational concepts are:
– quantities (of francs)
– time (in days)

The information has two components, and the drawing must thus utilize two visual variables: the two dimensions of the plane (figure 1).

Example 2: Comparison of the trends of stocks X and Y.
INVARIANT –*quotation in francs, cash payment, closing price, on the Paris exchange*
COMPONENTS –*quantities of francs, according to*
  –*the date*
  –*different stocks (X and Y)*

This information has three components; the drawing must employ three visual variables. A size variation can be used to distinguish X from Y, as in figure 2.

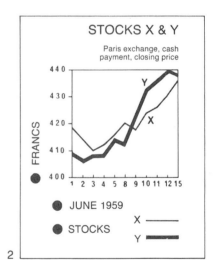

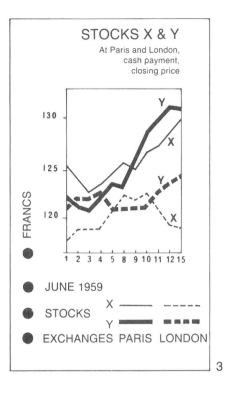

3

Example 3: Comparison of the trends of stocks X and Y in London and Paris.

INVARIANT —*quotation, cash payment, closing price*
COMPONENTS —*quantities (indexed), according to*
 —*the date*
 —*different stocks (X and Y)*
 —*different cities (London, Paris)*

Here the information has four components, and the drawing must employ four visual variables. In figure 3, for example, texture is used in addition to the three preceding variables.

Note that the definition of the invariant simplifies as the number of components increases.

In these three cases, the information is sufficiently familiar to be designated by a specific word: TREND. This word can serve as a title, since it summarizes the informational situation, and its immediate comprehension takes the place of a more lengthy analysis. However, modern information often correlates components of great diversity, which means that the invariant will be less familiar, its identification more difficult.

Example 4: Population residing in the Paris area, by department of birth (not including departments constituting the Paris area), given in absolute quantities.

Example 5: Distribution per 100 persons born outside of the Paris area but residing there in 1962, according to their department of birth. There is no single word to capsulize either of these examples; there is no known title, since the situations are too unusual. A thorough analysis becomes imperative if the designer is to understand what must be expressed and the reader to understand what is being represented; if not, each risks serious error. For instance, what is the difference between the data in examples 4 and 5?

Further analysis reveals that there is none. These are simply two verbal formulae which express the same content, as shown in figure 4. Both cases are, in fact, constituted by:

INVARIANT —*a person, living in the Paris area, born in the provinces and counted in the department of birth*
COMPONENTS —*population according to*
 —*departments*

The data are simply expressed in one case by the observed numbers, and the total corresponds to the number of observations (3 034 700); in the other case by numbers calculated per 100. All the numbers from the first example have been multiplied by the fraction 100/3 034 700 to produce the second. This merely involves changing the numerical scale, which in no way modifies the observed correspondences.

In such cases, it is only by seeking the precise definition of the invariant and the components that we can come to understand the true nature of the information.

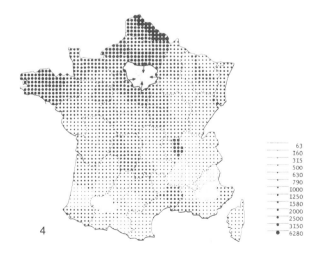

4

Example 6: The Cuban Missile Crisis—Principal elements of decision-making during the missile crisis of 1962 (see page 264). This is a nonquantitative problem; analysis alone permits a clear representation of the information.

INVARIANT —*a decision at the summit (made by a chief of state). The decisions are differentiated according to*
COMPONENTS —*nationality (American or Russian)*
— *(potential or actual)*
—*danger (increasing risk of war)*
—*date (made on such and such a date)*
—*nature (of such and such a nature)*

This information has five components, necessitating at least five visual variables. It cannot be perceived in its totality in one immediate image, as we note from observing figure 1.

No word exists for identifying the second component; certain components are characterized only by listing their categories. Nor does any word exist for the whole of the information. The "title" is merely a paraphrase which orients the reader toward the subject but does not specify which components are involved. We are far from the word "trend" in the first examples.

THE ORDER OF THE COMPONENTS

When the data involve percentages, or when we decide to derive percentages from the raw data, this must be reflected in the analytic description. For example, in comparing the extent of the three main sectors of the work force in several countries:

Figure 2:
INVARIANT —*working persons (1960)*
COMPONENTS —*different countries*
—*number (Q) per 100 working persons per country according to*
—*three main employment sectors*

In this case, all the countries are considered as similar and equal to 100.

This situation is reflected in the analytic description by placing those components not affected by the quantities first, and by having the quantities followed by only those components which they govern.

This rule is also exemplified in the following analytic description:

Figure 3:
INVARIANT —*working persons (1960)*
COMPONENTS —*absolute Q according to*
—*different countries*
—*Q per 100 working persons per country according to*
—*different employment sectors*

The first application of the notion of components is the following:

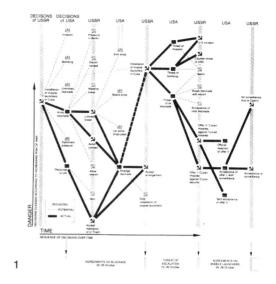

1

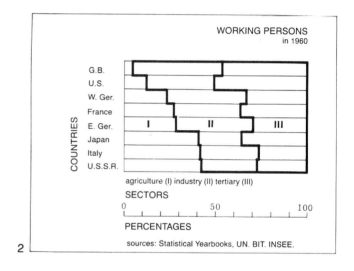

2

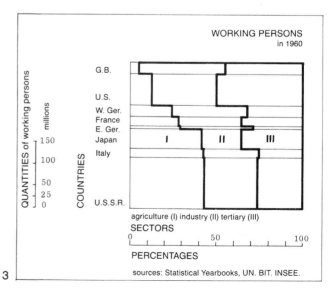

3

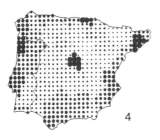

4

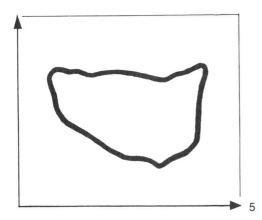

5

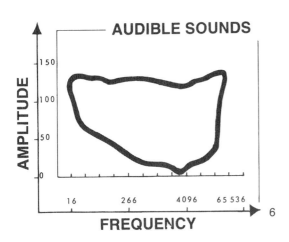

6

Wording of titles and legends

The two terms, "title" and "legend," are sometimes used interchangeably. A "legend," placed under a figure, will function as a title, while specialists will often seek the real "title" of a map in what some designers call the "legend."

Such confusion must be avoided. The headings (legend and title material) in a graphic image have two functions:
(1) to permit the reader to identify, *in the mind*, the invariant and components involved. This can be called EXTERNAL IDENTIFICATION, in the sense that it is independent from the graphic image itself.
(2) to identify, *in the drawing*, the visual variables corresponding to the components. This will be referred to as INTERNAL IDENTIFICATION.

EXTERNAL IDENTIFICATION

External identification is independent of the graphic representation, because the drawing, in itself, cannot furnish all the elements necessary for identification. Figures 4 and 5 are not identifiable. Written or oral statements are indispensable for their external identification. To identify the content of figure 5 we must know

– the invariant: audible sounds
– the first component: the frequency of the sound (cycles per second)
– the second component: the amplitude of the sound (auditory level in decibels).

These factors are clearly identified in figure 6.

One cannot study a graphic intelligently without knowing the invariant and the components displayed in it. Titling a drawing speeds the acquisition of this knowledge and dispels potential ambiguity.

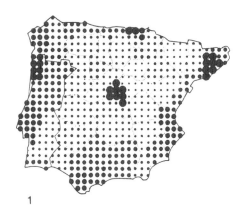

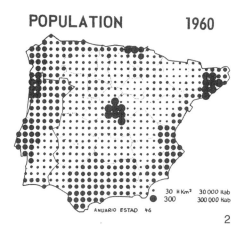

POPULATION 1960

1

2

Identification of the components

In numerous cases, however, identification of the components can result from the drawing itself. For a given audience, figure 1 clearly involves:
- a map of the Iberian peninsula: a geographic component
- quantities: a quantitative component.

All that remains is to specify the invariant, as in figure 2.

In certain cases the drawing itself can provide the means of identifying the components due to the familiarity of the subject.

For the same reason, a simple expression describing the invariant can sometimes be sufficient to define the components. Terms such as trend, price, temperature of X, barometric pressure, can define the two components of the diagram as well. The reader can understand these verbal cues because the images involved are of relatively current and common usage.

But modern scientific research multiplies the occurrence of innumerable combinations which cannot be named concisely or where the specific term employed is familiar only to a limited number of individuals. The designer can thus encounter three types of cases:
- common figurations, where a simple expression suggests the components or where the image itself is sufficient (particularly in cartography) to make them recognizable.
- new combinations, involving recent terms which are not very familiar to the average reader (agroclimatic diagrams, concentration curves, "stemmas" [see page 279]...). Here, the term serves its purpose only for a limited number of "experts" in the given field.
- new combinations, which have no precise term to characterize them.

In the last two cases, a written description of the components is indispensable to external identification (see figure 6 on the preceding page). Thus:

In most cases, the written designation of the various components is included in the title.

Identification of the invariant

An image (e.g., figure 1) or a term (e.g., "trend") can serve to define the two components. But quantity of what? Trend of what?

In all cases, a written term is needed to define the invariant.

Trend of what? Of stock X, cash payment at closing price, in new francs, on the Paris exchange.

This involves category X of the component "different stocks."

It involves the category "cash payment" of the component "different types of purchase."

It involves the category "Paris" of the component "different cities." . . .

The invariant specifies the common ground for different components that extend throughout the larger informational set being investigated or the work being consulted.

It can therefore be worded as a function of the information immediately related to that being considered. Thus, in a grouping of trends "on the Paris exchange," this term need not be repeated for each diagram; in a chapter on "cash payments" this term could be eliminated. . . .

But omissions are often dangerous, since graphic information ought to be detachable from its immediate context in order to be related to all data having a common element. This is the case in analytic documentation.

20

3

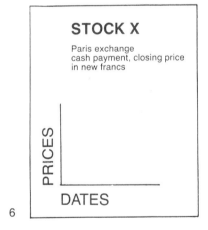

4

5

6

External identification: title composition

Consequently, in most cases external identification consists of writing, in visible characters and in a standard arrangement, (a) the invariant, and (b) all the components of the information. This is illustrated in figure 3.

The title will generally be set up as in figure 4, the order of the components following the rule outlined on page 18.

But the rigor and precision of this formula entails a relatively long title, which no longer fulfills the conditions of brevity and rapid comprehension, particularly necessary when studying a large number of drawings.

In order to avoid permanent omissions, it would seem that the most logical solution, the one meeting all the requisite conditions of external identification, is:

(1) to word the title according to the general formula outlined in figure 4;

(2) to cap it with a heading (see figure 5) whose wording depends on the larger informational set.

As we see in figure 6, the heading generally includes the name of the category (stock X) which belongs to the next higher informational set (different stocks). This solution affords the means of immediately perceiving the unique features of each representation, or, for the reader already acquainted with the documents, the means of rapidly locating a given representation. The following examples apply these principles and illustrate to what degree the current use of "title phrases" can be confusing. Indeed, there are a large number of title phrases which could fit a given set of information, while a given phrase could also apply to very different data.

TYPICAL TITLE PHRASES	SUGGESTED TITLES
Population residing in the Paris area, by department of birth (not including departments constituting the Paris area). Quantities in thousands. Percentage distribution of persons born outside the Paris area but residing there in 1962, according to the department of birth.	MIGRATION TO PARIS Residents of the Paris area born in the provinces – absolute quantities according to – department of birth
Number of persons residing in the Paris area (in 1962) per 100 persons born in each department (not including the departments of the Paris area).	RATE OF MIGRATION TO PARIS Residents of the Paris area born in the provinces – by department of birth – Q per 100 persons born in the department
Percentage of population living in rural communes [administrative subdivision of France] where 20–39.9% of the population is agricultural.	RURAL COMMUNES, 20–40% AGRICULTURAL Population living in rural communes where 20–39.9% of the population is agricultural – by department – Q per 100 persons living in all rural communes
Map of the distribution by ward of Parisian parents of students at the École Polytechnique	STUDENTS OF THE ÉCOLE POLYTECHNIQUE Residence of Parisian parents of students at the Ecole Polytechnique – quantity – by ward
Principal elements of decision-making during the Cuban "hot" crisis in 1962.	THE CUBAN CRISIS (1962) Decisions at the summit (made by chiefs of state) according to – whether the decisions were actual or only potential – date – danger of war – nationality (American or Russian) – nature of decision

TYPICAL TITLE PHRASES	POSSIBLE TITLES
Percentage of students enrolled in seventh grade, in private schools, by canton [administrative subdivision of France].	**PRIVATE SCHOOLING IN THE SEVENTH GRADE** Students enrolled in seventh grade, private schools – by canton – Q per 100 students in seventh grade, all schools **SEVENTH GRADERS IN PRIVATE SCHOOLS** Students enrolled in seventh grade, private schools – by canton – Q per 100 students enrolled in private schools
Distribution of the three main sectors (agriculture, industry, tertiary [other] of the work force, in percentage, by department.	**MAIN SECTORS OF THE WORK FORCE** Work force in 1954 – by department – Q per 100 working persons according to – three main sectors (agriculture, industry, tertiary) **MAIN SECTORS OF THE WORK FORCE** Work force in 1954 – by department – Q per 100 inhabitants according to – three main sectors (agriculture, industry, tertiary)
Changes in the population, aged 20 to 64, in France, between 1954 and 1962.	**GROWTH OF ADULT POPULATION (1954–1962)** Population aged 20 to 64. Difference between 1954 and 1962 – in absolute quantities – by department **EVOLUTION OF ADULT POPULATION (1954–1962)** Population aged 20 to 64 in 1962 – by department – Q per 100 persons aged 20 to 64 in 1954 **VARIATION IN THE PROPORTION OF ADULTS (1954–1962)** Difference between the 1954 percentage (Q of population aged 20 to 64 per 100 inhabitants) and the 1962 percentage – in quantities – by canton

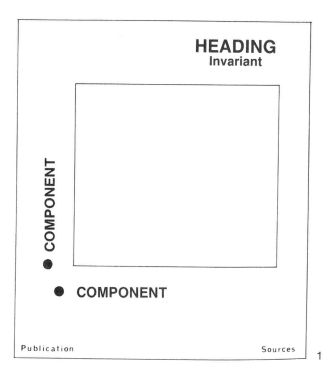

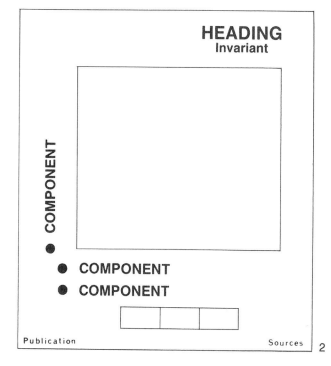

INTERNAL IDENTIFICATION

Having identified the invariant and the components, the reader must still recognize which visual variables are being used to represent these components.

An "ombrothermal" diagram becomes meaningful when it is known that this word signifies:

INVARIANT —*planted areas (for a given type of vegetation) according to*

COMPONENTS —*the precipitation recorded in these areas (annual total)*

—*the temperature recorded in these areas (annual average)*

But we must still be able to recognize that on the drawing precipitation increases as we read, say, from bottom to top (ordinate), while temperatures increase from left to right (abscissa).

Constructions involving two components

Each of the two planar dimensions must be named.

To avoid repeating terms in the diagram, external and internal identification of the components can be combined. This leads to the standard arrangement displayed in figure 1. However, a different situation arises when there is a third component, and, in cartography, where the geographic order occupies the two planar dimensions.

Constructions involving more than two components

When three or more components are involved, we must utilize "retinal" variations: variations in the size of points or lines, in the value or color of the marks, etc. These variations are independent of position on the plane.

For each retinal variable, we must therefore draw a standard variation, indicate the name of the component which it represents, and relate its "steps" to the categories of the component. This is what we mean by the "legend."

To avoid repetition, we arrive at the standard arrangement depicted in figure 2.

Maps

Here, two planar dimensions are occupied by the geographic entity, but any internal geographic categorization must still be specified. The enumeration areas can be departments, "agricultural zones," wards, cantons, communes, etc.

Furthermore, a map often involves several additional components.

This results in the standard arrangement shown in figure 3. It can also serve as a model for those diagrams in which it is useful to group all the elements of identification (provided that the terms affiliated with the planar dimensions are repeated, as in figure 4).

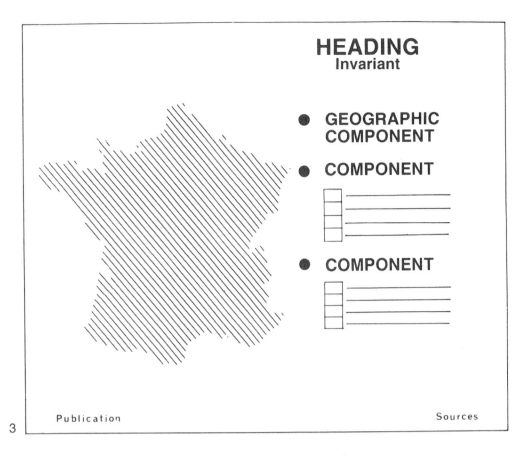

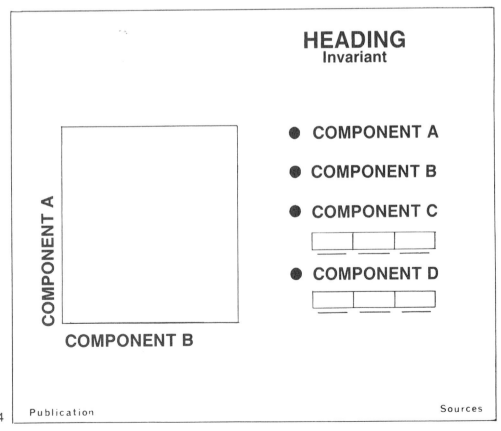

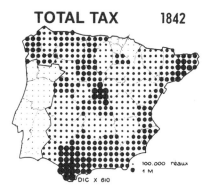
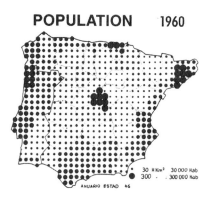

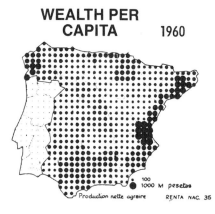

1

IDENTIFICATION OF A HOMOGENEOUS SERIES

In a homogeneous series such as the collection of information on Spain portrayed in figure 1 (see also page 398) or the series of phonic analyses of folk songs in figure 2 (see also page 267), each map or diagram is identified by the invariant. It specifies a particular category from each of the components involved in the entire series of images:
– the category "1848" of the component "time"
– the category "population" in the component "different studies"
or, with the folk songs,
– the category "United States" from the geographic component
– the category "ballad" of the component "type"
– the category "first stanza" of the component "stanza and refrain."

The information itself is constituted by the entire series of maps or the whole collection of diagrams. The components extending throughout the series are constituents of the information, even though they are transcribed only by the writing, not by the images themselves. These components are the basis for the operation of classing and grouping, which constitutes the real objective of such representations. It is therefore important that:
– **The components extending throughout a homogeneous series are always transcribed in the same place.**
– **The categories of these components have maximum visibility and, if possible, are written in bold face and CAPITALS.**

Although these recommendations may appear relatively unimportant, they are nonetheless essential. The investigator who knows how to utilize all the properties of graphic representation and to derive the most from information processing will follow these rules and reserve sufficient space for an organized and efficient identification.

IDENTIFICATION OF SOURCES

From the perspective of generalized documentation, each graphic must be capable of being extracted from its original source and incorporated into any other set of information. However, the reader must be able easily to find the documentary sources and identify the original author (an expert whose field is similar or related to that of the reader).

As a general rule, in or under each image or homogeneous group of images, the source, author, work, publisher, place of publication, and date should be indicated in such a way that these elements can be photographed with the drawing, as in figure 3.

COUNTRY	Region or Language	Type	Stanza or Refrain

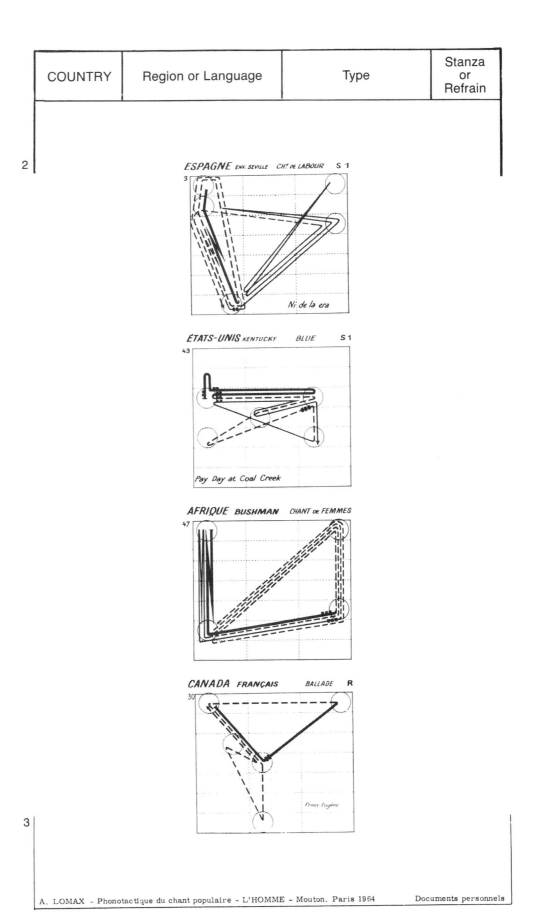

A. LOMAX - Phonotactique du chant populaire - L'HOMME - Mouton, Paris 1964   Documents personnels

B. The number of components

Consequences of this notion
– The number of visual variables necessary for the representation is at least equal to the number of components in the information.
– With three components, the information can be perceived as a single image. Beyond that, the perception of several successive images is often necessary.
– There are at least as many types of possible questions as there are components.
– The number of components is the best basis for a classification of graphic constructions.

Take the following information:
INVARIANT   –*expenditures by the British population per item*
COMPONENTS   –*twenty-eight different items of expenditure*
             –*Q per 100 expenditures per item according to*
             –*three income groups (upper, middle, lower)*

Because the information involves only three components, it is possible to replace figure 1, necessitating the mental addition of numerous images, by figure 2, which presents a memorizable image.

It is a feeling of uniformity, of nonvariation, which results from figure 1, yet the distribution of expenditures could

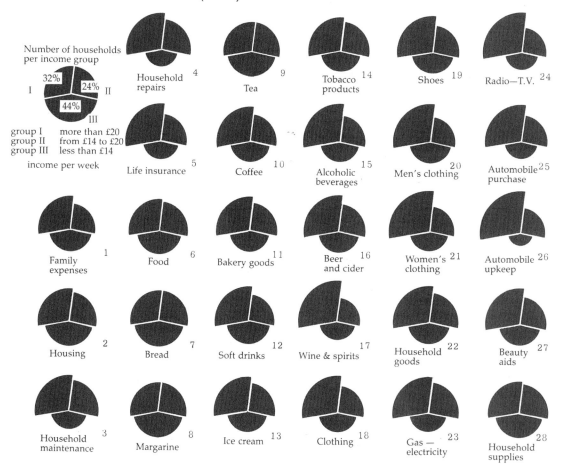

COMPARISON OF EXPENDITURES ACCORDING TO INCOME GROUPS IN THE UNITED KINGDOM (1960)

group I   more than £20
group II from £14 to £20
group III less than £14
income per week

from Harry HENRY. Thomson Organisation Ltd.
Sources: Central Statistical Office. London 1961

1

hardly be similar for all three income groups.

In figure 2 the reader is struck and guided by the visible differences (underscored, incidentally, by the use of black) and is able to concentrate on them. The reader can rapidly perceive the logical order of the image: From left to right are the groups, from top to bottom are the items, whose order constitutes the very purpose of the information. The reader can pose questions about the characteristics of a group or an item, or about the order of these items, and feel confident of obtaining an answer.

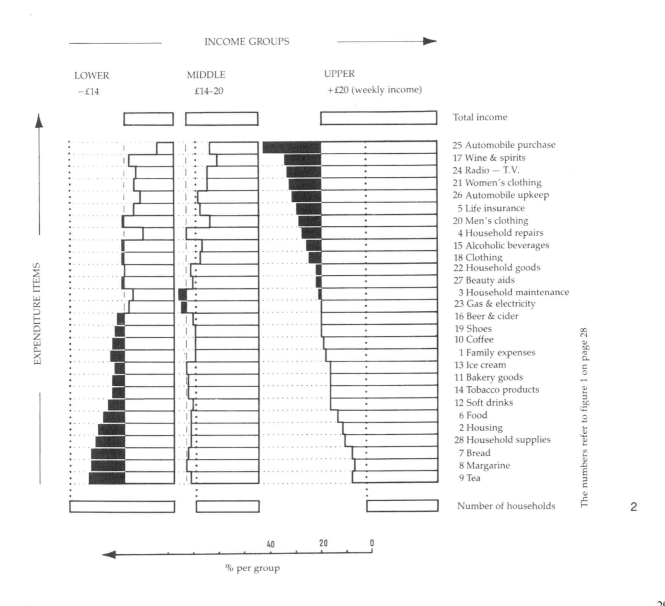

2

TRAFFIC ACCIDENT VICTIMS, in France, in 1958

VEHICLE (or pedestrian)	Pedestrians	Bicycles	Motorcycles	Four-wheeled vehicles
QUANTITIES	28 951	17 247	74 887	63 071

VEHICLE	Pedestrians		Bicycles		Motorcycles		Four-wheeled vehicles	
SEX	M	F	M	F	M	F	M	F
QUANTITIES	16 702	12 249	13 009	4 238	61 609	13 278	39 732	23 339

VEHICLE		Pedestrians		Bicycles		Motorcycles		Four-wheeled vehicles	
SEX		M	F	M	F	M	F	M	F
CONSEQUENCES (dead, injured)	d	1 232	570	701	126	2 664	322	1 817	694
QUANTITIES	i	15 470	11 679	12 308	4 112	58 945	12 956	37 915	22 645

VEHICLE			Pedestrians		Bicycles		Motorcycles		Four-wheeled vehicles	
SEX			M	F	M	F	M	F	M	F
AGE	C									
50	d		704	378	396	56	742	78	513	253
50	i		5 206	5 449	3 863	1 030	8 597	1 387	7 423	5 552
30	d		223	49	146	24	889	98	720	199
30	i		3 778	1 814	3 024	1 118	18 909	3 664	15 086	7 712
20	d		78	24	55	10	660	82	353	107
20	i		1 521	864	1 565	609	18 558	4 010	9 084	4 361
10	d		70	28	76	31	362	54	150	61
10	i		1 827	1 495	3 407	1 218	12 311	3 587	3 543	2 593
	d		150	89	26	5	6	6	70	65
	i		3 341	1 967	378	126	181	131	1 593	1 362

Source: Ministère des Travaux Publics

1

Information can have 2, 3, ... n, components, and n can be quite large. It is sufficient that one of the components or the invariant be common to all the data.
Consider the following example:
Analysis of highway accidents in France.
INVARIANT —*an accident victim*

This example can involve numerous components. As we see in figure 1 each additional component will generate new information. Furthermore, each additional component will also require a new visual variable leading to a different construction, as illustrated in figure 2 on the opposite page.

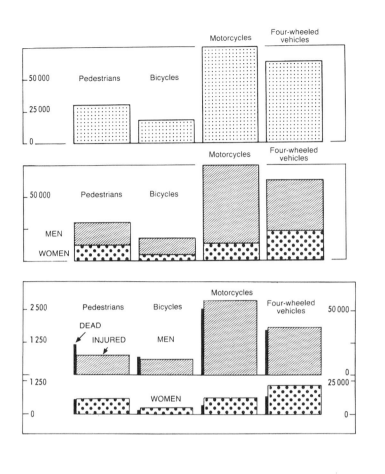
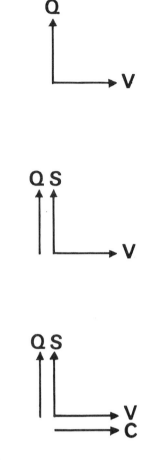
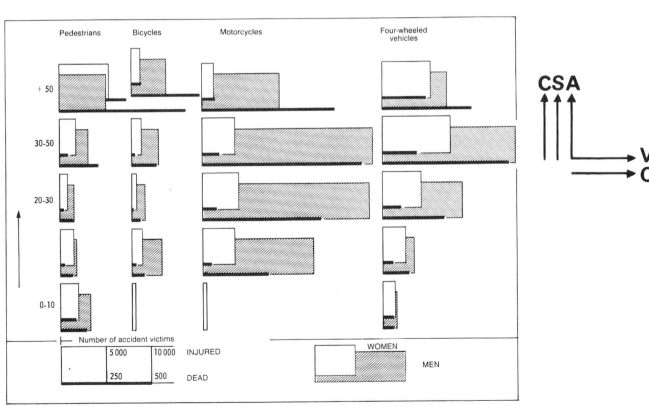

A set of information

Consider the distribution of the work force according to numerous concepts:

INVARIANT —*working persons in France*
COMPONENTS —*quantity according to*
—*geographic categories*
—*time*
—*age*
—*socioprofessional categories*
—*religious categories*
—*political categories*
—*various rates (mortality, fertility, birth, wealth, education ...)*

When components are numerous, we will speak of a *set of information*. It is useful to analyze it as a whole and to consider the finite set of components in order to determine the most efficient and most economical processing system.

Regional studies, which have a defined geographic space as their common component; sociological surveys, whose common component is a group of individuals; historic research, whose common component is a period of time, are finite sets. To reduce them to their minimum constituents different systems of "information processing" are used. Depending upon which system is adopted, the graphic representation will lead to a series of diagrams, double-entry tables (see figure 1, page 30), maps, etc.

Likewise, at the moment of publication, the entire set of drawings must be conceived as a demonstrational unit. The layout must be considered a scientific problem, linked to the imperatives of reading and comparison, before it can be treated as an aesthetic problem (see, for example, page 401).

Information can have a single component

Figure 1: Conversational relationships among several individuals.

INVARIANT —*an exchange of conversation*
COMPONENT —*a series of individuals*

Figure 2: Map of developed areas in Poland, taken from F. Uhorczak (see also page 318).

INVARIANT —*a site occupied by a building, whether a house, a factory, or a barn*
COMPONENT —*geographic space*

Definition of the number of components

In order to define the number of components in the information, the most convenient means consists of transcribing the data into a double-entry table (as in figure 1, page 30), before making any attempt at representing it graphically. There are as many components as entry-categories, plus one for quantities, if applicable.

Care must be taken, since such a table defines and clarifies the information but does not determine the most efficient graphic construction. This can sometimes be quite different from the layout of the table.

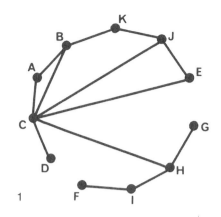

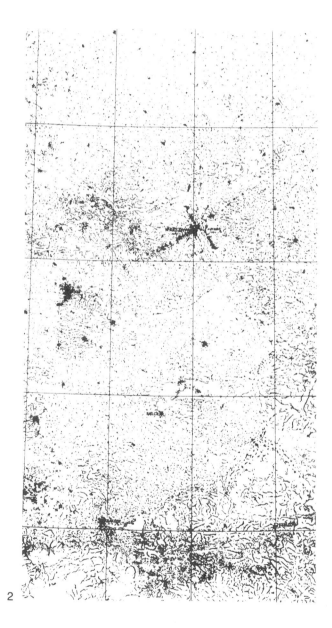

C. The length of the components

Consequences of this notion
- "Long" (extensive) components lead to basic or "standard" constructions.
- "Short" (limited) components lead to "special" constructions.
- The visual variables utilized must have at least the same length as the components which they are meant to represent.
- In a problem involving more than three components, the minimum number of images necessary is a function of the length of the components.

As variational concepts, components are, by definition, divisible. Their divisions bear different names according to circumstance and level.

One may speak of *elements* in the component "different persons forming a genealogical tree" or of *objects* in the component "different objects to be classed."

One may also speak of socioprofessional, geographic, or linguistic *categories*, and, in general, this term can be applied to all components.

One may speak of time, age, or income *classes*, and, in general, this term can be applied to all components which are ordered or quantitative. Finally, one may speak of *steps* of value, texture, or size for the visual variables (the components of the graphic sign-system). Each variable, each combination of variables, has a given length, which is most often quite limited. In fact, all these terms cover the same phenomenon, the useful and separable divisions of a component.

The term LENGTH of a component will be used to refer to the number of divisions that it enables us to identify.

The full significance of this notion can be appreciated when one comes to realize that the number of categories a person can grasp during the course of perception is quite small. This means that the visual variables can produce only a limited number of perceptible steps.

"Short" or limited components
The term *short components* will be used when length does not exceed four. Binary components have a length of two (sex, living or deceased, a decision which is actual or potential, etc.). Age is often divided into three main categories—youth, adult, elderly—as are the main employment sectors (agriculture, industry, tertiary). Short components are noteworthy in graphic problems. They simplify visual selection and enable us to use "special" constructions, which differ from the standard constructions.

"Long" or extensive components
The term *long components* will be used when length exceeds some fifteen divisions. Long components necessarily lead to "standard" constructions (see page 172).

The term *length* can be applied to a quantitative series when the latter is divided into steps or classes, or when it involves a "discrete" component (such as numbers of objects, inventories, or francs, which cannot be divided below a certain basic unit).

On the other hand, a series of numbers can be infinitely divisible; the term *length* is no longer applicable when the phenomenon is considered as continuous (such as speeds, altitudes, or temperature). Note, however, that the number of useful decimals is a finite number.

A phenomenon considered as continuous can nevertheless be expressed by graphic means, since the plane itself is continuous.

THE RANGE OF A QUANTITATIVE SERIES is the ratio of the smallest to the largest number.

This notion is quite different from that of length. A series whose extremes are 0.07 and 32 has the same range as a series whose extremes are 22 and 10 054; they both range from 1 to 457.

The practical range of a visual variation in size is limited; it cannot decrease below a ratio of 1 to 10 without losing the greater part of its efficacy. On the other hand, quantitative information can range from 1 to 1.2 (the size of individuals) as well as from 1 to 10 million (population maps). Thus it is easy to understand the importance of this notion and of the "range adjustments" which must often be introduced into quantitative representations in order to adapt the information to the faculties of visual perception (see page 357).

D. The level of organization of the components

The components of the information do not all involve the same intellectual approach. For purposes of information processing and/or display, the researcher will often attempt to order qualitative (nominal) categories such as trades, to compare ordered categories such as heat sensitivities, and to group neighboring quantitative values such as population densities.

A component can thus be characterized as *qualitative*, *ordered*, or *quantitative*; these are the three levels of organization. The visual variables which represent each component must permit parallel perceptual approaches. But, just like the components, the visual variables each have their own level of organization. An order will not be perceptible if the variable is not ordered; a ratio will not be perceptible if the variable is not quantitative. The notion of level of organization is thus of fundamental importance.

Consequences of this notion

– The ordering of qualitative data, the comparison of ordered components, the groupings resulting from a quantitative component are the basis for the graphic processing of information.
– The visual variables must have a level of organization at least equal to that of the components which they represent.
– The three levels of organization lead to the first subclassification of graphic constructions.

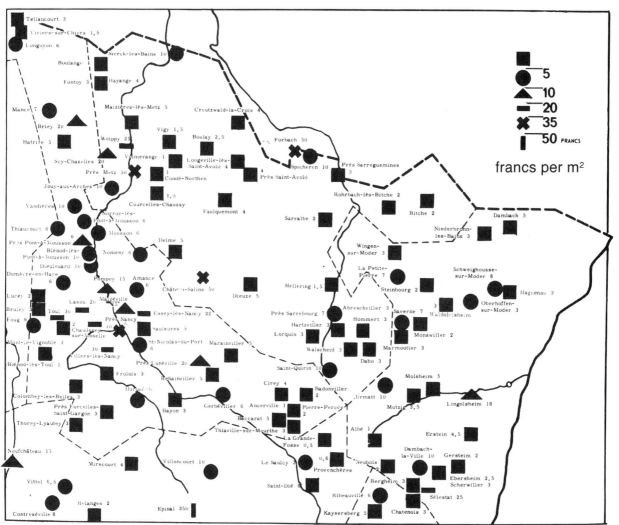

LAND VALUE IN EASTERN FRANCE, from the weekly magazine, "Elle." Paris, 1959.

It is because the level of the visual variable utilized does not correspond to the level of the component represented by it that the map in figure 1 is inefficient and necessitates the burdensome reading of several successive images. When the levels correspond, as in figure 2, the map is visually retainable. It necessitates only one immediately perceived image.

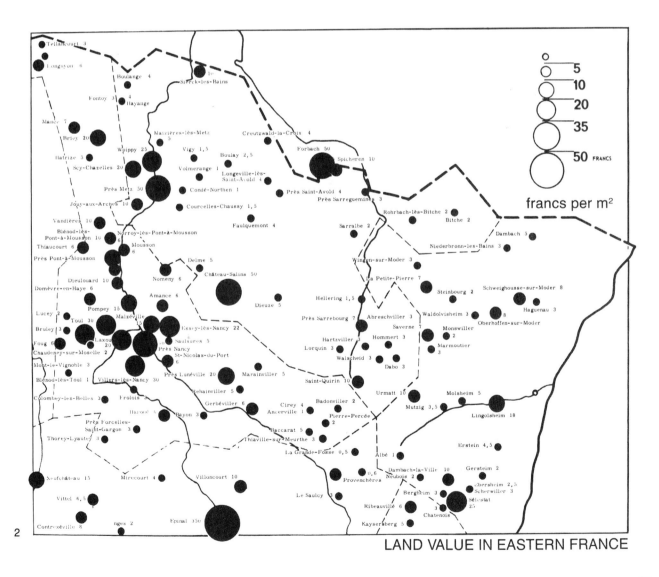

2

LAND VALUE IN EASTERN FRANCE

THE QUALITATIVE LEVEL (OR NOMINAL LEVEL)

This notion includes all the innumerable concepts of simple differentiation: professions, products, languages, races, religions, leisure activities, diseases, colors, forms, social, ethnic, cultural or political traits . . .

A component is qualitative when its categories are not ordered in a universal manner. As a result, they can be reordered arbitrarily, for purposes of information processing.

Qualitative categories are reorderable

In maritime commerce, for example, the categories coal, oil, wheat, wool, cotton, wine, wood, . . . of the component "merchandise" can be ordered in different ways, according to weight, total value, price per kilogram, volume, revenue production, fragility. . . .

Geographic groupings are also reorderable. Departments are commonly classed by alphabetic order, countries by their population, their production, their standard of living, their birth rate. . . .

The reciprocal ordering of two qualitative components (figure 1) or of a qualitative component in relation to an ordered component (figure 2) simplifies the images without diminishing the number of observed correspondences; indeed these operations of "permuting" and classing are the basis for graphic information-processing.

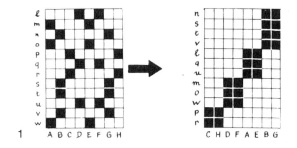

Qualitative categories are equidistant

As with ordered categories (discussed later), qualitative categories are, by definition, of equal importance, that is, "equidistant." Their graphic representation must not disturb this quality by highlighting a particular category or creating a priori groups of categories.

Two perceptual approaches

Faced with any qualitative concept, the observer can adopt two perceptual approaches:

This is different from that—a beech tree is different from an oak.

This is similar to that—beeches and oaks are similar—they are trees.

A *selective approach (difference)* is engendered by questions of an elementary or intermediate reading level. Where is a given category—the beech trees? When these questions are pertinent, it is important that the component be represented by a selective variable.

An *associative approach (similarity)* is engendered by questions of an intermediate or overall reading level. Where is a given component—the forest—all categories of trees combined? In order to reply to this question, the variable must permit equalizing and grouping all the categories during perception; it must be associative.

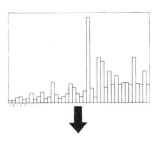
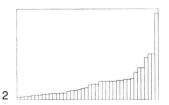

THE ORDERED LEVEL
(ORDERING AND REORDERING)

This level groups all the concepts which are capable of ordering categories in a universally acknowledged manner. Each person will agree in the same way that this is more than that and less than the other.

Ordered concepts are always defined, more or less directly, in relation to:
– a temporal order: age, generation, matrimonial status, geologic era
– an order of sensory discrimination: heat, vision (black–gray–white, large–medium–small, here–near–far), weight (heavy–medium–light), health
– an order of intellectual or moral discrimination (good–mediocre–bad)
– certain social structures, such as military or administrative hierarchies.

A component is ordered, and only ordered:
– **when its categories are ordered in a single and universal manner**
– **when its categories are defined as equidistant.**

Ordered categories cannot be reordered

More precisely we can say that the reclassification of ordered categories is generally a source of confusion in the process of communication.

Consider the following example:

Propensity to theft according to age and amount of theft (based on V. V. Stanciu, Theft in Department Stores, unpublished study).

INVARIANT —*theft in department stores*
COMPONENTS —*age groups*
—*quantities (per 100 persons per age group) according to*
—*classes of amount of theft*

The components "age" and "amount of theft" are ordered, producing figure 3: In black is the highest percentage for each column, that is, the age group with the strongest observed tendency.

It can be of interest to reorder the component "amount of theft" for the purposes of constructing a linear relationship (figure 4), which permits one to reflect on theft psychology. Interpretation is more delicate, however, because reading from left to right no longer has an ordered meaning.

These examples demonstrate that the amount of theft is not ordered in a direct way by age, nor age by amount of theft.

Ordered categories are defined as equidistant

This characteristic distinguishes an ordered component from a quantitative component. The series: bachelor, husband, widower, deceased, constitutes a universal order. But there is no reason, a priori, to bring together any two categories or to form groups. These categories are ordered and at equal distances from each other.

The same is true for the component "amount of theft," even though its categories are defined by *numbers*. These are *ordinal numbers* which merely serve to rank the categories.

In any graphic transcription involving an ordered component, particularly when using "retinal" visual variables, the designer must try to preserve this equidistance. A priori visual groups must be avoided, since the very purpose of graphic processing is to discover, a posteriori, the groupings which result from the information.

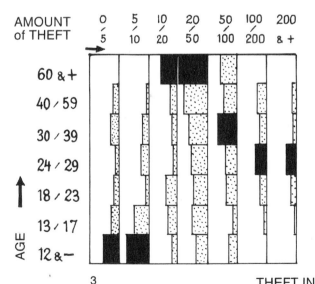
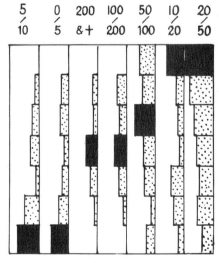

3 THEFT IN DEPARTMENT STORES 4
Distribution of delinquents according to age and amount of theft

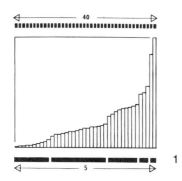 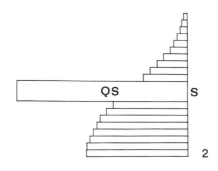 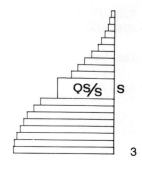

THE QUANTITATIVE LEVEL (INTERVAL-RATIO LEVEL)

This level is attained when there are countable units, leading us to say: this is double, half, four times that. . . .

A series of numbers is quantitative when its object is to specify the variation in distance among the categories.

With a quantitative series of numbers, Q, it is possible to represent a variation in the length of columns, as in figure 1, and from this to derive groups, characterized by slight differences in length (slight "distances").

Relations among quantities and enumeration units

Before representing given quantities, any graphic must first depict the units (geographical areas, time periods, age groups, etc.) within which these quantities are being enumerated. A population map by commune is, first of all, a map of communes.

When the enumeration units are unequal:
– the representation of these units on the plane can result in figures which are equal (points on a diagram) or unequal (a map of communes);
– the representation of the quantities can be independent of the inequality of these figures (a single point per area) or dependent on it (color over the entire area).

But the quantities themselves can be independent of the inequality of the enumeration units (death rate in a commune) or dependent on it (total population). Therefore, graphic representation necessarily leads to an initial analysis of any quantitative series in terms of these relations.

Quantities dependent on the enumeration units (or QS)

In comparing communes, the geographer must calculate population density in order to take the unequal areas of the communes into account. This calculation is necessary because the quantities of population *are not independent of the unequal areas* (S)* of the communes. The same is true for the historian who uses quantities of immigrants enumerated over unequal periods (S) of time or for the demographer who uses quantities of persons counted within unequal age groups (S). These quantities are not independent of the dimension (S) of the enumeration units.

*S will be used frequently to signify area, since it comes from the French equivalent *surface* (translator's note).

Quantities of the form QS are:

absolute quantities (Q) counted according to variable units (S), whether these quantities are expressed:

by the observed numbers:
Q of tons of milk per department (S), Q of persons per period (S) . . .

in hundredths (or in thousandths of the total):†
Q of milk per department (S), Q of persons per age-group, expressed in hundredths (or in thousandths) of the total of the series, that is, QS × 100/total of the series.

Test: For a hundred what? For a total of the series equal to a hundred;‡

by an index:†
Q of milk consumed per period (S), per 100 liters consumed in 1950 (Qi), Q of milk produced per department (S), per 100 liters produced in the department of Calvados (Qi), that is, QS × 100/Qi.

Test: For a hundred what? For a hundred liters produced in Calvados;‡

by a ratio based on a variable independent of S:
Q is monthly average amount of milk produced per department (S), or Q is communal average of expenditures per period (S), that is, QS total/number of units (months, communes). When S is represented by lines or by areas, the graphic transcription of QS can lead to serious errors (page 45). It is generally necessary to transform the data by making the calculation QS/S, as illustrated in figures 2 and 3.

Quantities independent of the enumeration units (or Q)

The geographer looking for quantities of population independent of the area of a commune must calculate densities: Q of population/area, that is, QS/S = Q.

The demographer will reduce a class variation in the same manner: Q of persons/length (in years) of the class,

†Hundredths and indices merely involve a simple change of numerical scale, useful in verbal communication to the degree that all the numbers become intelligible when it is understood what one hundred or one thousand represents. Graphically, absolute Q, hundredths, or indices produce the same image of a series.

‡The test, for a hundred what?, is indispensable. It permits the elimination of false percentages and the comprehension of what is in question. It obliges one to furnish the elements of an answer and it reveals, all too frequently, a series for which an answer is impossible, so that clarification or elimination of the series is required.

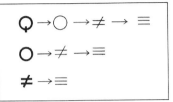

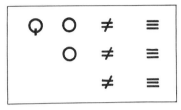

that is, QS/S = Q. But quantities Q are not all of this nature.

Quantities of the form Q are:
samples, altitude, temperature, commodity prices, number of workers per factory, etc. These are measurements or real values sampled at a point which is by definition without length or area. They thereby characterize an invariable enumeration unit;
reductions to a unitary class, densities, "absolute" frequencies, such as the examples cited earlier, which result from the operation QS/S = Q;
simple ratios, in which the variable unit (S) relates the two terms of the ratio:
Q of wheat produced per commune (S)/Q of hectares sowed per commune (S),
Q of emigrants per period (S)/Q of boats per period (S), that is, $QaS/QbS = Qa/Qb = Q'$;
"percentages" and *"rates"* which multiply the simple ratios by a hundred (or a thousand):
Q of deaths per commune (S) × 1000/Q of persons per commune (S),
Q of working persons per age group (S) × 100/Q of persons per age group (S), that is, $QaS \times 100/QbS = Qa/Qb \times 100 = Q'$ per hundred.
 Test: For a hundred what? For a total of a hundred persons per commune.‡
The graphic transcription of absolute quantities (Q) is simpler. However, one must know how to avoid a confusion with graphic solutions suitable only to QS (figure 16, page 45).

INCLUSIVENESS OF THE LEVELS OF ORGANIZATION

Graphic conventions
In order to designate a component and, at the same time, specify its level of organization, the following signs will be used:
Q—a quantitative series measuring variations in distance among ordered categories.
O—a component whose categories are equidistant and inscribed in a single, universally acknowledged order.
≠—a qualitative component whose categories are defined and equidistant.
≡—a qualitative component whose differential characteristics can be disregarded (i.e., which can be approached "all categories combined").

Inclusiveness of perceptual approaches
The level of organization determines the perceptual approaches that can be adopted toward a component. These approaches are ordered and inclusive. In effect, *for a quantitative component*, it is possible to adopt:
a quantitative perceptual approach and ask the question: What is the ratio between the two lengths, between the two populations, between the two areas...?
an ordered perceptual approach and ask the questions: In what order are the lengths given? Does the order of the departmental population quantities correspond to the alphabetic order of the departments...?
a selective perceptual approach and ask the question: Where are all the cities of 15 000 inhabitants?
an associative perceptual approach and ask the questions: What is the distribution of the "cities," disregarding any differentiation among them? Where is the forest, disregarding any differentiation of age, size, or kind of trees?

Thus:
– All quantitative series can be considered as merely ordered.
– All the categories of an ordered series can be considered as merely differentiated.
– All the categories of a qualitative series can be considered as similar.

But:
– A series which is only qualitative is not ordered (although we can reorder it arbitrarily, page 36).
– A series which is only ordered is not quantitative (although it may be defined by ordinal numbers, page 37).

The system of inclusion resulting from these statements is expressed in figure 4, a more readable version of which is given in figure 5. This enables us to identify the perceptual approaches which each component can generate, to choose a visual representation of at least an equal level, and to classify the visual variables among themselves.

II. The Properties of the graphic system

In order to utilize graphic representation, we must consider the scope of the system: that is, the visual variables which are available, their lengths, and their levels of organization.

A. The scope of the graphic system
B. The plane
C. The retinal variables

A. The scope of the graphic system

ITS LIMITS

A sign-system cannot be analyzed without a strict demarcation of its limits. This study does not include all types of visual perception, and real movement is specifically excluded from it. An incursion into cinematographic expression very quickly reveals that most of its laws are substantially different from the laws of atemporal drawing. Although movement introduces only one additional variable, it is an overwhelming one; it so dominates perception that it severely limits the attention which can be given to the meaning of the other variables. Furthermore, it is almost certain that real time is not quantitative; it is "elastic." The temporal unit seems to lengthen during immobility and contract during activity, though we are not yet able to determine all the factors of this variation.

Actual relief representation (the physical third dimension) has no place here either and will be referred to only for purposes of comparison.

In this study, we will consider only that which is:
– representable or printable
– on a sheet of white paper
– of a standard size, visible at a "glance"
– at a distance of vision corresponding to the reading of a book or an atlas
– under normal and constant lighting (but taking into account, when applicable, the difference between daylight and artificial light)
– utilizing readily available graphic means.

Consequently, we will exclude:
– variations of distance and illumination
– actual relief (thicknesses, anaglyphs, stereoscopics)
– actual movement (flickering of the image, animated drawings, film).

Within these limits, what is at the designer's disposal? MARKS!

In order to be visible a mark must have a power to reflect light which is different from that of the paper. The larger the mark, the less pronounced the difference need be. A black mark of minimum visibility and discriminability must have a diameter of 2/10 mm. But this is not absolute, since a constellation of smaller marks is perfectly visible.

THE VISUAL VARIABLES

A visible mark can vary in position on a sheet of paper. In figure 1 on the opposite page, for example, the black rectangle is at the *bottom* and toward the *right* of the white square. It could just as well be at the bottom and toward the left, or at the top and toward the right.

A mark can thus express a correspondence between the two series constituted by the
TWO PLANAR DIMENSIONS
Fixed at a given point on the plane, the mark, provided it has a certain dimension, can be drawn in different modes. It can vary in
SIZE
VALUE
TEXTURE
COLOR
ORIENTATION
SHAPE
and can also express a correspondence between its planar position and its position in the series constituting each variable.

The designer thus has eight variables to work with. They are the components of the graphic system and will be called the *"visual variables."* They form the world of images. With them the designer suggests perspective, the painter reality, the graphic draftsman ordered relationships, and the cartographer space.

This analysis of a temporal visual perception in eight factors does not exclude other approaches. But, combined with the notion of "implantation," it has the advantage of being more systematic, while remaining applicable to the practical problems encountered in graphic construction.

These variables have different properties and different capacities for portraying given types of information. As with all components, each variable is characterized by its level of organization and its length. We will first study the properties of the PLANE, then those of the RETINAL VARIABLES which can be "elevated" above the plane.

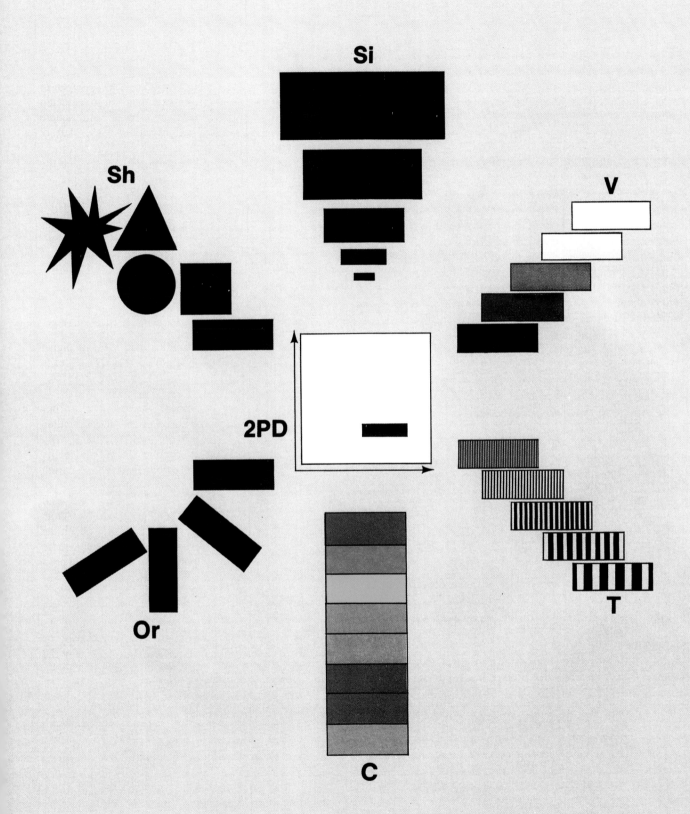

B. The plane

The plane is the mainstay of all graphic representation. It is so familiar that its properties seem self-evident, but the most familiar things are often the most poorly understood. The plane is homogeneous and has two dimensions. The visual consequences of these properties must be fully explored.

(1) Implantation (classes of representation)

The three types of signification—point, line, and area—which can be assigned to a mark on the plane will be termed "IMPLANTATIONS."* **They constitute the three elementary figures of plane geometry.**

Along a line, one can consider a point or a line segment. *On the plane,* one can consider a point, a line, or an area. Failure to grasp the various ramifications of this fundamental notion is a frequent source of error in graphics. Confusion stems from the fact that points and lines have no theoretical area, yet the marks representing them require a certain amount of "area" to be visible.

Consequences of distinguishing classes of representation
– The length (number of available steps) of the retinal variables and their use vary with the class of representation involved.
– The representation of quantities varies according to whether a point, a line, or an area is utilized.
– Differences in classes of representation are selective.
– In a single image, the same concept cannot be represented by different "implantations."

THE POINT

A straight line on a sheet of paper has a certain length which can be measured. But, at the moment of measuring, its extremities are considered not to have length on the line. These are POINTS. However, they do have a position on the line.

A point 51 mm from the horizontal edge of the paper and 34.5 mm from the vertical edge has a position on the plane. Whether it is made visible by a "pin prick" 1/10 mm in diameter, or by a "preprinted circle" 5 mm in diameter, its center has a precise position, but the mark itself is not meant to signify either length or area on the plane.

A POINT represents a location on the plane that has no theoretical length or area. This signification is independent of the size and character of the mark which renders it visible.

Consequently, a point can vary in position, but will never signify a line or area on the plane of the image. *By way of contrast, the mark which renders it visible can vary in size, value, texture, color, orientation, and shape,* but it cannot vary in position. Positional meaning naturally applies to the visual center of the mark. Any other usage must he made explicit.

Numerous examples can serve to illustrate this idea: geodetic or confluent points, a crossroads, the "corner" of a forest, the position of an airplane, or a transmitter are points in the planar space, without theoretical length or area. Their graphic representation nonetheless requires the presence of marks having sufficient size to render them visible. Represented cartographically, these phenomena are said to have a **point representation**.

THE LINE

Parallel reasoning permits describing a line as essentially the boundary between two areas. It has a length and a position on the plane, but has no theoretical area.

A LINE signifies a phenomenon on the plane which has measurable length but no area. This signification is independent of the width and characteristics of the mark which renders it visible.

Consequently, a line can vary in position but will never signify an area on the plane of the image. *However, the mark which renders it visible can vary according to all the variables other than those involving position on the plane: in width, value, texture, color, orientation of its constituents, and shape of detail.* Positional meaning naturally applies to the linear axis of the mark. The boundary of a continent, a nation, or a property, a ship's course, or a bus route, are linear phenomena without theoretical area. In cartography, they will be **represented by lines**.

THE AREA

A mark can, however, signify an area on the plane.

AN AREA signifies something on the plane that has a measurable size. This signification applies to the entire area covered by the visible mark.

An area can vary in position, but *the mark representing it cannot vary in size, shape, nor orientation* without causing the area itself to vary in meaning. *However, the mark can vary in value, texture, and color.*

If the area is visually represented by a constellation of points or lines, these constituent points and lines can vary in size, shape, or orientation without causing the area to vary in meaning. In cartography, phenomena such as lakes, islands, land, urban areas, and countries will be **represented by areas**.

---
*See translator's note, page 7.

ANALYSIS OF THE QUANTITIES TO BE REPRESENTED

When enumeration units have variable dimensions, the representation of the quantities associated with these units must take into account:
(1) the point, line, or area representation of the units;
(2) the nature Q or QS of the quantities to be represented (see page 38).

Take the following information, concerning four communes (units) A B C D:

Units (communes)	A B C D
Areas (S)	4 4 1 1 (tens of km²)
Quantity of pop. (QS)	4 8 2 4 (thousands of persons)
Density of pop. (Q)	1 2 2 4 (%)

In figure 1 the communes have a **point representation**. They are points in a scatter plot (distribution of the communes according to the percentage of agricultural [I] and industrial [II] population).

For each point a third factor can be added, either as quantities QS (figure 2), or quantities Q (figure 3) which will be perceived correctly.

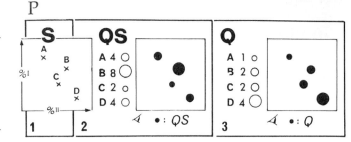

In figure 4 the communes are represented by **vertical line segments**, whose lengths are proportional to S. If one constructs the quantities QS on the other dimension of the plane (figure 5), the eye perceives the QS horizontally, but it especially sees the constructed area, that is, QS^2, which it interprets as being the population QS. The area and the general outline are erroneous. One must therefore construct QS/S (i.e., Q) along the horizontal axis, as in figure 6, and this gives an exact image of the quantity QS, in area, and an exact image of the density Q, horizontally.

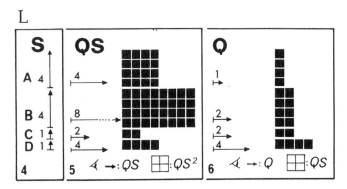

In figure 7 the communes are represented by **areas** proportional to S. The QS and the Q are distributed as in figures 9 and 10. The most simple representation (one point per 1000 inhabitants) produces figure 8, which is correct.

One can easily judge the visual confusion engendered by figures 11 and 13, which extend the value QS to the entire area.

The eye sees there, as in figure 5, QS multiplied by the area, that is, QS^2 (see also page 77, figures 5 and 6). The representation of the QS can, however, be useful (for example, in measuring the responsibility of the different mayors). In this case, figure 15, that is, a point QS per area, avoids the preceding visual confusion.

In contrast, constructing a point Q per area leads to an erroneous representation (figure 16), whereas figures 12 and 14 are correct.

It is interesting to note that the perceptual error seen in figure 5 is well known to statisticians and is nearly always avoided, whereas the erroneous perceptions produced by figures 11 and 13, where the error is expressed in a similar mathematical manner (perception of QS^2), are still found. Control of perception "above" the plane is less obvious than control of perception on the plane. It is no less important, since it is of concern to all cartography.

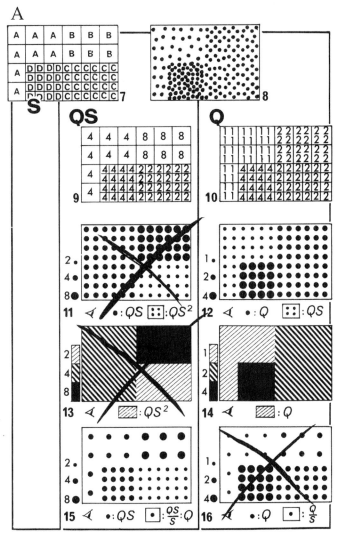

(2) The plane is continuous and homogeneous

The plane is capable of infinite subdivision. It is continuous. Its divisibility is limited only by the thresholds of perception and the limits of graphic differentiation. A line one centimeter in length easily supports ten identifiable divisions. Next to shape variation, *the plane offers the longest visual variable*. Therefore, it is to the plane that one will usually entrust the representation of the longest components.

Since it has no breaks in continuity, no "gaps," the plane will not admit informational *lacunae*. As a result, it is very difficult to evaluate fragmentary information within the plane. Even though it calls attention to missing data, the map in figure 1 gives an ambiguous impression of the distribution.
*It is very difficult to disregard a part of the signifying plane.**

The certainty of the uniformity of the plane entails a presumption of uniformity in the conventions adopted within the signifying space.

Consequently, **in a signifying space absence of signs signifies absence of phenomena**. Within a visible frame or limit, the space signifies something at every point. Any absence of signal signifies absence of phenomenon. This is the impression created by Figure 2. Therefore information must be applied to the entire area in such a way that the empty spaces signify *absence* of phenomena and not *missing* data.

In a signifying space any visual variable appears as meaningful; the introduction, for example, of a color variation whose only aim is aesthetic or decorative will lead to confusion if the color differences do not correspond to a component. This same property precludes spontaneity in the perception of logarithmic diagrams. Visible differences in length can only be disregarded after considerable training in perception.

In figure 3, for example, one must learn that only the differences in the slopes of the curves are meaningful.

Likewise, different lengths cannot immediately signify equal lengths. However, such a change in convention is sometimes necessary, particularly with networks (page 275), and the reader must be rigorously informed of it.

---

*By this term Bertin means that part of the figure which is intended to convey meaning. This excludes, for example, the space lying outside a geographic border or the frame of the drawing (translator's note).

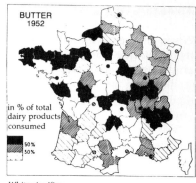

1 White signifies missing data

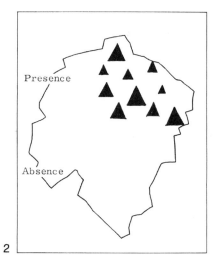

2

3

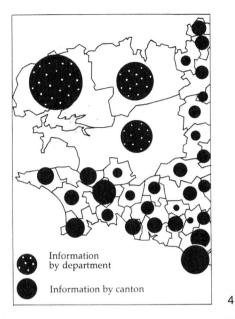

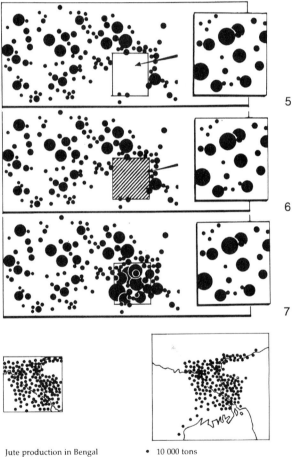

Jute production in Bengal   • 10 000 tons

8   9

In a signifying space a convention is invariable, and any change in the convention is naturally interpreted as a meaningful transformation in the structure of the distribution. Using a sign per department in one part of the image (figure 4), a sign per canton in another, will be interpreted as a meaningful change. The reader can correct this interpretation only by considering the two parts of the figure "separately."

The designer using an "inset" should be aware of these facts. In figure 5, however, the absence of signs immediately signifies an absence of phenomena, and in figure 6 the change in signs signifies a change in phenomena. Even if one understands that the sign refers to the inset, it is still impossible, in figures 5 and 6, to grasp the pertinent information, which only figure 7 enables us to comprehend and retain. Figure 7 creates the only homogeneous image, and the inset furnishes a clarification of it.

An "inset" is a supplementary image, which can never replace the main image (see, for example, page 188).

The frame of a graphic delimits the signifying space, but it does not necessarily delimit the phenomenon. A naturally circumscribed phenomenon, for example, the area of jute production in Bengal (figure 8), only appears spatially concentrated when it is circumscribed by a margin where the "absence" of the phenomenon is visible (figure 9). If the frame is too near, there is the presumption of an extension of the phenomenon beyond the frame (figure 8), and the demonstration of the spatial concentration of the phenomenon is not accomplished.

(3) The level of organization of the plane

THE LEVEL OF A VARIABLE

The perceptual properties of a variable determine its level.

Any individual will immediately class a series of values, ranging from black to white as in figure 1, in a constant order: A, B, C, D, or D, C, B, A, but never in another order. Value is ordered. But each individual will class a series of shapes, such as those in figure 2, differently: A, B, C, D, E, or B, A, D, C, E, or C, D, B, A, E. No visual classing asserts itself a priori. Shape is not ordered.

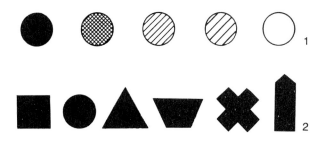

A value variation is thus capable of representing an ordered component: that is, of providing an easy visual response for any question implying an ordered perceptual approach. A shape variation, on the other hand, cannot represent an ordered component. If one adopts this variation, a question involving an ordered component will have no immediate visual response (see page 34).

The level of each visual variable can be defined as follows (see also page 65):

A variable is SELECTIVE (≠) when it enables us to immediately isolate all the correspondences belonging to the same category (of this variable).

These correspondences form "a family": the family of red signs, that of green signs; the family of light signs, that of dark signs; the family of signs on the right, that of signs on the left of the plane.

A variable is ASSOCIATIVE (≡) when it permits the immediate grouping of all the correspondences differentiated by this variable.

These correspondences are perceived "all categories combined." Squares, triangles, and circles which are black and of the same size can be seen as similar signs. "Shape" is associative. White, gray or black circles of the same size will not be seen as similar. "Value" is not associative. A non-associative variable will be termed dissociative (≠).

A variable is ORDERED (O) when the visual classing of its categories, of its steps, is immediate and universal.

A gray is perceived as intermediate between a white and a black, a medium size as intermediate between a small and a large size; the same is not true for, say, a blue, a green, and a red, which, at equal value, do not immediately produce an order.

A variable is QUANTITATIVE (Q) when the visual distance between two categories of an ordered component can be immediately expressed by a numerical ratio.

One length is perceived as equal to three times another length; one area is a fourth that of another area. Note that quantitative visual perception does not have the precision of numerical measurement (if it did, numbers would, no doubt, not have been invented). However, faced with two lengths in an approximate ratio of 1 to 4, unaided immediate visual perception permits us to affirm that the ratio being signified is neither 1/2 nor 1/10. Quantitative perception is based on the presence of a unit which can be compared to all the categories in the variable. Since white cannot provide a measuring unit for gray or black, quantitative relationships cannot be translated by a value variation. Value can only translate an order.

THE LEVEL OF ORGANIZATION OF THE PLANE

Among the visual variables, the plane provides the only variables possessing all four perceptual properties. The two planar dimensions have the highest level of organization and can thus represent any component of the information.

A variation in planar position is selective (≠)
Two similar marks, differing only in position on the plane, can be seen as different (figure 3), and we can immediately isolate all the correspondences, all the marks belonging to a given part of the plane. The best visual selection is obtained by the construction of separate images, juxtaposed on the plane (see, for example, figure 2, page 67).

A variation in position is associative (≡)
Two similar marks, differing in position, can also be seen as similar (figure 3), and, as a result, it is possible to perceive a whole group of points, lines, or areas, all positional characteristics combined.

A variation in position is ordered (O)
Marks A, B, C, when aligned as in figure 4, are ordered along the line, and this order will be universally perceived in the same manner: A, B, C, or C, B, A, but never B, C, A. Two examples of this are shown in figure 4. Thus for three aligned points, one is between the two others, and for three converging straight lines, one is between the two others.

This order can have a direction. Thus a point runs along a straight line in one or the other direction (figure 5), and a straight line rotates around a point in one or the other direction (figure 5).

Consequently, the plane permits representing an ordered collection, a ranking, or, in fact, any ordered component.

A difference in position is quantitative (Q)
This involves the interval and ratio properties of the plane. Anyone can evaluate the relationships displayed in figure 6 with a certain degree of accuracy:

$A > C > B$ $A = 2C$ $B = C/2$

The plane permits us to define equal segments or angles (superimposed) as well as to add segments (end to end; see figure 7), or angles (adjacent; see figure 8). This addition has all the properties of the addition of positive or negative numbers, once an orientation has been defined. As a result, the plane enables us to perceive ratios of length (figure 9), angle (figure 8), or area (figure 13); to measure (figure 10) or add (figure 13) areas; and to represent variable distances among categories, when they are represented by lengths (figure 11), angles (figure 12), or areas (figure 13).

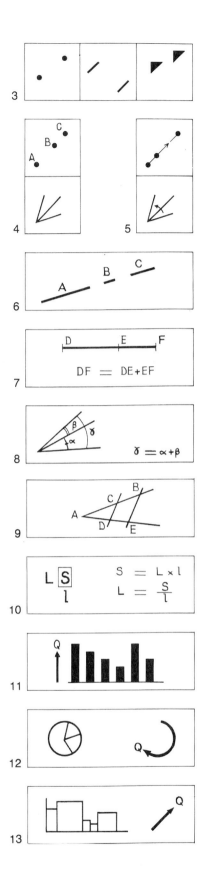

(4) Imposition (groups of representation)

The utilization of the two planar dimensions will be called "imposition." It depends primarily on the nature of the correspondences expressed on the plane, which enables us to divide graphic representation into four groups: diagrams, networks, maps, and symbols.

FIRST GROUP: DIAGRAMS

When the correspondences on the plane can be established between:
- all the divisions of one component
- and all the divisions of another component, the construction is a DIAGRAM.

Consider the information: trend of stock X on the Paris exchange. As shown in figure 1, the designer must first ensure that any date (component, time) can be correlated with any price (component, quantities). After that he or she will record the observed correspondences constituting the given information (figure 2). But the designer need not ensure a correspondence between two dates nor between two prices.

The process of constructing a diagram is as follows:
(1) defining a representation for the components;
(2) recording the correspondences.

SECOND GROUP: NETWORKS

When the correspondences on the plane can be established:
- among all the divisions of the same component, the construction is a NETWORK.

Consider the information: verbal relationships among different individuals A, B, C, D ... of a group.
INVARIANT –a verbal exchange between two individuals
COMPONENT –different individuals A, B, C, D ...

The designer must first ensure that any individual (of the component "different individuals") is capable of conversing with any other individual (of the same component). This is accomplished in figure 3. After that he or she will record the observed correspondences constituting the given information (figure 4). In the present case, the designer can then try to simplify the image by ordering the elements in such a way as to obtain the fewest possible intersections (figure 5).

The process of constructing a *network* is the opposite of that outlined for a diagram:
(1) recording the correspondences in an initial manner;
(2) deducing from them the representation of the component which will produce the simplest structure (the fewest intersections).

To construct a *diagram* from the above information, it would be necessary to add a component. One could consider, for example, that the correspondences are between one series of people who speak and another series of people who listen, the two series being composed of the same elements. This can be represented as in figures 6 and 7.

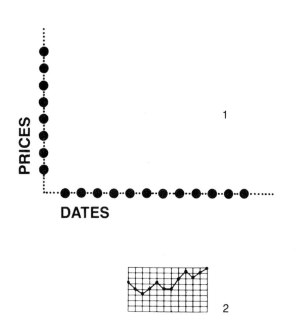

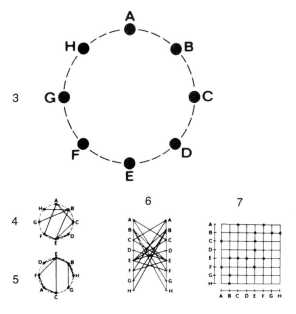

THIRD GROUP: MAPS

When the correspondences on the plane can be established:
– among all the divisions of the same component
– arranged according to a geographic order, the network traces out a GEOGRAPHIC MAP.

A geographic inventory of highways, for example, is constituted by the set of correspondences established among the elements of a geographic series, usually a series of towns arranged in a geographic order (figure 8).

Since a geographic network cannot be reordered arbitrarily, the image can only be simplified by eliminating certain correspondences.

The process of constructing a map is the simplest of all:
(1) reproducing the geographic order;
(2) recording the given correspondences.

It excludes any problem of choice between the two planar dimensions.

But a series of towns can obviously be arranged according to a reorderable *network*, a circular one, for example. After appropriate simplification, as in figure 9, the network provides another way of highlighting nodes and clusters, while displaying the function of each element. A series of towns can also be constructed in the form of a *diagram*, provided the series is represented twice; this permits orienting the correspondences and indicating, as in figure 10, that one can go from C to D, or B to A, for example, but not from D to C, nor from A to B.

FOURTH GROUP: SYMBOLS

When the correspondence is not established on the plane, but between a single element of the plane and the reader, the correspondence is exterior to the graphic. It is a problem involving SYMBOLISM, which is generally based upon figurative analogies of shape or color.

These are merely the result of acquired habits and can never claim to be universal (unlike fundamental analogies of differentiation, resemblance, order, or quantity).

Such is the case for road or railway signs . . . , conventional codes utilized in topography, agriculture, geology, or industry . . . , codes involving shape or color (safety signs, military symbols . . .). They are meaningful only if one recognizes a previously seen shape (figure 11) or has learned the signification of a conventional shape (figure 12).

Diagrams, networks and maps permit us to reduce information to its essential elements, by *internal processing*; whereas symbolism, like language, seeks only to resolve the problem of *external identification*, through immediate recognition.

Generally speaking, any construction within the graphic sign-system, whatever the group to which it belongs, will be termed a *"representation"* or a *"graphic."*

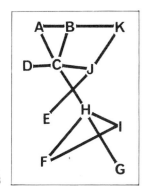
8

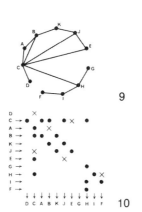
9
10

11
12

IMPOSITION	TYPES OF IMPOSITION				
GROUPS OF IMPOSITION	ARRANGEMENT	RECTILINEAR	CIRCULAR	ORTHOGONAL	POLAR
DIAGRAMS		⇒⇒ / ▭▭▭	↺ / ⊕	⌐→ / ▬▬	↺↑ / ◣
NETWORKS	S / ⧖	→ / ⌒⌒	↺ / ✦	⌐→ / ∷∷∷	
MAPS	GEO ⌐↑→ / 🇫🇷				
SYMBOLS	⚠				

1

GROUPS OF IMPOSITION AND TYPES OF IMPOSITION

With diagrams and networks, imposition is varied; the plane can be utilized in many different ways. The components can be inscribed:
- according to an ARRANGEMENT dispersed over the entire plane
- or according to a construction which is
- RECTILINEAR
- CIRCULAR
- ORTHOGONAL (rectilinear)
- POLAR (circular and orthogonal)

These will be called *types of imposition*. Our notion of IMPOSITION thus involves a first stage, the division of graphic representation into four GROUPS, and a second stage, the division of diagrams and networks into TYPES OF IMPOSITION (this is all shown in figure 1).

The use of retinal variables, either to represent a third component or to replace one of the planar dimensions, produces "ELEVATIONS," which can he combined with all the types of imposition in order to form TYPES OF CONSTRUCTION.

Note the wide variety of constructions possible with a diagram or a network; this poses a problem of choice of construction which does not occur in cartography.

The principal types of construction are expressed in figure 1 by SCHEMAS OF CONSTRUCTION, which will be developed later (see pages 172 and 270) to form a system of conventions capable of defining or analyzing any graphic construction.

PRINCIPAL TYPES OF CONSTRUCTION

Consider the following information:
Distribution of traffic accident victims according to type of vehicle:

INVARIANT — *victim of a traffic accident in France in 1958*
COMPONENTS — *Q of persons according to*
   ≠ four categories (pedestrians, bicyclists, motorcycles, four-wheeled vehicles

The data are as follows:
pedestrians   28 951   motorcycles   74 887
bicycles   17 247   four-wheeled vehicles 63 071

All the representations in the opposite margin portray this information.

The figures differ in the size of the lines, the arrangement of the text, the form of the letters, the precision of the drawing, its geometric or figurative style, the amount of black, and the shape of the whole. They could be further differentiated by the size of the figure, by the use of shading or value, by their colors, etc.

In fact, there is an infinite number of possible figures. But we know that they are alike in two ways:
– The pertinent correspondences are the same;
– The construction is a diagram utilizing at least two visual variables.

The manner in which the two planar dimensions are employed permits us to classify them and to define types of construction.

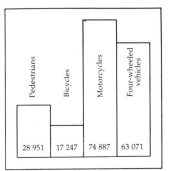
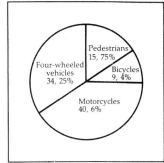
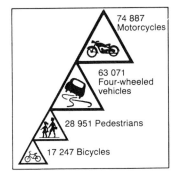
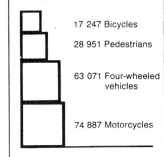
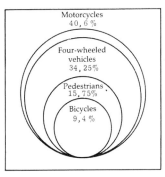
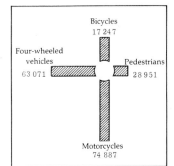
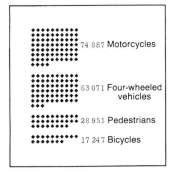
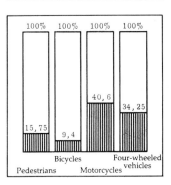
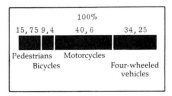
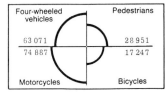

2

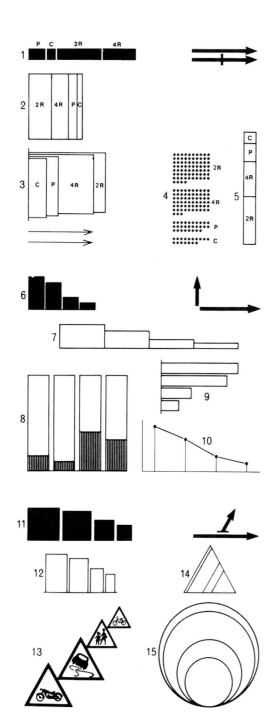

PRINCIPAL TYPES OF CONSTRUCTION FOR A DIAGRAM INVOLVING TWO COMPONENTS

Rectilinear (or linear) construction

In figure 1, a straight line represents the total number of accident victims. It is divided into parts proportional to the quantities in each category. Thus, the component **Q** and the component ≠ are portrayed on the same axis.

In figures 2, 4, and 5, the qualitative component "different vehicles" can be reordered by using the quantities in each category. The width of the straight line has no numerical meaning; it is simply the means for rendering the straight line visible.

In these examples the total is portrayed, which we indicate schematically by putting a bar through the arrow. The second dimension of the plane is not used; it remains available for representing any further component introduced into the information.

Orthogonal construction

If, as in figure 3, the partial quantities are not added but are related to the same base, we must employ a means of differentiation which will permit identification of the parts. The simplest way is to juxtapose them (figures 6–10). This juxtaposition forms an orthogonal construction, in which each dimension of the plane represents a component.

In these examples the total is not portrayed, but the different parts are easily comparable.

Rectilinear elevation

In figure 11, the areas are proportional to the quantities.

The signs are similar (homologous sides in a constant ratio).

The linear dimensions are proportional to $\sqrt{Q}$. The second dimension of the plane does not, therefore, represent the quantities. These are depicted by the amount of area, the amount of "black"; that is, the component **Q** is represented by a retinal variation (a variation in "size"). We indicate this by using an inclined arrow.

The quantities could also be juxtaposed along a straight line, as in figures 11, 12, and 13, or superimposed, as in figures 14 and 15. However, the total is not portrayed, and comparison of the parts is difficult.

Circular construction

By curving the construction in figure 1 we obtain a figure such as figure 18. This construction is a circular version of the rectilinear construction. The total is portrayed.

When the quantities making up a circular area are given equal radii, the amounts are designated both by their lengths on the circumference and by their angle at the center (figure 16).

The eye has acquired a great precision in judging this angle (figures 17 and 19), and this is easier to grasp than the circular length (figure 18).

Polar construction

By curving the construction in figure 6, we obtain figure 20. The polar construction is a circular version of the orthogonal construction. The total is not portrayed, and the parts are not easily comparable (figures 21 and 22).

Circular elevation

By curving the construction in figure 11 we obtain figure 24. The difference between this construction and the polar construction can be illustrated by comparing figures 23 and 26, or figures 22 and 27. The circular construction often appears as in figure 25. Circles are used to facilitate identification of the parts, whose areas are proportional to the **Q**.

These principal types of construction permit classing all the drawings on page 53 and, in fact, all planar constructions. Their diversity poses a problem of choice, which can only be solved by the notion of efficiency and by the rules of construction resulting from it.

We will discuss these constructions in later sections on diagrams (classed according to their perceptual properties on page 195) and networks (page 270).

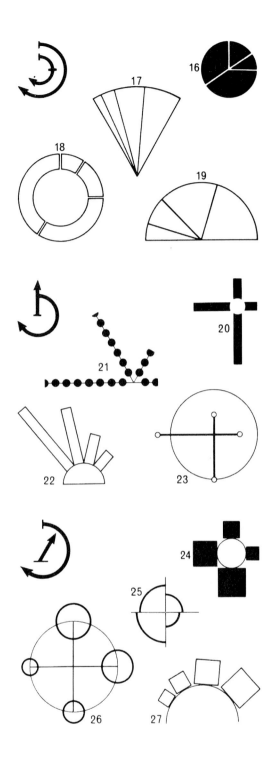

TABLE OF LEVELS AND IMPOSITIONS

To define a graphic construction, we will use the conventional signs below. They enable us to analyze all imaginable constructions and to indicate *schemas of construction* for them. When applied to the most efficient constructions, these signs denote *"standard" schemas*.

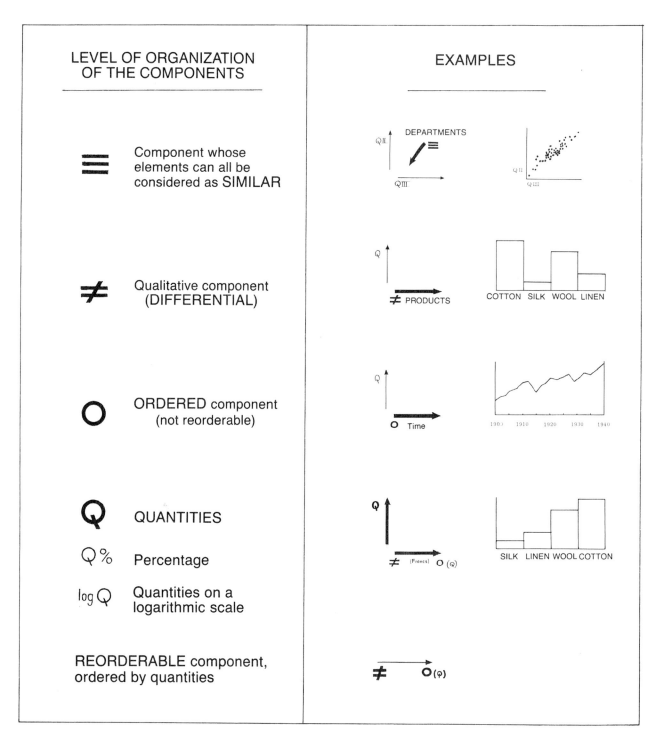

UTILIZATION OF THE DIMENSIONS OF THE PLANE

RECTILINEAR UTILIZATION

 Dimension of the plane utilized in a HOMOGENEOUS manner (the categories are established once and for all)

 Dimension of the plane utilized in a HETEROGENEOUS manner (the categories are repeated several times)

 n indicates the number of images or figures

Dimension of the plane representing CUMULATIVE QUANTITIES

 CIRCULAR UTILIZATION of the plane

ARRANGEMENT, TREE

EXAMPLES

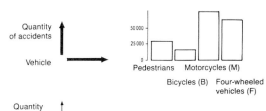

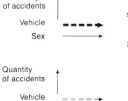 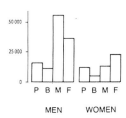

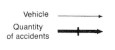 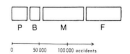

 RETINAL VARIABLE (read as an "elevation" above the plane)

 POINTS, LINES or AREAS (not differentiated)

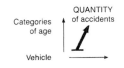

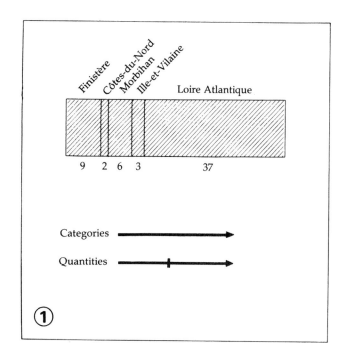

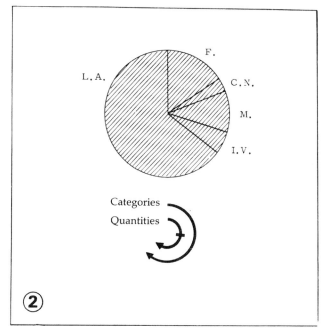

UTILIZATION OF THE PLANE

In cartography, the geographic component occupies the two planar dimensions. Consider the following information:

INVARIANT —*salaried workers in establishments with more than 500 employees*

COMPONENTS —*Q (in thousands of salaried workers), according to*
≠ *five departments in Brittany*

This information has two components. Its graphic representation must utilize at least two variables, and, depending on the type of construction, will result in *diagrams* (figures 1, 2, or 3).

However, the qualitative component ≠ is geographic in nature. The various categories are spatially defined—they are departments—and the information can also produce a *map* (figure 4). In this representation, the reader is invited to superimpose on the natural map, as seen from an airplane, elements which are invisible but nonetheless "real."

The reader is invited to perceive the sheet of paper, not as a medium, but as a geographic space. The surface of the paper signifies the surface of the earth; an excellent analogy, since space is utilized to signify space.

This is more natural, more readily comprehensible than the analogies used initially in figures 1, 2 and 3, or, say, the correspondence of a planar dimension with time. Perhaps this explains why figurative representation and cartography were used several millennia earlier than the diagram, whose analogies imply a higher degree of abstraction.

However, this natural analogy is obtained at the price of the complete utilization of the two planar dimensions, and it leaves no dimension of the plane available to represent the quantities.

They must become secondary to the geographic

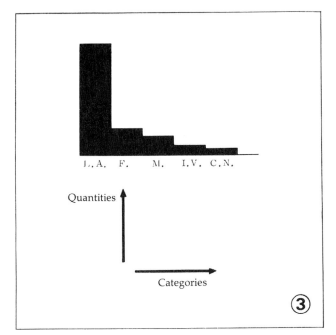 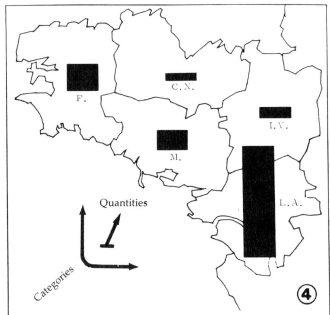

arrangement. The perception of the quantities can no longer be based on the comparison of the juxtaposed elements of a whole, as in figures 1 and 2, nor on the differences in the length of the elements aligned along a base, as in figure 3. Their perception must call upon other visual variables, upon new "stimuli" whose utility was not considered as long as the planar dimensions were sufficient for the representation. In figure 4, it is not so much the height of the column as the amount of "black" which permits perceiving the quantities. This becomes all the more evident as the number of correspondences increases (pages 360 and 374).

When two components occupy the plane, we must seek new variables to represent additional components. These are the "elevated" or "retinal" variables.

C. The retinal variables

With the introduction of a third component into the information (or a second component in cartography), the graphic representation must utilize the retinal variables.

THE VISUAL VARIATIONS AVAILABLE "ABOVE" THE PLANE

Experimental psychology defines depth perception as the result of multiple factors:
– binocular vision, within a limit of several meters
– the apparent movement of objects when the observer moves
– a decrease in the size of a known object
– a decrease in the values of a known contrast
– a reduction in the known texture of an object
– a decrease in the saturation of the colors of known objects
– deformations of orientation and shape (perspective).

All these variations, with the exception of the first two, are at the disposal of the graphic designer, who can use them to add a third component to those of figure 1, for example. The designer can relate the categories of the additional component with any one of these variables:

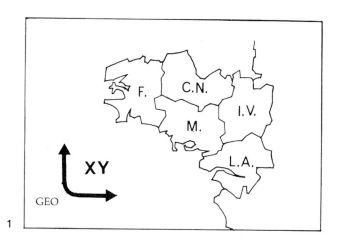

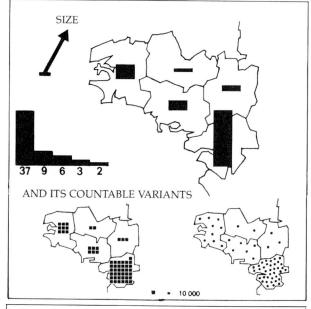

– categories of **SIZE**: height of a column, area of a sign, number of equal signs (figure 2)

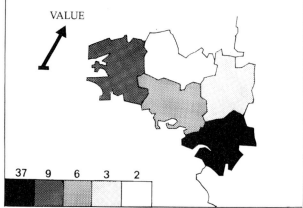

– categories of **VALUE**, the various degrees between white and black (figure 3)

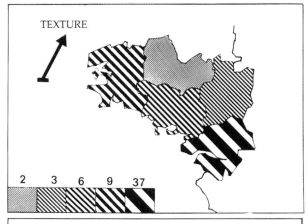

— categories of **TEXTURE**, that is, with a variation in the fineness or coarseness of the constituents of an area having a given value (figure 4). This variation can be obtained by enlarging or reducing a ruled photographic screen

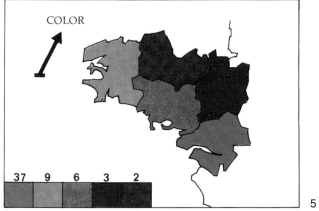

— categories of **COLOR** (hue), using the repertoire of colored sensations which can be produced at equal value (figure 5)

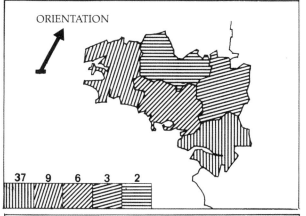

— categories of **ORIENTATION**, various orientations of a line or line pattern, ranging from the vertical to the horizontal in a distinct direction (figure 6)

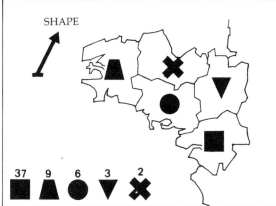

— categories of **SHAPE**, since a mark with a constant size can nonetheless have an infinite number of different shapes (figure 7).

Thus any retinal variable can be used in the representation of any component. But it is obvious that each variable is not suited to every component. It is the notion of level of organization which provides the key to solving this problem.

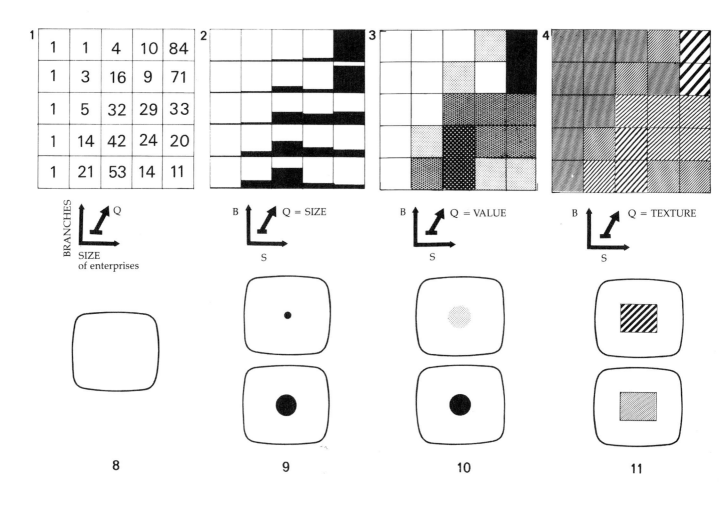

PLANAR DIMENSIONS AND "RETINAL" VARIABLES

The use of retinal visual variables is not required by cartography alone. It is necessary in all graphic problems involving three or more components, when the two dimensions of the plane are already being utilized.

Consider the information: amount of salaries, distributed according to branches of the economy and size of enterprise.

INVARIANT —amount of salaries distributed by enterprises
COMPONENTS —≠ five branches of the economy (commerce, energy, transportation, industry, service)
—Q (salaries) in % per branch of the economy, according to
—O five, business enterprise size-categories (0, 1–5, 6–100, 101–500, more than 500 employees)

The quantities are given in figure 1. As in the map of Brittany (figure 1, page 60), the two dimensions of the plane are utilized; the branches on one axis, the size of the enterprises on the other. Retinal variables must be called upon once again to represent the quantities, as illustrated in figures 2–7. In order to choose the best representation, we must determine what distinguishes the planar dimensions from these variables and what characterizes the different retinal variables.

When the planar dimensions represent two components of the information, they constitute an image, whose organization and basic form are established once and for all. They lend the plane a meaning which translates into quantities, categories, time (in diagrams), or space (in maps). They also define the field of vision. Beyond its frame the plane once again becomes a sheet of paper; it no longer has a meaning or else it changes in meaning to support another image. Visual "scanning" is thus involved; the reader perceives the planar dimensions through the intermediary of eye movement. Overall perception of the plane depends on "muscular" reactions of the optic system.

The retinal variables are inscribed "above" the plane and are independent from it. The eye can perceive their variation without requiring movement.

One could thus imagine a frame (figure 8) in which two

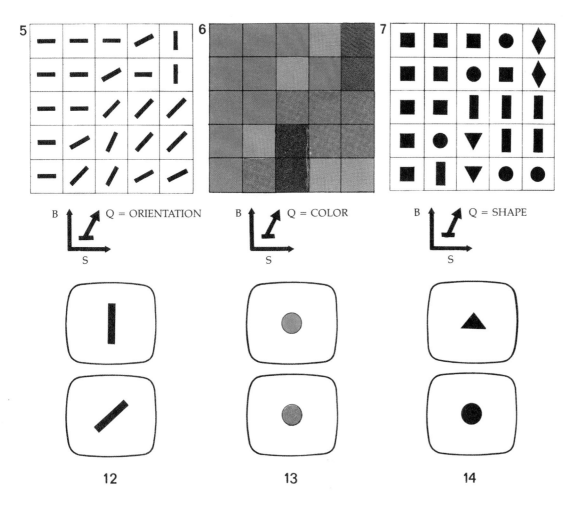

different examples of each variable would appear successively, in the same place. This is shown in figures 9–14. No muscular movement is required in order to distinguish between the two examples. These variables rely upon other visual reactions in which scanning does not seem to intervene in a significant manner.

In order to distinguish them from "muscular" responses, we will speak here of "retinal" responses and consequently of retinal variables.

On the scale of ordinary perceptions, which alone interest us here, the retinal variables are physiologically different from the planar dimensions. However, with a very large point, for example, there exists a limit beyond which it is no longer visible as a point. Perception must then call upon "muscular" movement, and the point becomes meaningless in terms of the retinal perception designated by the legend (and reinforced by the other signs). We will now examine the perceptual properties of each of these retinal variables.

1. The level of organization of the retinal variables

While the plane is at once selective, associative, ordered, and quantitative, the preceding pages show that the retinal variables possess only some of these properties. Their levels of organization differ. The correct representation of a quantitative component, for example, can only be accomplished by a variation in size.

Along with the notion of "imposition" (see page 52) in diagrams, that of "level" is probably the greatest potential source of graphic error. Level of organization assumes particular importance in cartography, because the two planar dimensions are committed to the geographic base, which means that retinal variables must be utilized whenever a second component appears in the data.

This notion could be studied variable by variable; however, it seems more useful here to proceed level by level.

The visual variables used in the following tests are "pure" variables; that is, they are considered with all other variation excluded. For example, color (hue) variation is considered for one given value. This precaution is indispensable in order to avoid confusion. In most graphic constructions several variables are combined. They must first be examined individually. This will permit analyzing and understanding each of the innumerable possible combinations. There are sixty-three basic combinations for differentiating two point signs! The level of organization of each one of these combinations, as we will see on page 186, corresponds to that of the individual variable having the highest level of organization.

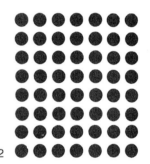

64

ASSOCIATIVE PERCEPTION (≡)

Associative perception is useful when one is seeking to equalize a variation, and to group correspondences with "all categories of this variation combined."

Example 1: What is the distribution of the density of the signs, and of the population density, in a map where each sign represents 500 inhabitants, but where the signs differ according to whether the inhabitants are farmers, herdsmen, or nomads? If the nomads are in black, the herdsmen in gray, and the farmers in white, only the density of the nomads will be perceived. A variation in value (black-gray-white) in not associative.

Example 2: Associativity is required when the representation combines two components, such as cephalic index and height of individuals, as in figure 1. The eye can easily isolate a given category or height by grouping the signs, with cephalic indices combined. Shape variation is associative. But we cannot immediately isolate all the dolichocephali, with all heights combined. Size variation, utilized here for representing height, is not associative. It is "dissociative." A dissociative variable dominates all combinations made with it and prohibits carrying out an immediate visual selection for the other variables.

Test for associativity. Since it is a question of disregarding a variation, the best test seems to be a series of undifferentiated points forming a uniform area, as shown in figure 2. If the eye can immediately reconstruct the uniformity of the area, in spite of a given visual variation, this variation is associative (≡). If not, it is dissociative (≠).

The tests given in figures 3–8 show that SIZE and VALUE are dissociative, while all other variables are associative. The same is true for line and area representations.

Visibility

All the signs in figure 2 appear to us with the same power. They have the same visual "weight" or "visibility."

An associative variable does not cause the visibility of the signs to vary.

Signs differentiated by size and value appear to us with different power, and our moving away from the images, for example, would cause the signs to disappear in succession. They do not have the same visibility.

A dissociative variable causes the visibility of the signs to vary.

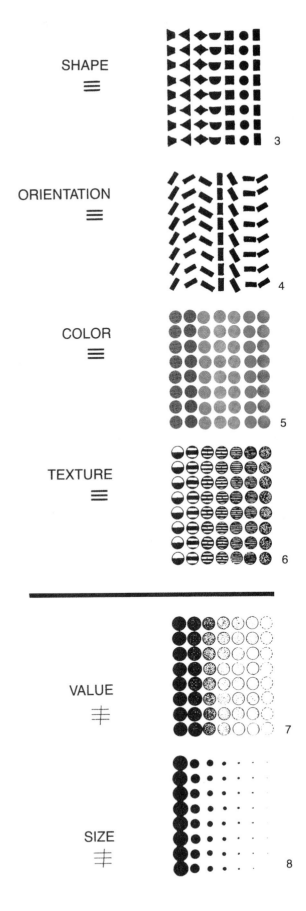

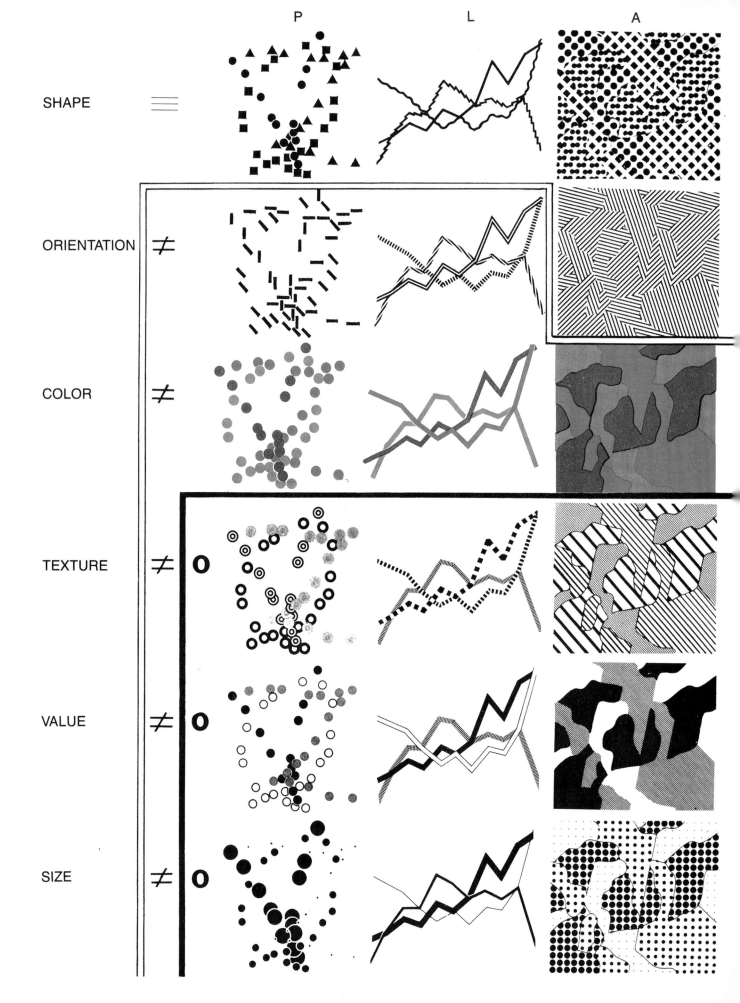

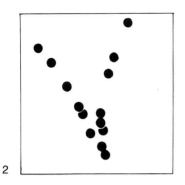 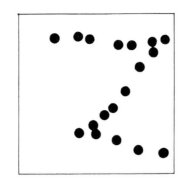 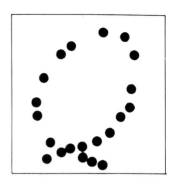

2

SELECTIVE PERCEPTION (≠)

Selective perception is utilized in obtaining an answer to the question: "Where is a given category?" The eye must be able to isolate *all* the elements of this category, disregard all the other signs, and perceive the image formed by the given category.

Such perception can be immediate, in which case the variable is selective, and each category forms a family. On the other hand, the perception can necessitate going through sign by sign, in which case the variable is not selective.

Test. In each of the images in figure 1 on the opposite page, we attempt to isolate all the signs in the same category, then recognize and retain the image which they form as a whole.

For all three implantations—point, line, and area—shape is not selective; nor is orientation when represented by area. The best visual selection is achieved by juxtaposing separate images on the plane (figure 2).

ORDERED PERCEPTION (O)

Ordered perception must be used in comparing two or several orders: "Is the ordering of geographic locations by birth rate similar to their ordering by death rate?" "Is the classing of departments according to their overall population similar to a classing according to tertiary population or agricultural population?"

Such a comparison can be *immediate*, in which case the variable is ordered. On the other hand, it can require a scrupulous analysis of all the correspondences, point by point, in which case the variable is not ordered.

Test. When a variable is ordered, it is not necessary to consult the legend to be able to order the categories. It is obvious that this is before that and after the other. The best test is to ask the reader to immediately reestablish the universal order of the signs for each variable.

In examining the graphics on the opposite page (figure 1), it is obvious that the shapes, the orientations, and the colors (value excluded) are not ordered. Each person can establish any order whatsoever; none asserts itself immediately.

Conversely, texture, value, and size impose an order which is universal and immediately perceptible. Texture, value, and size are ordered for all three types of implantation.

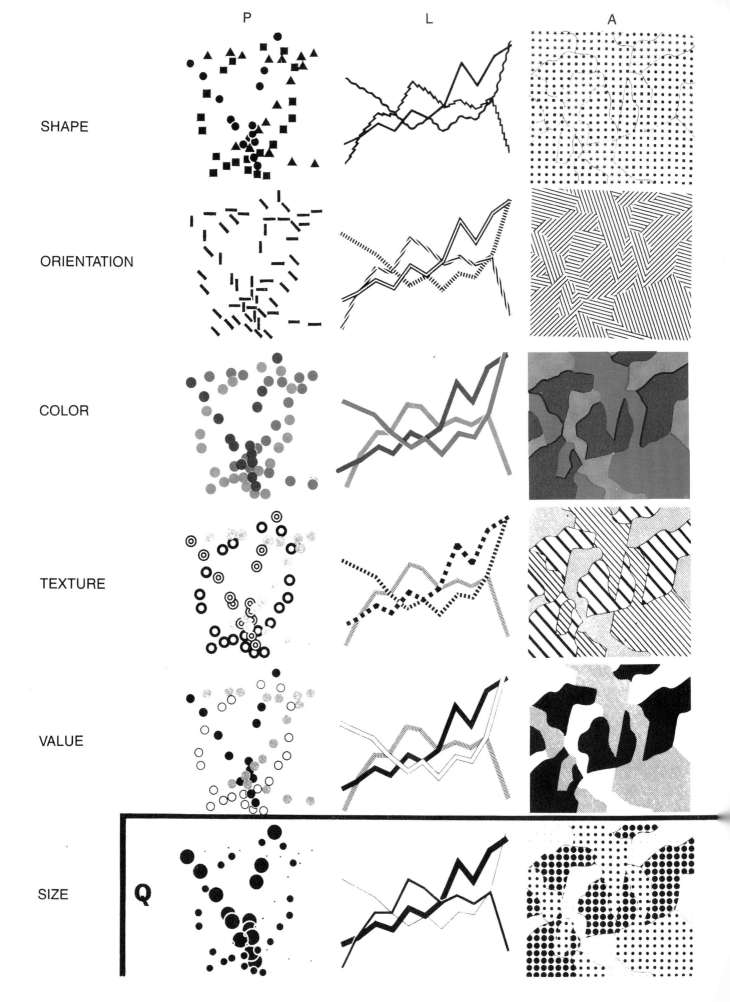

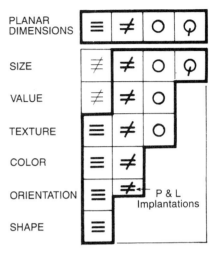

2

QUANTITATIVE PERCEPTION (Q)

Quantitative perception is involved:
(1) when we seek to define numerically the ratio between two signs;
(2) when we seek to group homogeneous signs, that is, ones involving small quantitative "distances," and thus define the natural steps resulting from a statistical study.

Test. When perception is quantitative, the numerical ratio between two signs is immediate and necessitates no recourse to the legend; it appears immediately to the reader that this is double that or is eight times that. The best perceptual test is to ask the reader the value of the larger sign if a value of one is attributed to the smaller sign.

It is readily apparent that *only size variation is quantitative* (see figure 1).

Value variation is not: White cannot serve as a unit for measuring gray, nor can the latter for measuring black.

Texture variation is not quantitative either; the absence of texture (or an invisible texture) cannot serve as a unit for measuring a coarse texture. However, between two coarse textures, a quantitative relationship can be discerned, since the spacing is more evident. We must remember that quantitative perception represents an accurate approximation but not a precise measurement.

A CLASSING OF THE VISUAL VARIABLES

The above observations are summarized and schematized in figure 2. The levels of organization and perceptual approaches order the visual variables in a necessary sequence: planar dimensions–size–value–texture–color–orientation–shape. We can thus identify "higher-level" variables, that is, those possessing a greater number of perceptual properties, which makes this classification of fundamental importance in the choice of a graphic representation.

Note, however, that, unlike the plane, no retinal variable possesses all four properties and that the inclusive nature of the properties is disrupted in the case of associativity, which is absent in size and value.

This table (figure 2) will be further developed on page 96, after we have studied each variable in terms of its length, which is a type of implantation and the intended perceptual level.

69

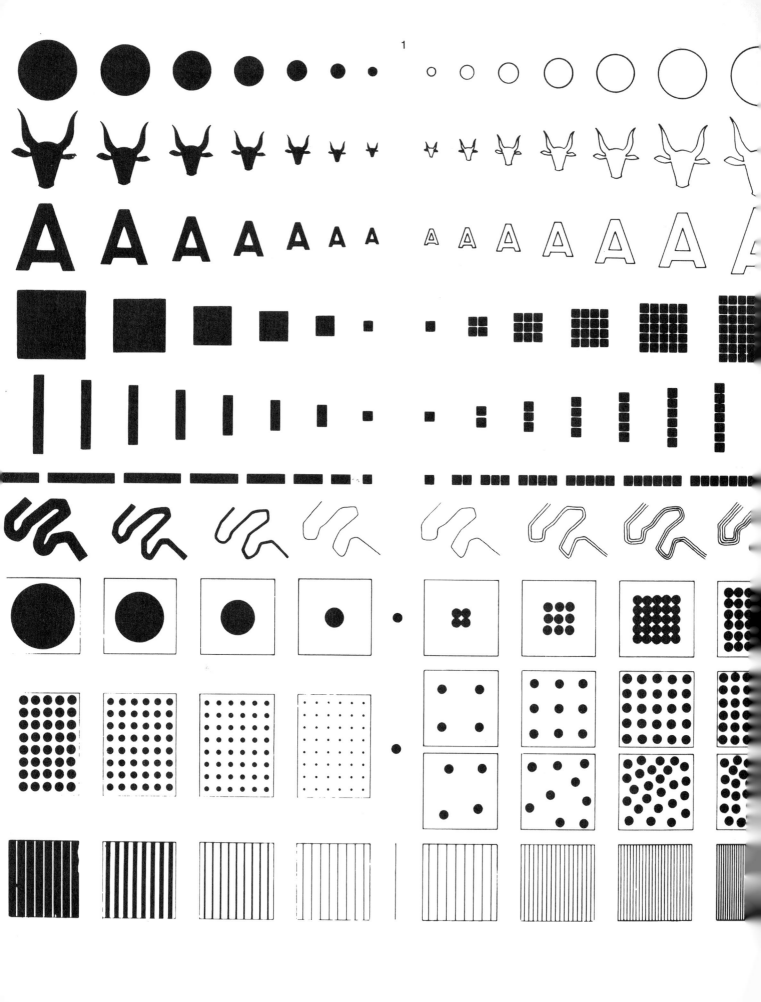

2. Characteristics and properties of the retinal variables

SIZE VARIATION

In point or line representation, any figure can vary in size without varying in position, value, texture, color, orientation, or shape.

It is a variation in the dimensions of the mark (area) which constitutes the perceptual stimulus for size variation.

Implantation

In *point representation*, the figure can assume many forms: It can be geometric or mimetic; it can be a column of proportional height; it can be formed by parts which are joined and countable. The extent of variation is very large in point representation; we can, for example, usefully construct two visible points, one of which has an area 10 000 times larger than that of the other (pages 182 and 363).

In *linear representation*, a line can be made to vary in thickness.

We can also juxtapose countable parallel lines. However, the extent of variation will be limited if the lines converge, as in a highway network.

In *area representation*, the area cannot vary in size. But its constituent points or lines can vary in size and number. Here, the extent of variation is limited by the size of the area involved. However, it is possible to construct several circles or figures which extend beyond the area (page 373).

Length

In quantitative perception and ordered perception, the number of possible steps is unlimited, but on the average the eye will not distinguish more than twenty steps between two points whose areas are in a ratio of 1 to 10. This limit has caused us to propose a "natural series of graduated sizes," indicating the steps which are necessary and sufficient for any quantitative representation (page 369).

Perception

With *selective perception*, size variation is very limited. In the circumstances that occur in an average drawing, it is not advisable to count on more than four or five selective steps (such that one can isolate and precisely define the form constituted by all the points of a given size). *Size variation is dissociative*, and it is not possible to disregard it visually. Any other variable combined with a size variation will be dominated by it, and its length will diminish accordingly. With very small points, color, for example, becomes practically imperceptible.

The combination of size and value. Since value is also dissociative, a size variation will be perceptible only for signs having a dark value. Black or white signs can even deprive the size variation of all of its immediate properties.

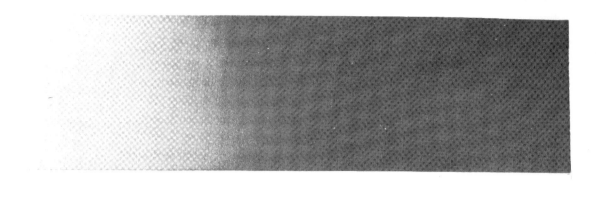
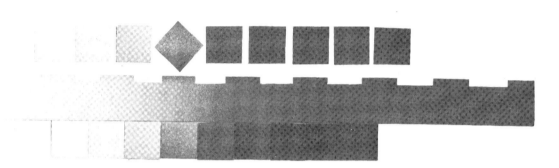

VALUE VARIATION

Value variation is the continuous progression which the eye perceives in a series of grays ranging from black to white (figure 1). On white paper the strongest value can be conceded to be black.

This progression is independent of color (hue); one can pass from black to white by grays, by blues, or by reds . . . (page 85). We will use the term "medium value" to designate one of the instances intermediate between black and white, whatever the color. However, a given color—blue, red, or green—no matter how dark, will always be lighter than black at the moment when it is detectable. Consequently, **the ratio between the total amounts of black and white perceived on a given surface will be called value**.

The value of a colored surface is designated by the gray whose value is equivalent to that of the surface. **But, in practice, a gray is designated by the ratio between a given surface of paper (not of white) and the printed surface; we speak of a ten percent gray, which means ten percent black and ninety percent white.**

A medium value in a given color or in black is obtained in three ways:
– by mixing white with the color as do painters and printers
– by "simulated-engraving," which transforms, by photomechanical processes, a flat gray tint into a regular pattern of points that are very small and generally invisible to the eye (in printed photos, for example)
– by "shading" or "dotting," which involves drawing fine but generally visible marks. This latter process is the simplest and most commonly used.

Length of value variation

A value variation is *ordered*, whatever the number of steps being constructed.

For *selective perception*, it is advisable not to exceed six or seven steps of value—black and white included.

Length obviously depends on the *distance* available between black and white. This will be even more limited if the "white" is not white. It is thus a serious error to use colored paper (gray, blue, green, red, . . .) for a graphic or map where value is to be used as a meaningful variable. The loss of legibility is considerable (see figure 2). Length also varies with the *size* of the marks. The smaller they are, the greater the reduction in the number of selective steps (see figure 1).

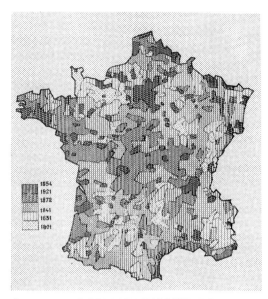

2   DATES OF MAXIMUM POPULATION

Furthermore, it is clear that the contrast between gray and white is greater as the marks become smaller (figure 1). In order to retain the appearance of a "medium" gray (equidistant from white and black) on a white sheet of paper, we must use a lighter gray for a smaller mark. However, this effect only applies where a large white surface is involved. In other cases, the phenomenon does not seem to occur.

Value variation is dissociative. It is not possible to disregard it visually. Any other variable combined with a value variation will be dominated by the latter, and its length will diminish accordingly. With very pale values, the identifiable number of sizes, colors, shapes, orientations, and textures diminishes and disappears at null value.*

TERMINOLOGY

*The same visual sensation can be provoked by different stimuli. Consequently, one distinguishes in experimental psychology between: (a) something that refers to the perceived sensation and permits defining a visual sensation (but is not measurable); and (b) something that provokes this sensation: the (measurable) stimulus.

The determination of a value can be considered in different ways:
– by measuring the quantity of reflected light. This is the "luminance" of the stimulus, which corresponds to the "brightness" of the sensation;
– by measuring the ratio between the luminance of a gray and that of a "perfect" diffuser (a white). This is the "luminance factor" to which corresponds the "luster" of a sensation.

It does not seem helpful to utilize these distinctions in this study. The term "intensity" could also apply. But we prefer to use VALUE since this is the only term that involves the adjectives, light and dark.

1

BLACK / WHITE	10/90	20/80	30/70	40/60	50/50	60/40	70/30	80/20	90/10
ratios		9/4	12/7	14/9	3/2	3/2	14/9	12/7	7/4
progression		125%	71%	56%	50%	50%	56%	71%	125%

BLACK / WHITE	6/94	11/89	20/80	33/66	50/50	66/30	80/20	89/11	94/6
ratios		2/1	2/1	2/1	2/1	2/1	2/1	2/1	2/1

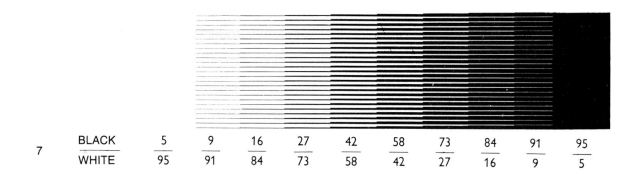

BLACK / WHITE	5/95	9/91	16/84	27/73	42/58	58/42	73/27	84/16	91/9	95/5

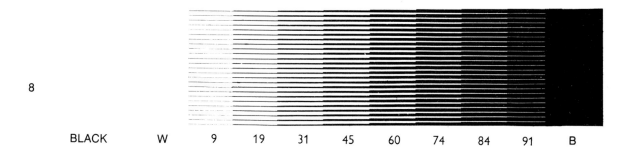

BLACK	W	9	19	31	45	60	74	84	91	B

CONSTRUCTION OF EQUIDISTANT STEPS OF VALUE

Dividing available values into equal parts is a common problem, whose solution is relatively complex.

In figure 1 the progression of black is arithmetic (see series 2). Note that equidistance is not obtained—the step limits are much more perceptible at the extremities of the scale than at the center. In effect, the eye sees ratios and not absolute quantities. Thus, the ratio between the first two grays is

$$\frac{20}{80} \bigg/ \frac{10}{90} \text{ that is, } \frac{9}{4}$$

and the ratio between the second and the third gray is

$$\frac{30}{70} \bigg/ \frac{20}{80} \text{ that is, } \frac{12}{7}.$$

The sequence of ratios (series 3) and progressions (series 4) indicates the variation in visual distance among the steps of the scale.

The values defined in series 5 are more regular. The sequence of ratios is constant (series 6); the ratio of the progression is 2, producing a regular increase of 100%. It is easy to determine the formula for this series, if we give white and black values other than 0 and 100, which, incidentally, corresponds to reality. If we call

W   the value of white
B   the value of black
r   the ratio of the progression
n   the number of steps

it is clear that the value of white must be multiplied by r (n − 1) in order to obtain the value of black, that is, W r (n − 1)=B, from which the progression is derived as a function of the number of steps:

$$r = \sqrt[n-1]{(B/W)}$$

If white is given the value 5/95 and black 95/5 we obtain:

$$r = \sqrt[n-1]{361}$$

and series 7 is derived from it. But note that the progression is not rigorous and that a slight adjustment must be made for the light values in order to obtain a good equidistance (see series 8). From this we derive figure 9, in which the logarithmic progression is adjusted, and also the table in figure 10. This table shows the various equidistant values that should be used, depending on how many steps are involved in the series.

Common cases. For a small number of steps the following principles can be applied:
Three steps: black–medium gray (slightly below rather than above 50/50)–white (figure 11)
Five steps: as for three steps, but with a black "broken up" by fine white lines or dots and a white "broken up" in the same way (figure 12)
Four steps: black–two grays of about 25/75 and 75/25–white (figure 13)

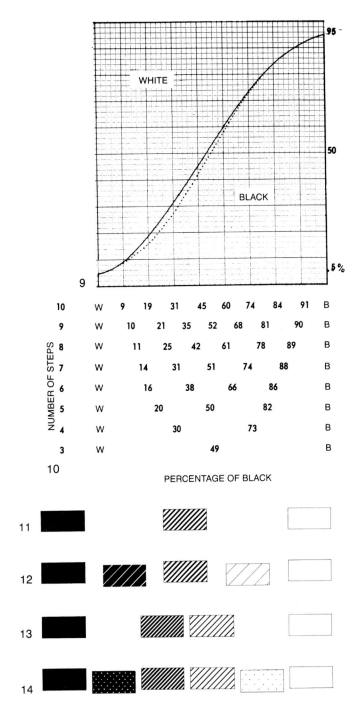

Six steps: as for four steps, but with a white and black which are "broken up" (figure 14).

In area representation value variation can also be obtained with prepared screens, in combination with other variables (page 340).

75

PRINCIPAL SOURCES OF CONFUSION IN THE USE OF VALUE VARIATION

Value is ordered, and we cannot reorder it

In figure 1 (death rates in Paris [number of annual deaths per 1000 inhabitants]), the designer was not careful to order the grays according to the order of the death rates. The figure must be read point by point, and it is only through considerable mental effort that the information can be grasped. Visual evaluation of the image would be erroneous.

When the two orders correspond (figure 2) the image becomes meaningful, and its retention useful.

Value is not quantitative

When oil consumption in Europe (1954) is portrayed by a value variation, as in figure 3, only the order in which the consumptions are classed is indicated. Even if it is known that Portugal consumes one million tons, the image by itself permits no evaluation of consumption in the other countries. The reader is dependent upon a legend. Visual means are underemployed.

However, if a size variation is used, ratios are perceptible and evaluations are possible (figure 4).

The population of the various sections of Paris, as represented in figure 5, indicates the order of the sections by population size. However, this image produces an erroneous impression of the concept—population—because an area representation suggests the notion of density. But we can see from figure 6 that the representation of density gives a very different image.

The representation of quantities by value obliges the designer to transform a series of numbers into a series of numerical classes. There are several rules for doing this. But the problem is precisely that there is no single rule; the reader never sees more than a certain choice of steps, among all the possible ones.

The maps in figures 8–11 all represent the same information that is given in figure 7, i.e., the working population in the industrial sector in 1954. Depending on the particular design, each reader will form a different image of the same information. The only image which does not involve a prior choice of class intervals is figure 12, where the quantities are represented by proportional circles.

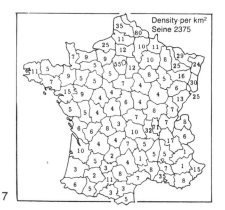
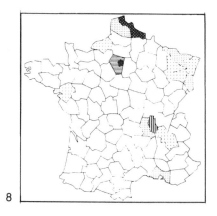
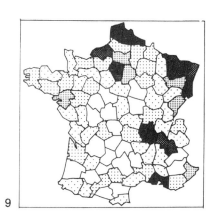

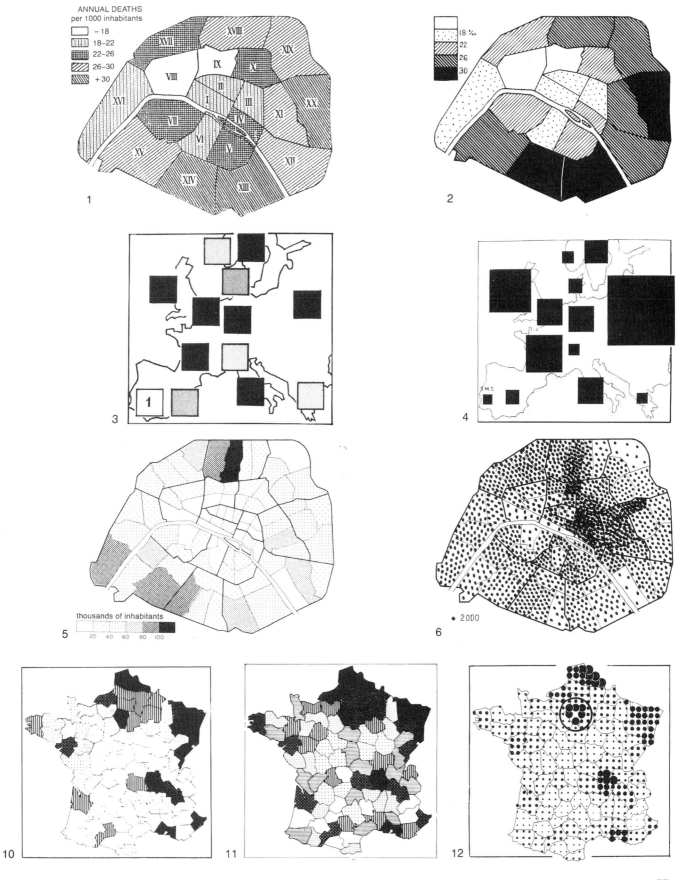

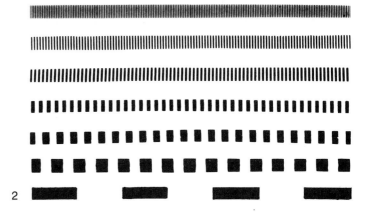

2

TEXTURE VARIATION*

Texture variation is the sensation resulting from a series of photographic reductions of a pattern of marks. For a given area and with a regular pattern, these reductions increase the number of marks, without causing the value of the area to vary. (In figure 1 opposite, texture is given horizontally; value and shape [pattern] vertically).

At a given value, the TEXTURE is the number of separable marks contained in a unitary area.

This variation (figure 2) begins at null texture, in which the elements are so numerous, but at the same time so small, that they are no longer identifiable to the naked eye.

From null texture, the variation extends to large textures (or "coarse" ones), which form the limit beyond which there is ambiguity concerning the implantation of the sign (in a pattern which is too coarse, the notion of area threatens to disappear).

The fineness of the texture is directly proportional to the number of marks; in photogravure a screen of 50 (that is 50 points to the inch) constitutes a coarse texture; a screen of 300 corresponds at the present time to the finest texture available by this technique.

If, in the above discussion, we have taken photographic reduction as an example, it is because it is directly linked to the notion of texture. However, many designers of scientific atlases are unaware of this link, since:

a texture which is too fine does not reduce and will disappear in microfilmed reproduction.

Since the scientific documentation of tomorrow will be primarily on microfilm, most of the fine-screen representations will disappear, carrying with them the information which they now contain (see, for example, page 365).

Value and texture

Any value can support a variation in texture (except, of course, for the two extreme values, black and white). In medium values steps based on texture are highly perceptible and can produce the "vibratory effect" shown in figure 1 on page 80.

Size and texture

The length of a given texture variation is directly linked to the size of the mark. The larger the mark, the greater the number of separable steps. Consequently, it is the type of implantation which controls the available length.

Implantation

It is obviously in *area representation* that texture will be associated with the largest marks, thus furnishing the greatest number of perceptible steps. Such steps will be quite numerous for ordered perception, but limited to four or five for selective perception.

In *linear representation* (for a line thickness of 1 mm) texture is also limited to three or four selective steps.

In *point representation* texture requires quite large signs but yields only two or three selective steps. Texture is often used to produce a vibratory effect. However, this only occurs when the marks are of a certain size and shape.

*For definition, see note on page 11 (translator's note).

1

THE VIBRATORY EFFECT OF TEXTURE

Black and white combined in a certain way over areas can create the uncomfortable sensation produced by figure 1. This visual effect constitutes a remarkable selective possibility when it is properly utilized.

When does it occur?

Figure 2 on the page opposite combines variations of texture and value. It is repeated for the purposes of highlighting the phenomenon (i.e., the six lines at the bottom are simply the inverse of the six lines at the top).

From left to right is the value variation; from top (or from bottom) toward the center is the texture variation.

The vibratory effect appears in the central region and consequently involves values bordering on 50% and constituent sizes of more than 1 mm. This "vibration" seems to result from the collusion of a physiological effect—the creation of a certain resonance at the retinal level—and a psychological effect—the hesitation between "figure and ground." In a given graphic representation this corresponds to an immediate ambiguity concerning the implantation: Is it an area, line, or point sign?

It is the designer's duty to make the most of this variation, to obtain the resonance without provoking an uncomfortable sensation, to flirt with ambiguity without succumbing to it.

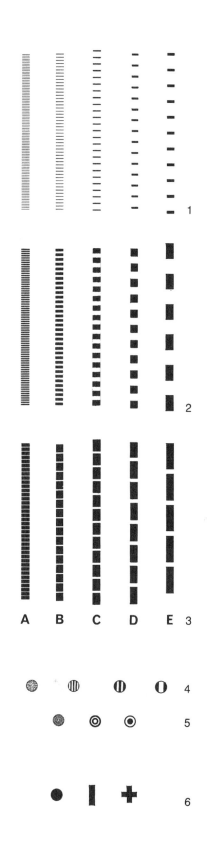

The vibratory effect in linear representation
This effect is easy to obtain (see figure 2B, C, and D), and it determines the selectivity of texture in linear representation. (The value variation runs from figure 1 to 3, and the texture variation from A to E.)

The vibratory effect in point representation
Providing the signs are large enough (larger than 2 mm, approximately), this vibratory effect can be produced in two ways:

(1) by a variation in *internal complexity* within a given shape, as in figures 4 or 5.

But among all the possibilities for construction of figure 5, what sizes of the white ring yield the best vibratory effect?

The table in figure 7 combines variations in the arrangement of the ring (from left to right) and variations in value (from top to bottom). A cyclical construction (figure 8) emphasizes the power of this effect, which dominates the value variation. This construction confirms that it is in values of 50% to 60% and around the designs marked by a cross in figure 8 that the effect is at a maximum.

(2) by a variation in the *external complexity* of a sign, that is, by contrasting the circle, the cross, and the dash (figure 6). In terms of perceptual effect, this variation approaches that of texture. It tends to create, within the limits of the sign, a zone of vibratory confusion and ambiguity. Owing to this effect, the circle, the cross, and the dash are the three *shapes* which, within certain limits, allow for a selective perception. However, these three signs are not visually ordered.

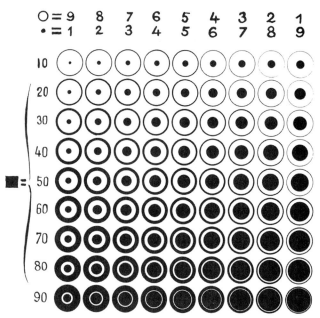

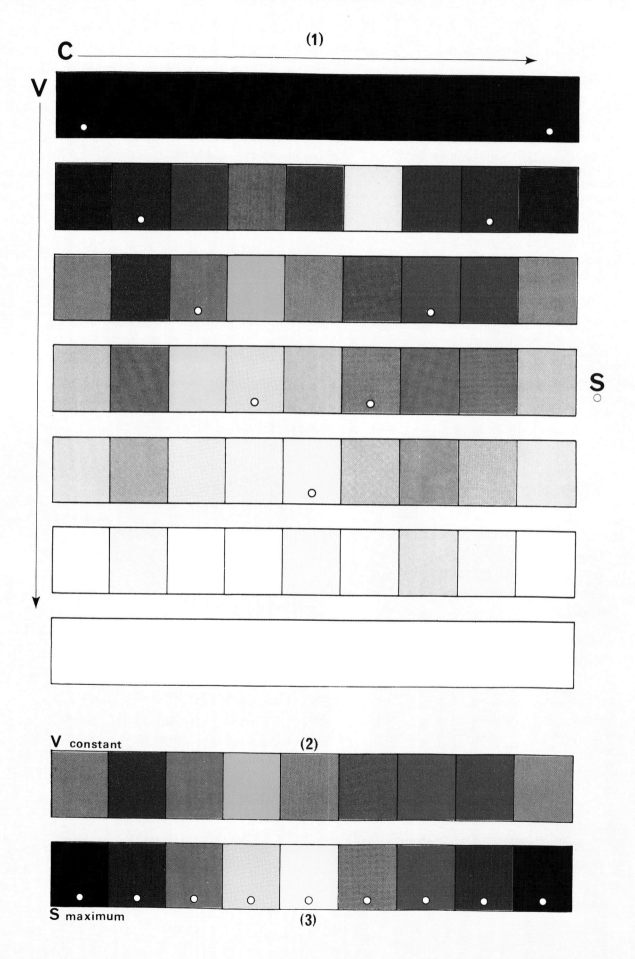

COLOR VARIATION

The use of color cannot be understood unless the notion of color (hue) is distinguished in a rigorous and definitive manner from the notion of value. They are two different sensations, which, by nature, overlap. We have already defined what we mean by value variation.

COLOR variation is simply the perceptible difference which can be perceived between uniform areas having the same value.

In the double-entry table (figure 1), which represents the color-value combination, it is clear that we can pass through the entire series of colors without changing in value (from left to right), just as we can pass through the entire series of values within each color (from top bottom).

Each of the boxes in figure 1 is a TONE. A tone placed on a sheet of paper is therefore defined by two parameters: color (hue)* with the categories violet, blue, green, yellow, orange, red, purple, gray (or neutral tint);† and value, defined by the percentage of black in the corresponding gray.

Color saturation

In order to construct the entire series of values, it is necessary, no matter what the color chosen, to add some white in order to obtain light values, or some black in order to obtain dark values. For each color, there is a central value which does not require the addition of white or black. In this privileged value the color, being neither "washed out" with white (desaturated), not "soiled" by black (darkened), appears as most brilliant. This is the "pure tone" in painting, printing, and colorimetrics. It is the "saturated tone" in experimental psychology.

White corresponds to the addition of all other colors, black to the diminution of all reflecting power. The saturated tone corresponds then to a color involving no mixture with other colors, that is, to a very thin band on the color spectrum.

Value and saturation

The saturated color is marked by a dot in figure 1 opposite. Note that the pure yellow is in row 5, the pure green and orange in row 4, the blue and the red in row 3, the violet and the purple in row 2.

The saturated tone is not of constant value (figure 2), but varies in value according to the color (figure 3).

It is this fact that leads to many of the main problems raised by the use of color. In graphic representation its ramifications are numerous.

TERMINOLOGY

*The word "tint" is too loaded with ambiguity in common language to help us in defining the notion of color. Indeed, we often speak of light or dark tints (value variation), warm or cool tints (color variation), flat or textured tints (texture variation).

†Color variation is the visual sensation resulting from a difference between uniform areas. Thus gray constitutes a color variation in relation to any other color. Conversely, a monochrome drawing in red contains no color variation. (It can be photographed in black and white without loss of information.)

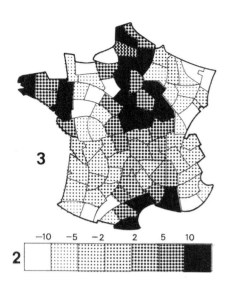
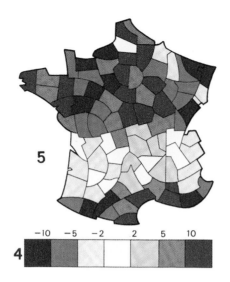
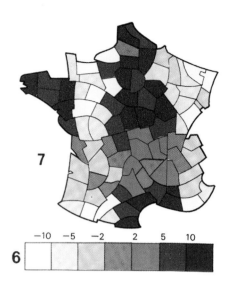

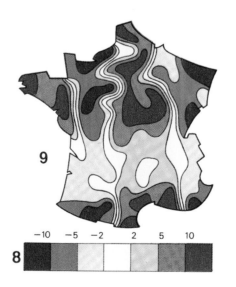
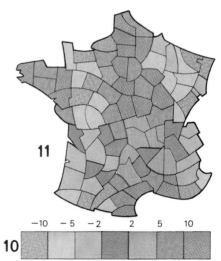

86

AT EQUAL SATURATION (MAXIMUM SATURATION)

The series of pure tones combines color and value
If we retain only the pure tones, which is natural since these are the most brilliant, we simultaneously construct a color variation and a value variation. This is the color spectrum represented in figure 4.

Each of the two regions of the spectrum, on either side of yellow, produces an ordered series
The lightest color, near the center of the spectrum, is yellow. But for all the other levels of value the eye encounters two different colors. The order of the values does not correspond with that of the spectral scale.

Immediate visual perception will follow the order of the values and combine the two extremities of the spectrum
Let us consider the information in figure 1. Represented by a value series (figure 2), it produces the map in figure 3, which is a "North–South" image. Represented by the color spectrum (figure 4), it produces the "East–West" image in figure 5, which is a misrepresentation of the information. Note that it is impossible to disregard this orientation. The eye seems to combine the two extremities of the spectrum into the same perceptual unit, which then contrasts with the unit formed by the central colors. This is because the central colors are "light," whereas the extreme colors are "darker."

Value perception dominates color perception
In figure 5, blue and red are immediately perceived as similar, rather than different. If this is true, then a legend ordering the colors according to their value (figure 6) will produce an image corresponding to the correct distribution of the information, as we saw it in figure 3. This is confirmed by figure 7.

Thus the combination of the "cool" series with the "warm" series is a source of visual confusion in the representation of an ordered component.

Such a combination is only possible with isarithms, where colors which are not adjacent on the spectral scale are never brought together on the plane. The total image is then based on a "warm–light–cool" series (figure 8). The resultant map (figure 9) indeed captures the "North–South" distribution of the information displayed in figures 3 and 7.

AT EQUAL VALUE

Color variation is not ordered
A color variation without value variation (figure 10) will produce a flat, meaningless image (figure 11). What is properly called color variation is only selective and associative; it cannot be used to represent an ordered component.

The choice of selective colors differs according to value
At a given value, that is, on a horizontal row in figure 1 on page 84, there are only two saturated colors (and a single one for row 5). The farther one gets horizontally from a saturation point, the more the other colors, "soiled" by black or "washed out" with white, tend to fuse into grayness.

Selectivity is at a maximum near the saturated color and diminishes as one moves away from it
As a result, with *light values*, steps should be chosen around yellow, that is, from green to orange; blue, violet, purple, and red at light value are grayish and not very selective.

Medium values offer the greatest number of selective color steps. The two saturated colors—blue and red—are diametrically opposed on the color circle, and the "grayish" sectors are reduced to a minimum.

When *dark values* are being used, the best colors for making the marks stand out well run from blue to red (through violet and purple). Dark green, dark yellow, and dark orange are dull and not highly differentiated.

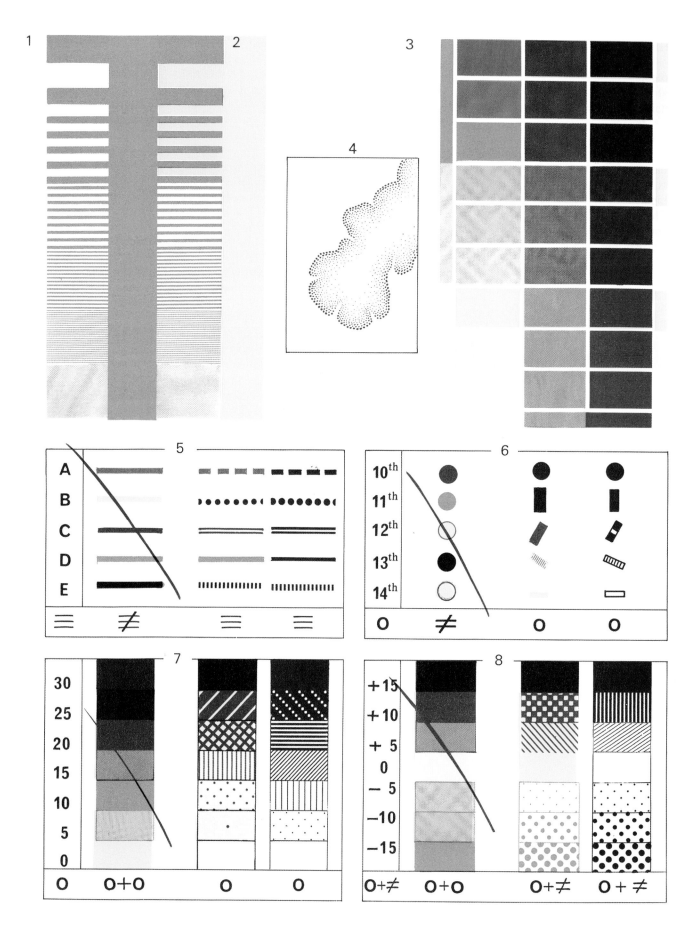

THE SIZE OF THE MARKS AND COLOR "FUSION"

No color sensation is perfectly pure. Nature mixes colors, and the problem is to separate them in space. In order to obtain a color sensation, that is, a perceptible color difference, there must be an area covered with a *uniform color* (similar over the entire area) and contrasting with another area (of *another uniform color*). Obviously, the smaller the areas, the more one approaches a natural mixture, and the less the problem is resolved. Any effort at coloration is a struggle against natural mixture, against color "fusion."

The smaller the mark, the less distinguishable are the colors; the length of a color variation is thus a function of the size of the marks.

The marks must be at least 1.5 mm in diameter to usefully support even a very few color variations. On the other hand, it is easy to see a difference between any two colors, no matter how similar, when they are applied over a large area. For very large areas, perceiving a million different separable tones is conceivable.

Color fusion permits us to create new colors from given colors

Once the problem of separation is more or less resolved, by the use of colored pigments, it is easy to move in the other direction and mix the colors. The eye naturally mixes separate colors, all the more easily when spatial separation is reduced by the arrangement of the marks.

A blue and a white produce, at certain points, a blue, and a light blue, and a white (figure 1). From one area to another, we progressively see the blue becoming altered, the white becoming colored. A blue and a yellow produce, at a certain point, a green (figure 2). This visual property is the basis for:
(1) trichromatic representation. From "primary" blue (cyan), yellow, and red (magenta), virtually the entire spectrum can be reconstructed in a manner sufficient for most common problems (figure 3).
(2) trichromatic analysis. The three primary colors can be used to represent three ordered components. The multiple combinations of the components can then be studied and any regions of the plane corresponding to similar or neighboring combinations can be discovered. They will have the same tone. This is trichromatic analysis (figure 3).
(3) the spatial division which transforms the color of a mark according to the sharpness of its boundaries. Blurred edges favor fusion. The most common graphic application is an arrangement of graded points; it extends its coloration well beyond the last point (figure 4).

PRINCIPAL SOURCES OF CONFUSION IN THE USE OF COLOR

The colors of the spectrum are dissociative ($\neq$)

The series of pure tones varies in value and, as such, is dissociative.

Yellow is a light color, very near to white. Consequently, line B of figure 5 is not very visible.

In linear (or point) representation, light colors should be avoided. Black can replace them, and the series will have sufficient visibility. Furthermore, it is necessary here and in the following examples to ensure that a black and white photograph of a color graphic would still be comprehensible (see page 90). This is achieved by combining color with other visual variables (shape, texture, and orientation). Finally, we note that in all these examples (figures 5–8), a monochrome legend is, practically speaking, as efficient as color.

Color is selective ($\neq$); it is not ordered

The series of points in column 2 of figure 6 is supposed to represent an order of centuries, but it would not produce an ordered image. This result could only be obtained by a series of values from one of the scales of the spectrum (here the "warm" scale), preferably combined with orientation (column 3, figure 6).

The eye mixes the two scales of the spectrum in a single series of values

In figures 7 and 8, it is not the order of the component which the eye will reconstruct, but the order of the values. This will lead to the confusion observed in figure 5 on page 86.

In figure 7 the component is homogeneous. We can go from white to black either by the "warm" scale or the "cool" one, each excluding the other.

In figure 8 the component is a dual one, differentiating a positive series from a negative one. The solution consists of representing the positive series by the "warm" scale, with quite fine screens (fine texture), and the negative series by a variation in the size of highly visible points (coarse texture) rendered in a single color borrowed from the "cool" scale (green or, preferably, blue).

Color is not indispensable

Since visual order derives from value perception, a monochrome series running from white to black is itself sufficient for the representation of an ordered component (see figure 2, page 86).

A difference in texture can be used to differentiate the negative part of a given series, as in figure 1, for area maps (isopleths), and figure 2, for contour maps (isarithms).

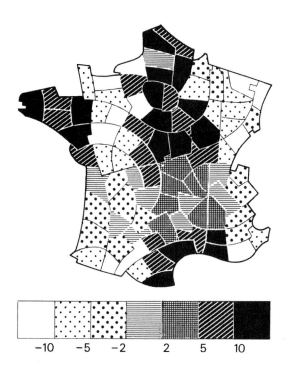

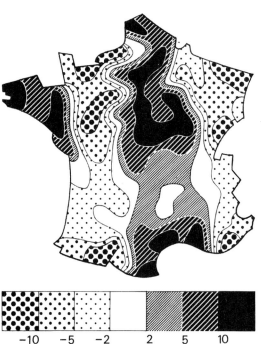

SYMBOLISM AND COLOR AESTHETICS

Color symbolism

Throughout the entire world, water, seas, and rivers are never red; fire, heat, and dryness are not generally accompanied by a blue sensation; vegetation is most often green. . . . At the same time, birth, marriage, and death are often associated with particular colors. Color symbolism results from the more or less universal nature of certain visual images.

But there are only eight distinct colors. Whatever the conceptual domain treated by graphic representation, color is efficient only if the entire length of the variation is utilized. Therefore, spectacular color effects which are irreconcilable with the universal nature of a concept being represented should be avoided.

Color harmonies

The efficient use of color is primarily governed by all the preceding observations. Within the margin of freedom which remains, it is necessary to draw upon accepted aesthetic rules. They often contradict the rules which assure a good color selection. We will call those compositions which utilize different values within a given color "value harmony" (or tone-on-tone harmony, or cameo). We will call compositions which utilize several neighboring colors on the color circle (for example, blue, blue-green, green) "hue harmony."

Combined with value harmony, hue harmony produces a highly "aesthetic" effect.

A "contrast" is created by adding to a hue harmony the color which is diametrically opposed on the circle, that is, the complementary color. The artistic effect depends on the size of the colored areas. The complementary area must be small in relation to the original area, which we will call the "dominant" area.

ADVANTAGES AND DISADVANTAGES OF COLOR

Advantages
Color is an excellent selective variable. It combines easily with other variables and is eminently "legible." Thanks to colored pens and inks, it will always remain the privileged variable of the graphic designer in problems involving differentiation, *particularly at the moment of inventory and classification*. However, remember that color is only selective, whereas information processing involves problems of ordering. Consequently, color is never strictly indispensable in the stages of research, analysis, and processing.

Above all, color exercises an undeniable *psychological attraction*. In contrast to black-and-white it is richer in cerebral stimulation, and in numerous cases where it can appear as a luxury, this luxury nonetheless "pays off." It captures and holds attention, multiplies the number of readers, assures better retention of the information, and, in short, increases the scope of the message. Color is particularly applicable to *graphic messages of a pedagogical nature*.

Disadvantages
Aside from its high cost when publication is involved, color has two major disadvantages:

– *Anomalies of chromatic perception* (color blindness). These anomalies are more frequent than is generally believed. They become serious for those who utilize the graphic document as a research tool. The color-blind person alleviates this disadvantage by looking, often unconsciously, for alternative signals which will permit restoring the meaning of the message. It is important that the drawing furnish these.
– *Reprographic problems*. The immediate reproduction of several copies is the basis for the scientific documentation of tomorrow. For practical reasons, color will be excluded from this. In contrast to monochrome reproduction, color can multiply the cost many times over, and increase the time factor considerably. And in spite of these tremendous differences, color reproduction on paper or film reduces color differentiation and thus destroys part of the information.

In order to alleviate these two disadvantages of using color, a color variation should only be applied to those problems where it appears indispensable: precision inventories, "trichromatic analysis," pedagogy. Furthermore, one must:
(1) make a potential monochrome photograph decipherable—by combining color with other variables (figures 5–8 on page 88)
(2) in any complex selective problem, add a small monochrome image for each of the components. These smaller images enable us to perceive the distribution of a given component.

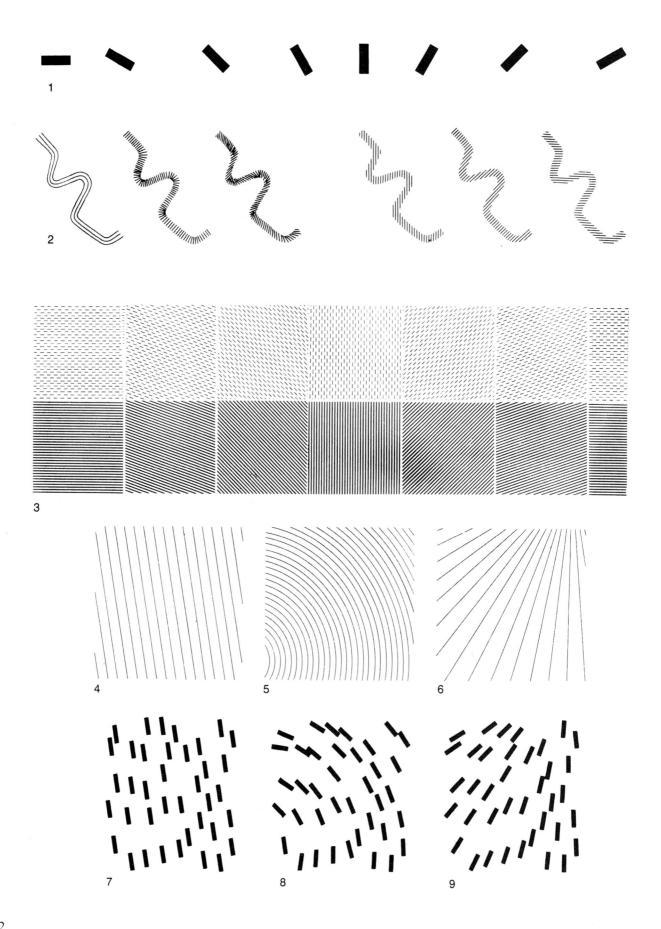

ORIENTATION VARIATION

In point representation, a mark can assume an infinite number of different orientations without changing the position of its center. However, we are sensitive to such a variation only when the mark has a linear aspect (figure 1). (The ratio height/base must be at least 4/1.) Furthermore, for this variation to be meaningful, we must be able to detect categories of orientation (figure 2); this obliges the designer to limit the number of categories in order to distinguish among them. Parallelism among the signs in a meaningful orientation is thus of fundamental importance, and we can further state that:

it is the difference in angle between fields created by several parallel signs which constitutes the perceptible "stimulus" for orientation variation.

Parallel circular systems (figure 5) and their rectilinear counterparts (figures 4 and 6) also involve variations in orientation. They create well-defined fields in area representation as well as in point representation (note that the centers of the corresponding signs are the same in figures 7, 8, and 9).

Implantations

Since shape variation is not selective, it is clear that, in *point representation, orientation is the only available variation which can differentiate signs of equal visibility* (see 323). (Color is another possibility, but this cannot be used for reasons mentioned earlier. Texture too is possible, but it has limited length and is difficult to construct.)

It is preferable to restrict oneself to four orientations (figure 10), constructing the oblique lines at 30 degrees and 60 degrees rather than at 45 degrees. Five orientations are possible, but selectivity diminishes. Note that all the oblique signs tend to form a family in contrast to the orthogonal signs.

Linear signs can vary in shape (figure 11). If we study a drawing of oriented signs, the axis of perception being as distant as possible from the perpendicular to the plane of the paper (figure 12), all the signs parallel to this axis stand out and isolate themselves from the other signs.

In *linear representation* we must limit ourselves to two orientations: the axis of the line and its perpendicular. Selectivity is reduced but still exists.

In *area representation* variation in orientation is the easiest to construct, but it is at the same time the least selective (figure 3). It can only be used at the inventory stage or in combination with a selective variable.

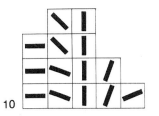

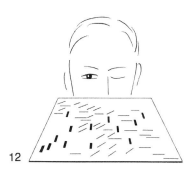

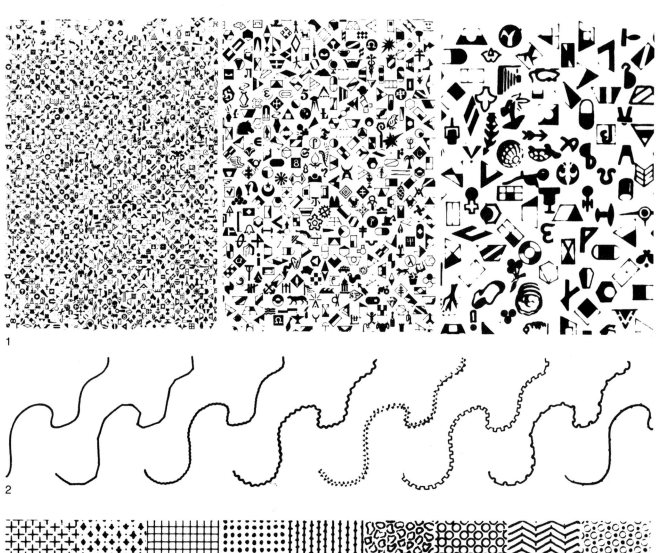

1

2

3

SHAPE VARIATION

The world of shapes is infinite

A mark with a constant area can assume an infinite number of different shapes. This variation has *unlimited length*, and it is tempting to abuse it.

Figure 1 contains several hundred different signs. *Three pairs are strictly similar.* In searching for them the reader will notice that the eye focuses only on the mimetic shapes and virtually ignores the geometric ones. An idea, not a figure, creates interest and facilitates retention. Note the difficulty experienced in attempting to study more than one sign at once and in searching for the same sign several moments after observing it. This is because identification of the sign absorbs all the reader's attention, obscuring the position of the sign on the page.

It is the element of "similarity" recognized in the shape which constitutes the stimulus for this variable and confers its principal characteristics upon it:

Shape variation is associative and can be used when the image of the density of the signs, "all shapes combined," is meaningful (page 157 and figure 1, page 322).

Shape variation is not selective. It does not permit answering the question: "Where is a given category (differentiated by shape)?"(See, for example, page 157.) All signs of like shape cannot be grouped at a single glance, since it is necessary to construct an image for each sign in order identify it. *Shape is not utilizable in problems involving regionalization.*

Shape variation is only applicable for elementary reading. It serves:
(1) to reveal similar elements, and therefore different elements
(2) to facilitate external identification (page 19), through shape symbolism.

Shape and implantation

In *point representation* (figure 1), when the forms are amorphous, two similar marks are difficult to identify. If they are geometric, we can easily redraw them and construct identifiable marks. Some are highly familiar, such as numbers and letters. They can also be mimetic (persons, animals, objects) and tend to evoke the same concept for the majority of readers. The innumerable "flourishes" which can be added to any sign also enter into this variation. Since selectivity depends on other variables, the designer tempted by shape variation should first draw a dash. By making it vary in value, texture, orientation, or color (if applicable), the designer can obtain a selective series which is generally sufficient for most problems (see figure 1, page 324, and page 325). Shape can be added subsequently, but must not disturb the differentiation just obtained.

In *linear representation* (figure 2), a line can vary in shape and translate different concepts by differences in angularity (page 329).

In *area representation* (figure 3), especially for large areas, which will accommodate large signs, a certain selectivity can be obtained by carefully contrasting points and lines. Optimum selectivity is produced by other variables—size, value, and texture—which come more naturally to the designer (figure 4). Certain patterns have achieved the status of symbols (figure 5).

Shape symbolism

"What is at a given place?" If the visual response is a triangle or a square, the reader must have recourse to the legend. On the other hand, a mimetic shape will often avoid this necessity and resolve the problem of *external identification*. How is this effect obtained?

It should be stated at the outset that the meaning of a shape is never obvious. The signs on page 157 do not enable us to dispense with the legend. Indeed, even the most recognizable shapes can suggest numerous meanings. A horse's head can just as easily correspond to a race track, a stable, a stud farm, a riding school, a bridal path, a horse butcher, a glue factory, a harness factory, a chess game, etc. The cross, "symbol" par excellence, allows students armed with bad maps to imagine New York as garnished with cemeteries: The fine black crosses of the cemeteries and the fine red crosses designating monuments are similar at first glance!

There is no universal shape signification. The meaning of a symbol becomes familiar to us only by habit; through the repetition of a similar situation. A shape can become a symbol only within a restricted domain, rigorously defined and previously familiar to the observer. However, we must recognize that modern information tends to mix different domains and hinder such familiarity!

Thus, creating an efficient code of conventional signs is less a problem of discovering effective shapes than a problem of defining the field in which they are to be used and in which *their meaning will be constant and sufficiently repeated to become established.* Within a given code, the efficiency of the signs will depend less on their evocative capacity than on the visual distances which can be obtained *among the shapes* in order to avoid ambiguity and confusion.

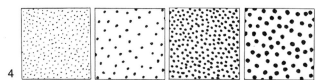

4

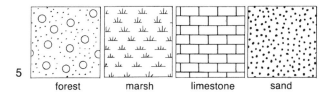

5   forest   marsh   limestone   sand

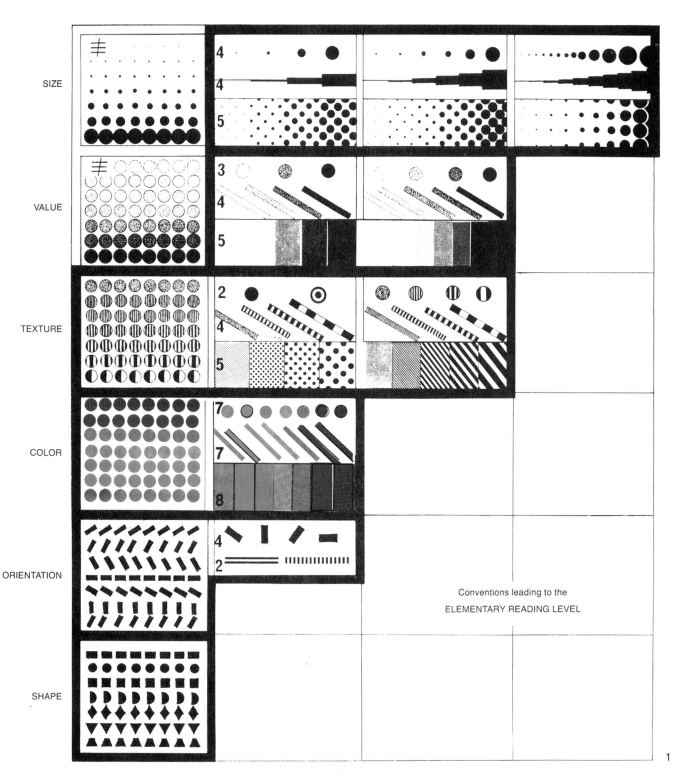

TABLE OF PROPERTIES OF THE RETINAL VARIABLES

The perceptual properties (level of organization, length of the retinal variables), which have been defined and illustrated in the preceding pages, are summarized in this table. It completes figure 2 on page 69.

Level of organization

In the column on the left (association), what is shown in point representation is also true for the other "implantations."

In the three columns on the right (selection, order, quantity) each box contains an example for each of the three implantations.

The level of a given variable is indicated by which boxes are filled in; a component of level O can only be represented by one of the planar dimensions, or by size, value, or texture. The use of any retinal variable falling into an empty box will lead to a representation which remains on the elementary reading level.

Length

The length of each component, unlimited in quantitative, ordered, or associative perception, is reduced to a small number of steps in selective perception. This number, marked by a numeral for each implantation, is calculated as a function of common graphic problems and needs. However, remember that the useful length of each variable is linked to the size and value of the marks. For very small marks and light values, the number of perceptible steps is reduced even further.

This table is indispensable for any problem involving the use of a retinal variable. It defines the "pure" variables, excluding all other variation, and it is the basis for all combinations of variables. As we will see on page 186, the properties of a combination are easily derived from this table. Note that the two planar dimensions are the only visual variables which display all the pertinent perceptual properties.

III. The rules of the graphic system

For a graphic to be "useful," it must be "efficient." The rules governing graphic efficiency stem from the properties of visual perception.

A. The basic graphic problem
B. Image theory
C. Three functions of graphic representation
D. General rules of construction
E. General rules of legibility

A. The basic graphic problem

Should a graphic be used?

The appropriateness of using a graphic to represent a given problem depends on several factors. The decision whether to transcribe the information graphically should be based on an evaluation of the specific properties and efficiency of each "language," each sign-system. Such a decision also depends on acquired habits, personal aptitudes, and even "fashion." A negative decision is often due to time considerations, since many investigators feel (perhaps erroneously) that utilizing graphics can involve a considerable amount of additional time.

The potential utility of graphics can only be assessed if we are able to answer rigorously a second question: What type of graphic should be used?

To utilize graphic representation is to relate the visual variables to the components of the information. With its eight independent variables, graphics offers an unlimited choice of constructions for any given information. When the information contains a geographic component, the problem of the graphic designer is to determine whether a diagram, a network, or a map should be used; and, for each case, what type of construction or which "retinal" variables should be chosen.

The spectrum of choices is probably wider than most graphic designers suspect. The following pages display a collection of the principal kinds of representation possible for a given set of information. Application of "image theory" will then enable us to determine, from within this collection, the particular graphic which should be used. The "basic problem" in graphics is thus to choose the most appropriate graphic for representing a given set of information.

A hundred different graphics for the same information

Consider the information given in the table in figure 1.
The work force, in France, in 1954.
≠ *by department*
Q *quantities (in thousands) according to*
≠ *three main sectors of the work force:*
   primary (agriculture)
   secondary (industry)
   tertiary (commerce, transportation, services)

Further calculation augments this information with:
(1) the total working population (in thousands) per department;
(2) the percentage of each sector of the work force for each department.

This information offers the possibility of constructing either diagrams or maps (see figure 2).

1

DEPARTMENTS		QUANTITIES (000)				PROPORTION		
		I	II	III	Total	I	II	III
1	AIN	67	43	40	150	45	28	27
2	AISNE	56	71	66	193	29	37	34
3	ALLIER	65	45	57	167	39	27	34
4	Bses ALPES	15	8	12	35	43	24	33
5	Htes ALPES	16	8	13	37	44	21	35
6	ALPES Mmes	31	61	122	214	14	29	57
7	ARDECHE	48	32	25	105	45	31	24
8	ARDENNES	25	53	35	113	22	47	31
9	ARIEGE	33	17	14	64	52	26	22
10	AUBE	28	48	36	112	25	43	32
11	AUDE	50	20	32	102	49	19	32
12	AVEYRON	70	32	29	131	54	24	22
13	BOUCHES DU RH.	42	143	226	412	10	35	55
14	CALVADOS	70	55	69	194	36	28	36
15	CANTAL	45	13	20	78	58	16	26
16	CHARENTE	65	36	38	140	47	26	27
17	CHARENTE Mme	79	39	65	183	43	21	36
18	CHER	43	41	36	120	36	34	30
19	CORREZE	64	23	29	116	55	20	25
21	COTE D'OR	43	41	59	143	30	29	41
22	COTES DU NORD	131	35	62	228	58	15	27
23	CREUSE	58	13	17	88	66	15	19
24	DORDOGNE	104	34	41	179	58	19	23
25	DOUBS	35	67	39	142	25	47	28
26	DROME	46	38	35	119	39	31	30
27	EURE	48	52	45	145	33	36	31
28	EURE & LOIR	44	27	38	110	41	25	34
29	FINISTERE	164	76	89	329	50	23	27
30	GARD	40	51	52	144	28	36	36
31	HAUTE GARONNE	64	67	84	216	30	31	39
32	GERS	63	10	16	89	71	11	18
33	GIRONDE	115	107	170	392	30	27	43
34	HERAULT	62	40	71	173	36	23	41
35	ILLE & V.	137	60	82	279	49	21	30
36	INDRE	54	30	32	116	46	26	28
37	INDRE & L.	61	41	55	157	39	26	35
38	ISERE	68	136	78	282	24	48	28
39	JURA	39	34	27	100	39	34	27
40	LANDES	70	25	28	123	57	20	23
41	LOIR & CHER	51	27	30	108	47	25	28
42	LOIRE	56	160	82	298	19	54	27
43	Hte LOIRE	52	23	22	97	54	24	22
44	LOIRE INF.	101	108	105	314	32	34	34
45	LOIRET	51	51	54	156	32	33	35
46	LOT	41	10	16	67	61	15	24
47	LOT & GAR.	70	24	30	124	57	19	24
48	LOZERE	22	5	7	34	64	15	21
49	MAINE & L.	104	65	65	234	44	28	28
50	MANCHE	116	42	56	214	54	20	26
51	MARNE	44	57	67	168	26	34	40
52	Hte MARNE	25	28	28	81	31	35	34
53	MAYENNE	74	23	28	125	59	19	22
54	MEURTHE & M.	23	127	91	241	9	53	38
55	MEUSE	24	31	27	82	30	38	32
56	MORBIHAN	132	47	59	238	55	20	25
57	MOSELLE	36	173	94	303	12	57	31
58	NIEVRE	34	27	33	94	36	29	35
59	NORD	81	483	296	860	9	56	35
60	OISE	40	69	55	164	24	42	34
61	ORNE	65	30	35	130	50	23	27
62	P.D.C.	94	242	137	473	20	51	29
63	PUY DE DOME	80	79	63	222	36	36	28
64	Bses PYRENEES	80	49	62	191	42	25	33
65	Htes PYRENEES	37	27	28	92	40	29	31
66	PYRENEES ORIENT.	35	20	33	88	40	23	37
67	BAS-RHIN	76	122	114	312	24	39	37
68	Ht-RHIN	40	121	74	235	17	51	32
69	RHONE	44	215	194	453	10	47	43
70	Hte SAONE	34	32	23	89	38	36	26
71	SAONE & L.	94	77	62	233	41	33	26
72	SARTHE	87	45	58	190	46	24	30
73	SAVOIE	44	38	35	117	38	32	30
74	Hte SAVOIE	52	42	45	139	37	30	33
P	PARIS	2	575	940	1517	0	38	62
75	SEINE	8	574	550	1132	1	51	48
76	SEINE INF.	75	152	174	401	19	38	43
77	SEINE & M.	37	72	76	185	20	39	41
78	SEINE & O.	46	328	356	730	6	45	49
79	DEUX-SEVRES	71	29	33	133	53	22	25
80	SOMME	57	68	61	186	31	36	33
81	TARN	55	47	33	135	41	35	24
82	TARN & G.	44	13	18	75	59	17	24
83	VAR	33	50	81	164	20	31	49
84	VAUCLUSE	40	30	41	111	36	27	37
85	VENDEE	110	38	40	188	59	20	21
86	VIENNE	60	29	39	128	47	23	30
87	Hte VIENNE	64	47	45	156	41	30	29
88	VOSGES	36	95	43	174	21	54	25
89	YONNE	41	28	37	106	39	26	35
90	BELFORT	3	25	13	41	8	60	32
		5212	6705	6905	18825	28	35	37

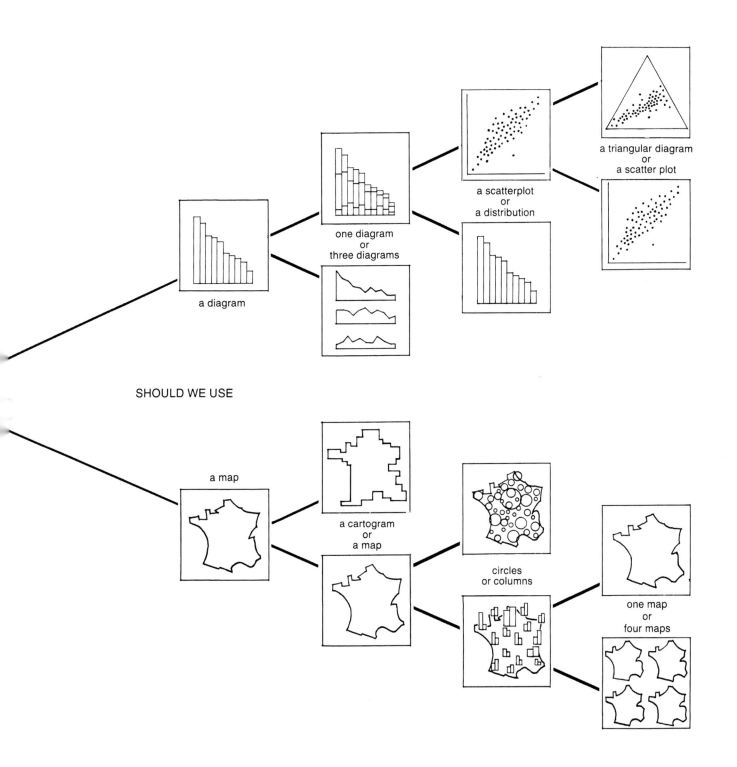

Diagrams

The most elementary construction (figure 1) represents the quantities by following the layout of the data table. This means adopting an alphabetic order for identification of the departments on one dimension of the plane and arranging the sectors and quantities on the other dimension. This form can vary; the quantities can be cumulative by department (figure 2), or be replaced by percentages (figure 3). In no case does the image yield useful information. It remains complex; its memorization is pointless; and reading the quantities at the elementary level is less efficient than reading them directly from the original data table (figure 1, page 100).*

*Note, however, that when the primary purpose of the graphic is information processing, rather than information display, an arrangement closely duplicating that of the data table is usually the best solution. See Jacques Bertin, *La Graphique et le traitement graphique de l'information* (Paris: Garnier Flammarion, 1977) [*Graphics and Graphic Information-Processing*, trans. William J. Berg and Paul Scott (New York and Berlin: Walter de Gruyter, 1981)], pp. 24–31. Hereafter *GIP* (translator's note).

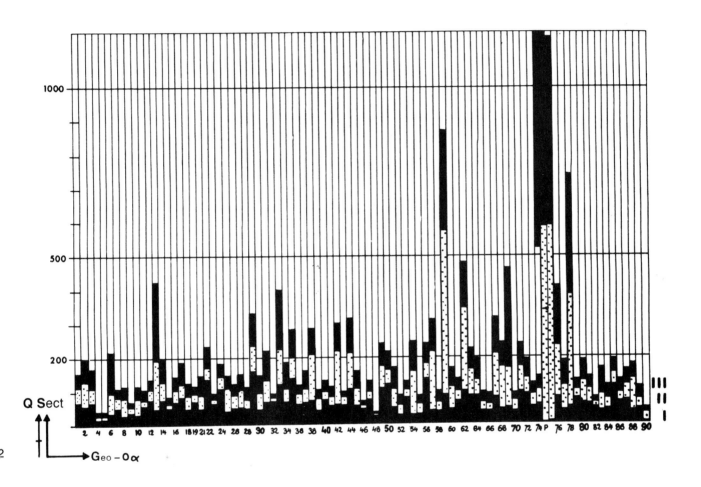

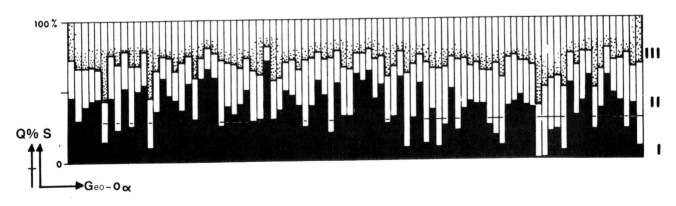

On these two pages (104 and 105) the images have been simplified. The geographic categories have been ordered in a quantitative series which tends toward a straight line. However, the information contains four different series— one for sector and another for the total quantities (Q_t)— and all are equally capable of being utilized to reorder the alphabetic list of departments. In figure 1, for example, the departments are ordered by total working population, and, based on this order, the respective quantities for each sector are represented cumulatively. In figures 3 and 4, the same is

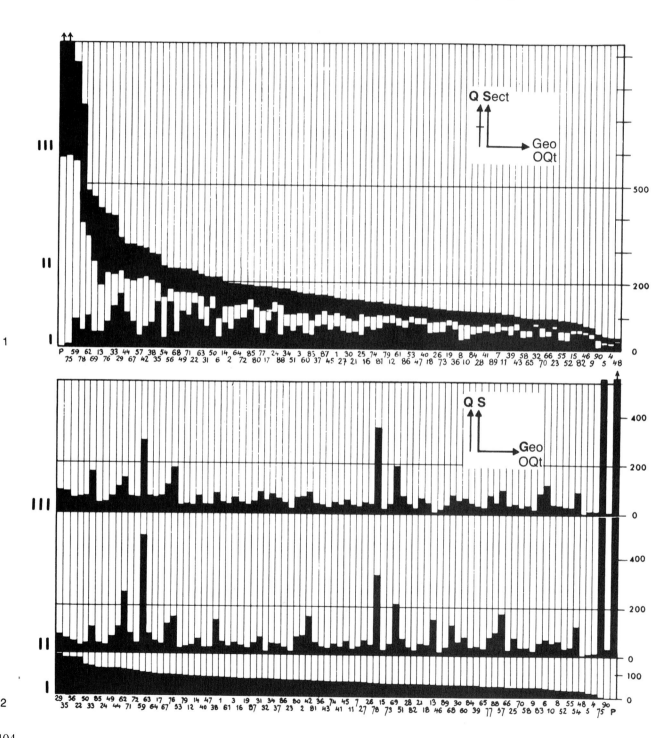

true, but the order of the departments is defined by the series of quantities in sectors II and III, respectively. In figure 2, the order is determined by the series of quantities in sector I, and the other quantities are not cumulative. Note, incidentally, the use of a schema within each figure to define the principles behind each construction. It is difficult to draw a useful conclusion from this group of images, even though figures 1, 3, and 4 are based on the same principle.

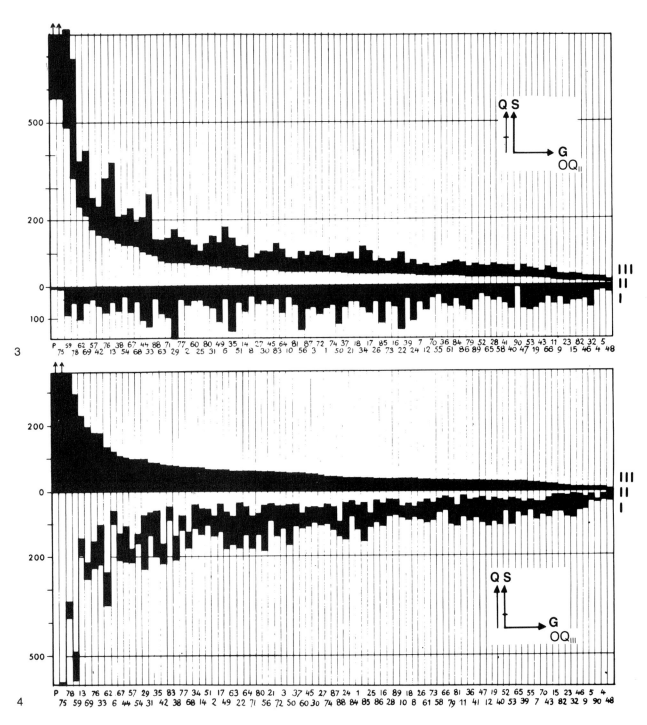

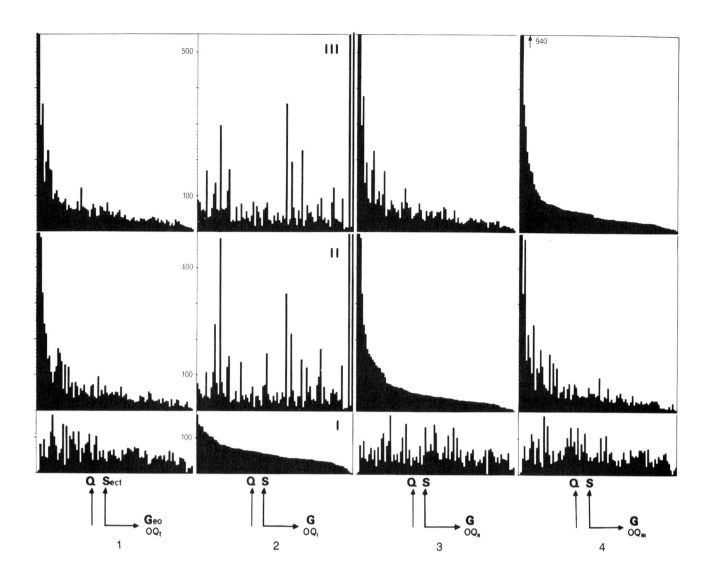

The use of cumulative quantities in figures 1, 3, and 4 on pages 104 and 105 does not aid interpretation. In fact, interpretation is easier when all the images are based on separate series, as in figures 1, 2, 3, and 4 above. This type of construction enables us to perceive a greater resemblance between series II and III than between these and series I.

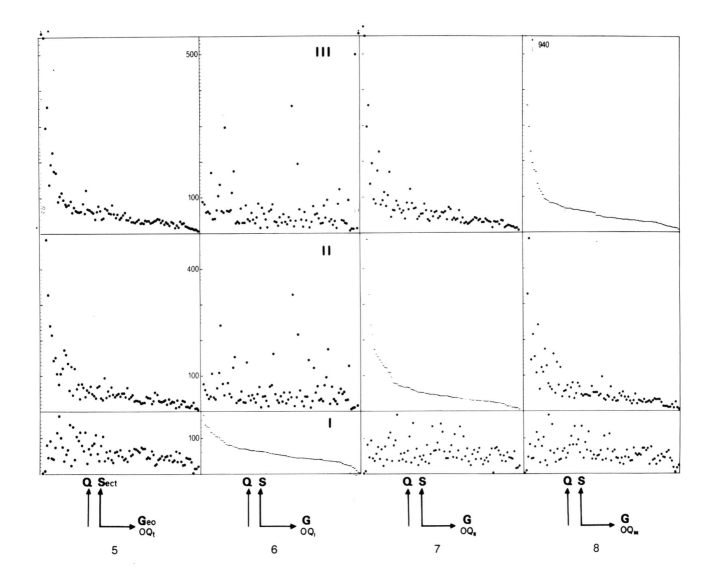

In fact, the very purpose of the diagrams is to bring this resemblance to light. It is even more obvious in the above drawings (figures 5, 6, 7, and 8). Their construction is similar to those on page 106, but here the quantities themselves are not depicted. Only the significant elements of the resemblance, the tops of the columns, appear. The complex "outline" displayed in the previous drawings is replaced by a linear form, which will be less dispersed as it bears a degree of resemblance with the series used for ordering the image. But drawing these constructions is a rather long and laborious process.

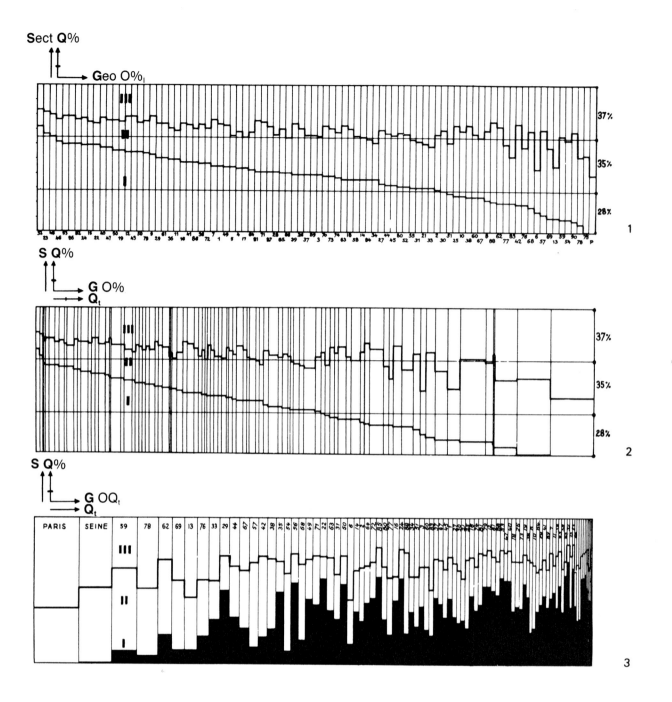

By using percentages and adopting the construction from figure 1 on page 104, much simpler images are obtained, since the total is constant and equal to 100. The resemblance referred to on page 107 appears in figure 1 here in the form of a progression, which is positive from left to right for sectors II and III, whereas it is negative for I.

Like figure 1, figure 2 is based on the order defined by the percentage of agricultural population, but here the *total quantities* per category are depicted by a variation in the width of the columns representing the different departments. Note that:
(1) the Q_t progress from left to right as with sectors II and III and in contrast to sector I;

(2) the national averages (27%, 35%, 37%), represented by the two horizontal lines, are meaningful and indeed correspond to the visual averaging of series III and I, which is not the case in figure 1.

The Q_t can be used to order the departments, as in figure 3, and we arrive at the same conclusions. However, reading is more difficult.

Given that we can choose categories ≠ (figure 1) or **Q** (figures 2 and 3) and that we can use four different orders and represent four different quantitative series, there are twenty possible graphic constructions of the form shown on page 109.

Discovering a degree of resemblance among the orders

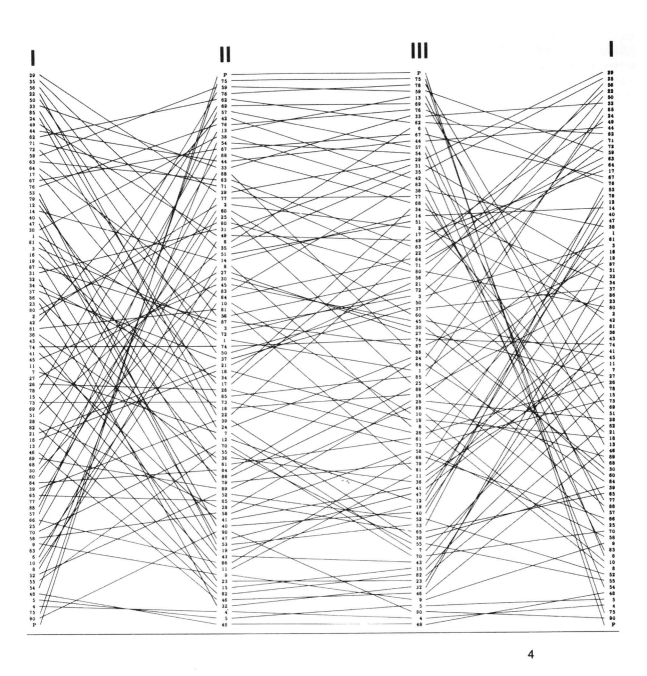

4

does not necessarily imply that the quantities, which are useful in the ordering process, must be represented. They are not portrayed in figure 4.

The departments, identified by a number, are ordered according to quantities in sector I (first column), II (second column), III (third column), and again I. The lines link the same department from one column to another. The nearer the order between two columns, the less numerous are the intersections. The resemblances and differences already discerned appear clearly here. Furthermore, a precise count of the number of intersections, although painstaking, enables us to "measure" the degree of resemblance (see page 248). This kind of construction is practical when the length (number of categories) of the component is limited. Beyond about thirty categories, construction becomes tedious and reading difficult, as is the case here.

Again, the search for similarity does not necessarily imply that departmental identifications must be represented.

Identification of the different departments, which is useful for construction of the curves, is not used, for practical reasons, in figures 1–14 opposite. Only the ordered quantities enter into the interplay of comparisons.

The curves in figures 1, 2, and 3 are derived from the graphics on page 106. They are **repartitions**. They can be superimposed, as in figure 4, provided their totals are equalized (reduced to 100, for example). (See also page 246.)

Another construction is possible. Thus a **distribution** curve can be utilized when the information is not known for each element, but only by classes (see page 206). The transition from a repartition to a distribution can be made graphically by following, sequentially, the operations shown in figure 5 (calculation of the number of departments per class of quantity), figure 6 (portrayal of this number), and figure 7 (this figure corresponds to figure 6, rotated 90 degrees). All these distributions can be superimposed, as in figure 10, provided they correspond to the same number of elements (here ninety departments plus Paris).

The **concentration** curve (see page 206) shows to what degree each of the populations (sectors) being considered approaches an equal dispersion across all the departments, which would give a straight line instead of the curves in figures 11, 12, and 13. *Conversely*, if each of the populations were concentrated in a single department, this would produce a line corresponding to the coordinates of the figure. The superimposition of the different curves (figure 14) shows that:

(1) population I is more dispersed than II and III or, if one prefers, that the latter are more "concentrated";

(2) from this perspective, II and III again resemble each other and are different from I.

In figures 4 and 10, this resemblance appears after careful study, whereas in figure 14, it is striking.

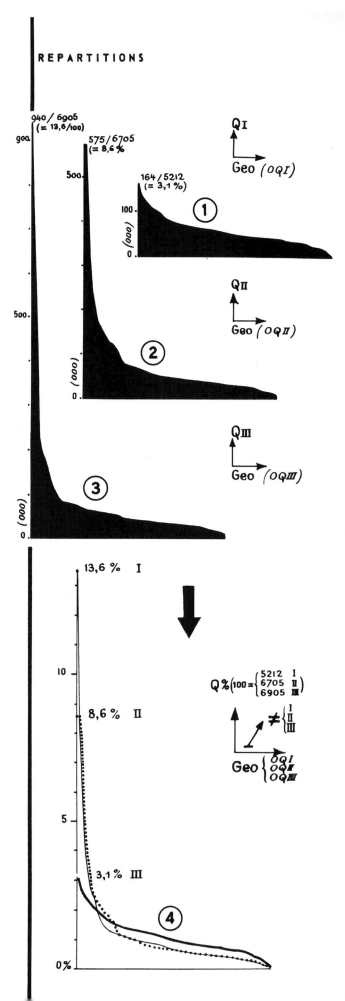

110

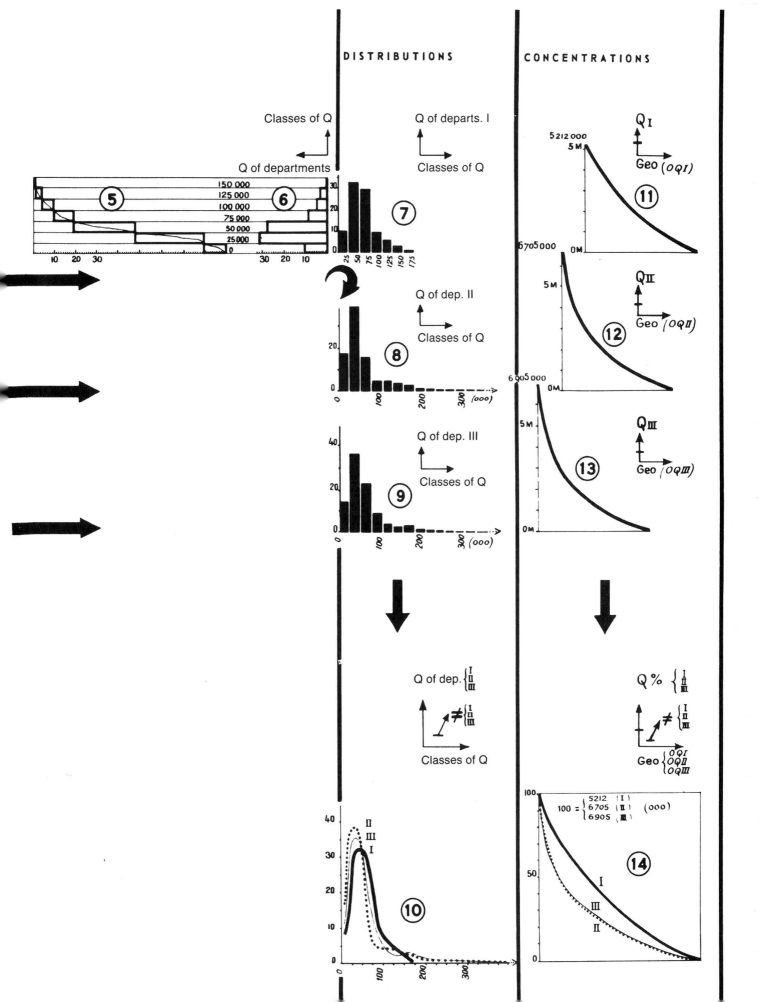

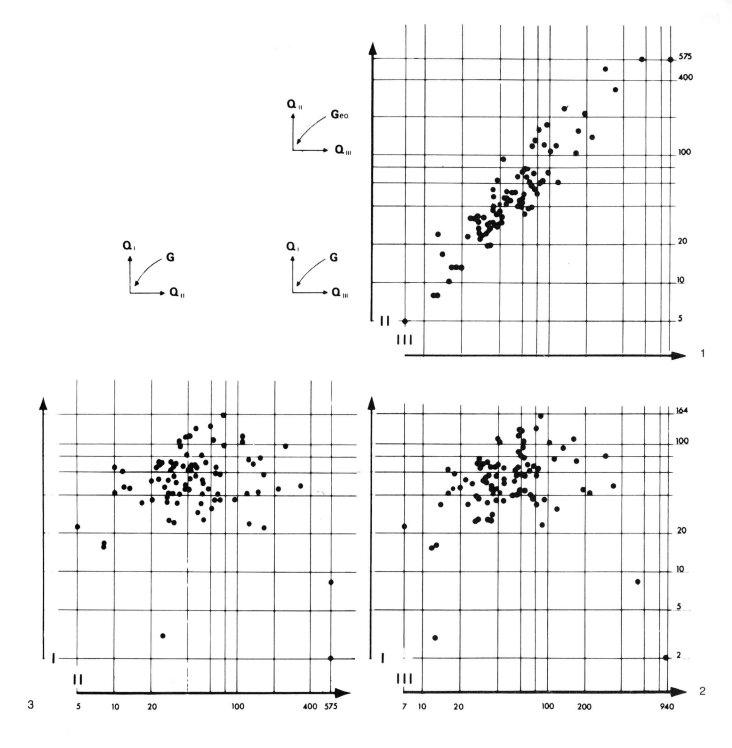

All the preceding constructions require long and tedious procedures: ordering the departments; graphically recording the quantities; calculating cumulative quantities, etc. However, if we are willing to sacrifice both identification of the departments and visualization of the quantities in the final image, as in figures 1, 2, and 3, the graphics can lead rapidly to the intended result: perception of similarities, which, we should stress, is independent from portrayal of the quantities and identification of the departments.

Figures 1 to 3 are based on the following visual principle:

In figure 1, for example, a department with a large population in sector III will be represented by a point to the right of the diagram, while a department having a large population in sector II will appear near the top.

In the case of a perfect correspondence between the two classings, the set of departments would be aligned on a straight diagonal line. Conversely, a lack of correspondence would be shown by a scattering of points over the entire figure. The alignment in figure 2 and the dispersion visible in figures 2 and 3 confirm the results obtained from the examples given on the preceding pages. Here the graphic

112

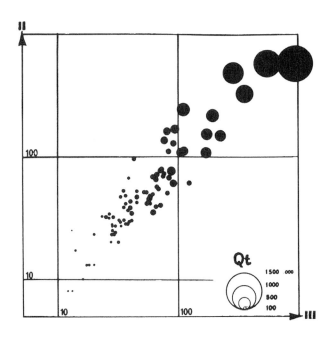
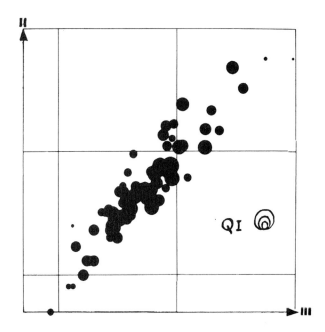

4

5

6

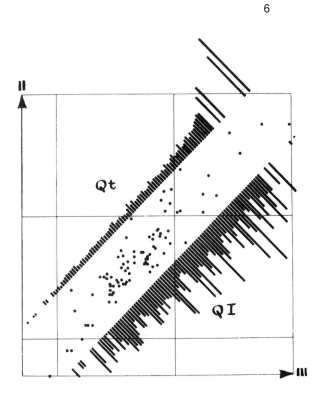

operation is much more rapid. However, comparison of the four orders would still require three additional figures relating total population to the populations of sectors I, II, and III, respectively. The complete picture would thus involve six constructions in all.

The proliferation of diagrams can be avoided by utilizing a third, ordered variable. For example, one can suggest an "elevation" above figure 1, which portrays the order resulting either from the quantities of total population (figure 4), or the quantities of population in sector I (figure 5).

Finally, as shown in figure 6, the four orders can even be represented in a single graphic. The useful conclusions are obviously still the same.

Note that:
(1) the constructions in figures 1, 2, and 3 require no ordering or calculating;
(2) the same is true for figures 4 and 5, if we use the "natural series of graduated circles" (see page 369);
(3) the double logarithmic scale is only useful here for separating points which would be too close on an arithmetic scale; this corresponds, in the final analysis, to a reduction in the size of the sheet of paper;
(4) any regular scale (provided it has the same progression for the two coordinates) would lead to the same conclusions.

113

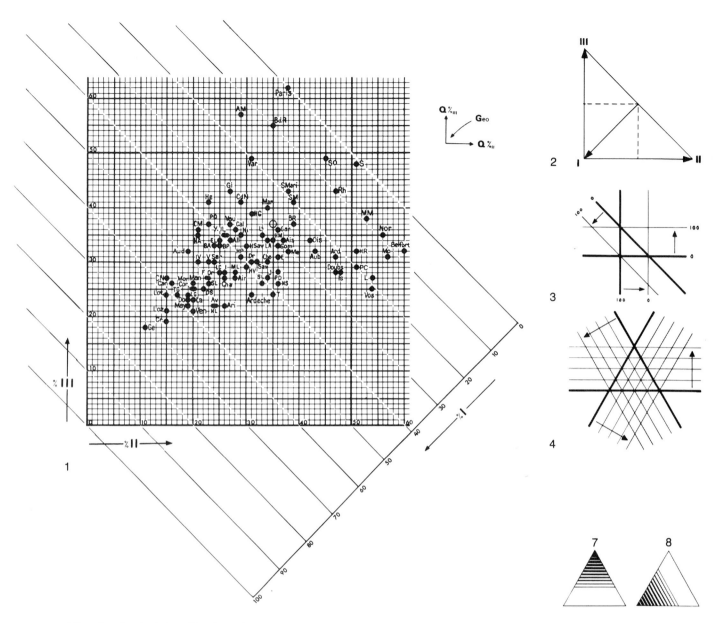

The visual principle utilized on the preceding pages can also be applied to percentages, as in figure 1 here. Since by definition the three sectors have a constant sum (100), it then follows that when we represent two sectors, the third will also be represented.

Obviously, a department having 50% of its working population employed in sector II and the remaining 50% in sector III will have a population in sector I of zero. Indeed, the line of 0 population in sector I passes through the intersection of 50% for sector II and 50% for III. Likewise, line 0 passes through the intersection of 40% and 60%, 30% and 70%, etc. Conversely, a department with a population in sector I equal to 100% will have populations in II and III equal to 0, which corresponds to point 100 for sector I. Thus the greater the working population of a department in sector II, the farther it will appear to the right of the diagram. The more employment in sector III, the nearer the department will be to the top; the more employment in sector I, the closer the department will approach the lower left-hand corner. All three orders are visible.

However, as figures 2 and 3 illustrate, the visual variation representing sector I is more limited than those representing sectors II and III.

By equalizing the three fields of variation, as in figure 4, we construct an equilateral triangle, which is the basis for all triangular graphic constructions. Such graphics are applicable to any information involving a component which has three categories (sectors here) and quantities whose total is meaningful.

Figure 5 shows a triangular graphic, constructed by ordering the percentages of sectors I, II, and III. This graphic is interpreted by first noting that the greater the percentage of employment in sector I, the more a department approaches angle I; the greater the percentage of II, the more the department approaches angle II; likewise for III. This means we must graphically identify the three

114

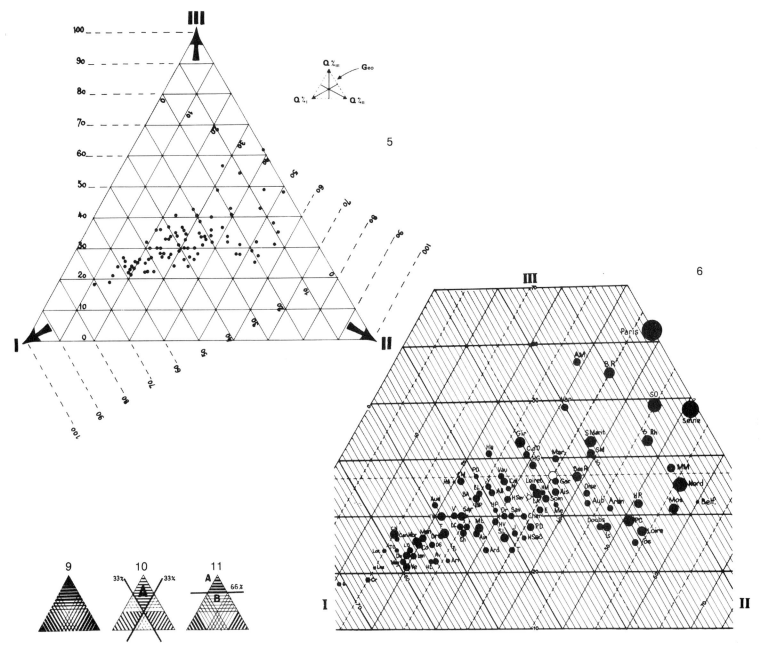

angles of the graphic and not the three sides, as habit would dictate.

In the information being considered here, one notices that whatever the percentage in sector I, the corresponding percentages of II and III have a tendency to be similar; the shape of the cluster of points reconfirms the preceding discoveries, though less strikingly. This is because the relationship is less pronounced for percentages than for absolute Q.

Finally, we can compare the three orders obtained to a fourth, that of the total population (figure 6), according to the procedure used on page 113. Here, however, the "retinal variable" (size) is based on the absolute quantities, not their square, as was the case with figure 4, page 113, for example. Legibility is reduced, but we perceive that the link between percentage of II and III and Q of total population is much weaker than the link between Q_{II} and Q_{III} and Q_t depicted earlier.

Note that the triangular graphic permits us to divide the collection of departments into different categories, defined by their place in the triangle, that is, by a quantitatively measurable characteristic resulting from the entire set of the three categories. For example, we can group in the same category all departments below 33% for two sectors and consequently between 33% and 100% in the third sector, as in figure 10, A. This category can then be divided in two, as in figure 11, A and B.

Finally, this hypothetical central value of 33% can be replaced by the observed central value—in this case the intersection of 28%, 35%, and 37% (see page 108)—which is represented by a white circle in the middle of figure 6. This point can serve as a center for defining the categories, as outlined above (see also page 232).

These categories can then be recorded on a map or a cartogram, as we shall see in the following pages.

115

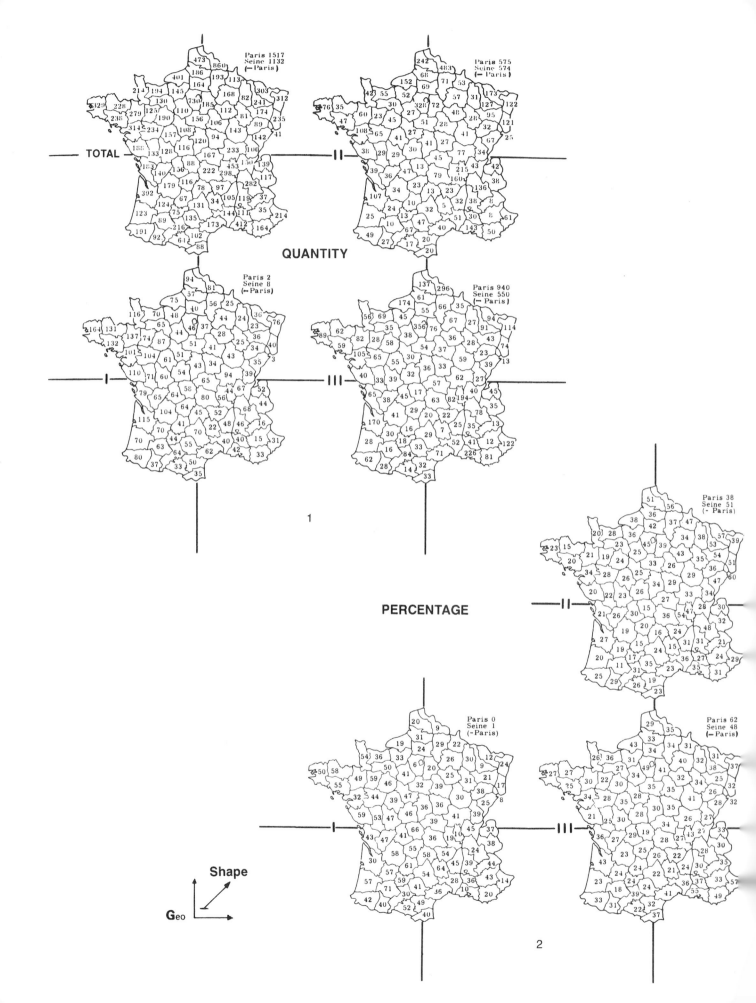

Maps

Since the information from the table on page 100 is geographic in nature, it can also be represented by maps.

The first operation is to record all numbers in their geographic areas (departments). We obtain the maps shown in figures 1 and 2. Obviously, this does not result in a single image, but in numerous figurations and a multitude of shapes, which would have to be *read* one by one in order to grasp the content of the information.

The first objective of graphic expression is to reduce these multiple shapes to the smallest possible number of meaningful shapes (images), in such a way that they can be retained and compared, without losing the least bit of information.

Cartographic representation, in fact, produces additional "new" information: the density, that is, the relationship of the numbers to geographic area.

It will be necessary to represent this information, either by special maps, as in figure 3, or by other constructions which make the concept of density visible through the representation of the quantities.

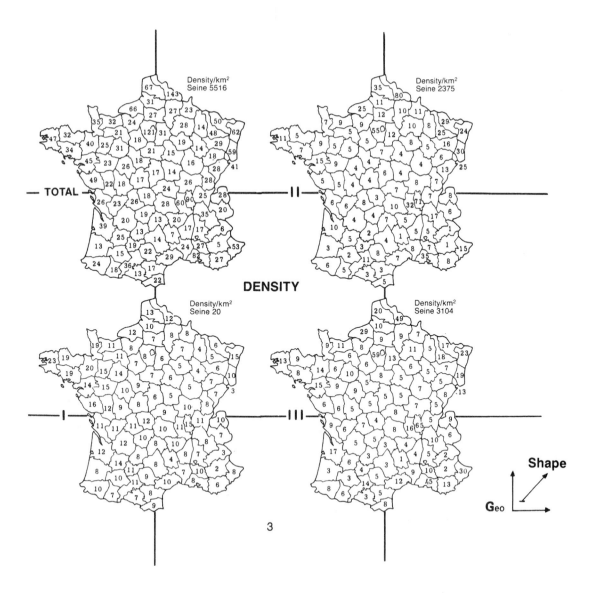

3

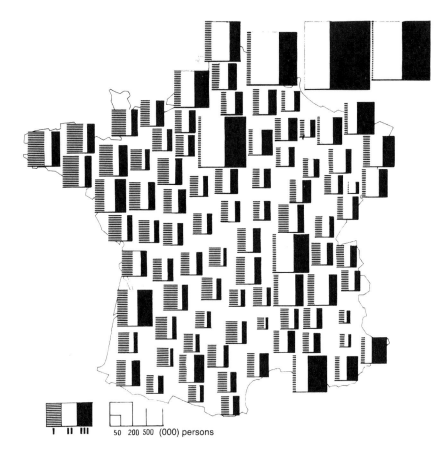
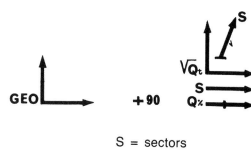

S = sectors

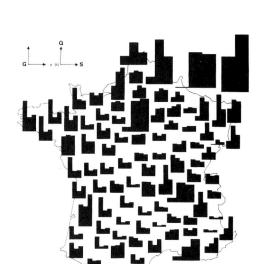

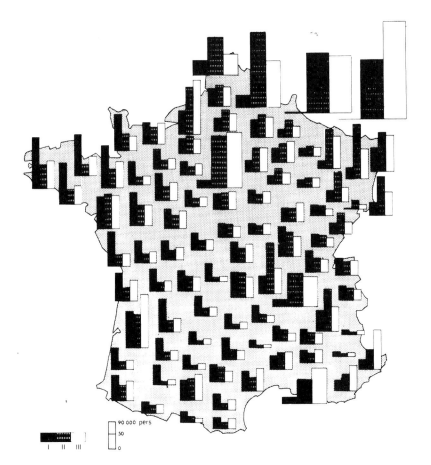
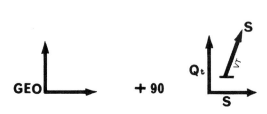

118

CHARTMAPS

It is obviously very simple to temporarily disregard the geographic component and construct a diagram for each department. Arranged later according to geographic order, the diagrams form a "chartmap," comprised, of course, of as many images as there are departments.

With this principle, "anything goes," and indeed the designer's imagination can go a long way, as is evidenced in the chartmap on page 394.

These constructions (figure 1, here) are so inefficient that they require the use of a retinal variable in order to differentiate the nongeographic categories (i.e. the sectors) and promote any comparison of individual diagrams. But even color is insufficient for resolving these problems.

Although the chartmap answers elementary questions introduced by the geographic component—"What is there at a given place?"—it is inefficient for any question of an intermediate level (regional evaluations), for overall conclusions, and, of course, for external comparisons.

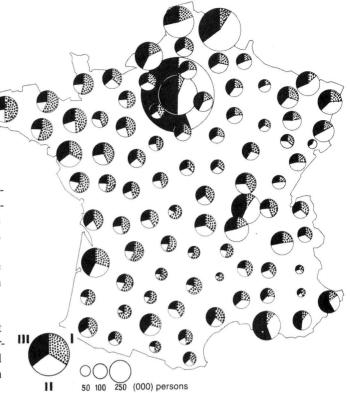

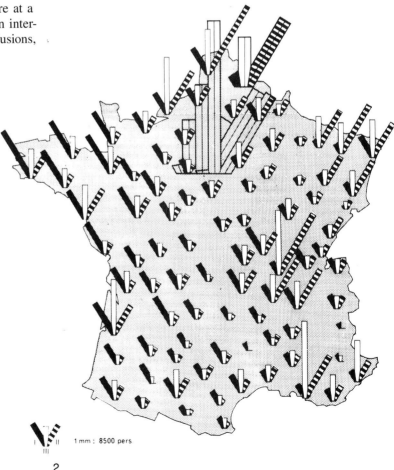

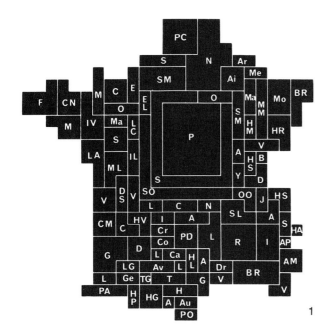
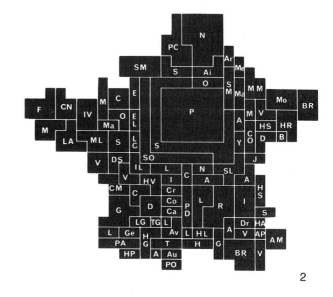

CARTOGRAMS

By relating the divisions of the map (departments), not to the actual geographic areas occupied by these departments, but to the total quantities of population per department, we adopt a very different system, THE CARTOGRAM, in which visible space changes its meaning.

The image is spectacular, and the observer is immediately struck by the gigantic proportions of the Parisian population. We evaluate these quantities of population by size differences, as well as by the distortions of geographic space which they entail. But it is precisely these distortions which prohibit the reader from identifying the departments by means of geographic familiarity. Acquired habit has become useless! We must either "read" the name or number of the department (figure 1), that is, return to the elementary level of reading, or else form a new habit of identification. Unfortunately, the principle of construction itself prohibits the forming of such a habit for two reasons!

(1) Each designer will construct a different cartogram from the same information. Thus, figure 2 differs from figure 1 because of the principle of transformation being used. The characteristic "outline" of France has been purposely deformed in figure 2 in order to emphasize the regions where the population is relatively large (Brittany, Nord) or small (the Aquitainian Basin). In figure 1, on the other hand, the designer has attempted to preserve the exterior geographic shape of France by allowing only internal changes.

(2) There are as many components able to serve as the basis for a cartogram as there are imaginable concepts. And for each component the place, and thus the shape and position of a given department, will be different.

One can compare several images based on the same cartogram. However, they will not facilitate geographic identification, which draws a large part of its effectiveness from the constancy of its component shapes. A cartogram, like a map, can accommodate several different retinal variables, as in figure 3, which is similar to figure 3 on page 123, or figure 4, which is similar to figure 5 on page 123. Note that the resemblances among the sectors are hardly visible and, in fact, are overwhelmed by the striking differences in total working population per department.

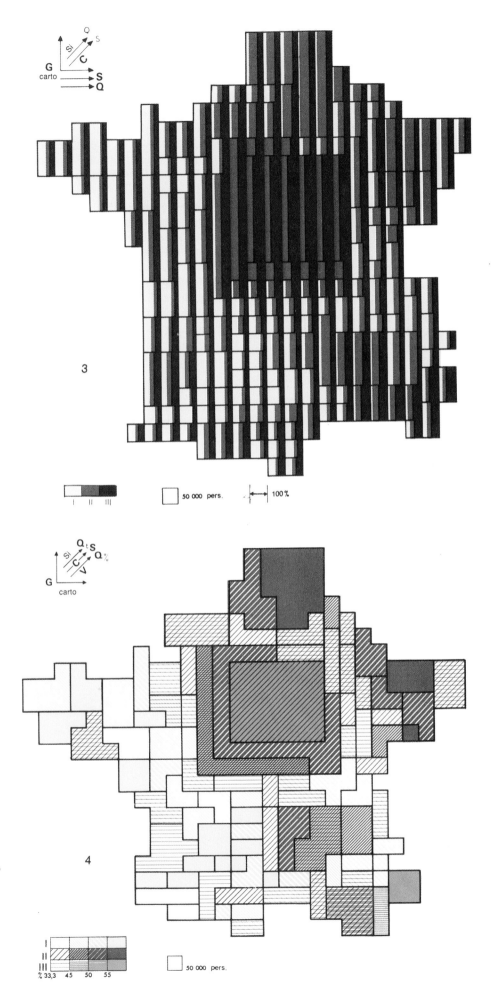

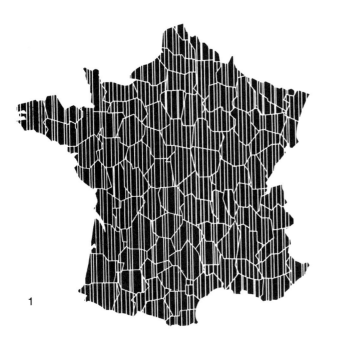

1

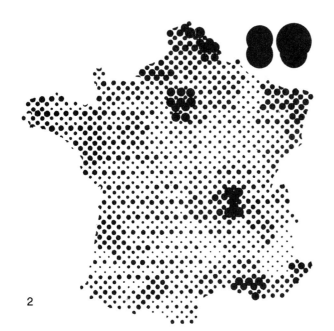

2

COMPREHENSIVE MAPS: FIGURES 3 AND 4

Here the plane is something other than a simple receptacle for two-dimensional diagrams. Furthermore, it is once again geographic.

These maps result from the superimposition of three images, one per sector, and the use of a retinal variable to try to differentiate them. This variable is indispensable in figure 3; without it the overall image (figure 1) displays only the totals, which are all equal (100%). In figure 4 the retinal variable is indispensable for differentiating the sectors, but the overall image (figure 2) denotes the absolute quantities, all sectors combined.

These constructions are *comprehensive* and represent the totality of each quantitative series. However, it is still difficult: (1) to obtain a rapid answer to the question, "What is the distribution for a given sector?"; (2) to compare this distribution to that of the other sectors; and (3) to retain the information visually in order to compare it with external information.

SIMPLIFIED MAPS: FIGURES 5 AND 6

The map in figure 5 is more legible than the previous ones, but this is achieved at the price of considerable *simplification*. No quantitative perception is possible. Four ordered steps, whose numerical meaning would be impossible to define without the legend, replace all the numbers.

This map imposes on us a certain regional perception of France, based on the definition of the steps (which could have been entirely different), and it prohibits us from making different evaluations and from criticizing the regionalization with full knowledge of the facts. To this area representation, one could, as in figure 6, add point signs (here denoting the total quantities). They quantify the populations, previously given only in percentages.

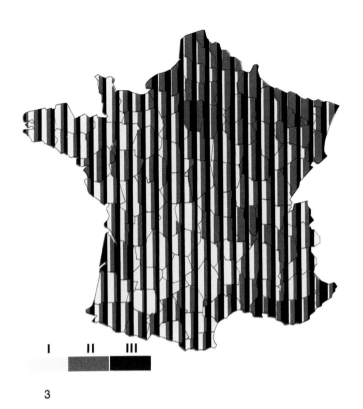

3

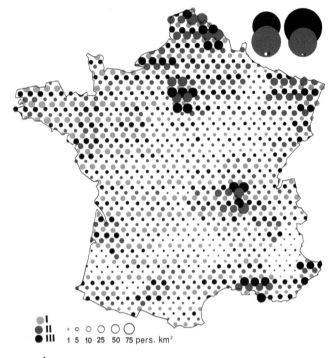

I
II
III

○ ○ ○ ○ ○
1 5 10 25 50 75 pers. km²

4

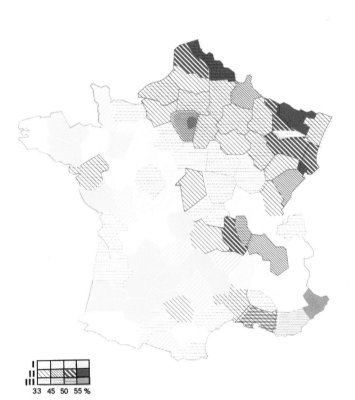

I
II
III
33 45 50 55 %

5

6

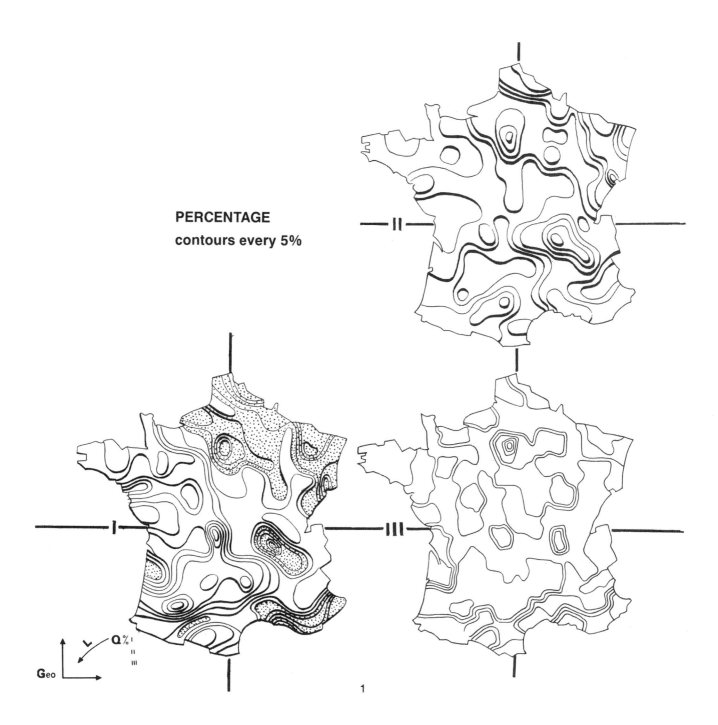

1

CONSTRUCTIONS INVOLVING SEVERAL MAPS

Isarithms (contours)

It is relatively easy to retain several separate images. We can therefore represent the information in three images, one per sector (sometimes with a fourth image for the total population). Here again, numerous graphic solutions are possible.

Isarithms, applied to the percentages, lead to the maps of sectors I, II, and III in figure 1 above. We see (III) that the contours are not sufficient for portraying the quantities. The immediately perceptible visual variation applies to the spacing of the curves, that is, to the gradient which separates two regions. But, we remain unaware of the "sense" (up or down) of this gradient and, as a result, uninformed about the quantities themselves. The sense can be suggested by shading (II), especially in combination with a discrete dot pattern (I).

In no case, however, can the notion of quantitative value be obtained from these images. Note that the contours can be traced either by following the department boundaries (III) or by utilizing a reference point placed at the visual center of the department (I and II). (For further discussion of isarithms, see page 385.)

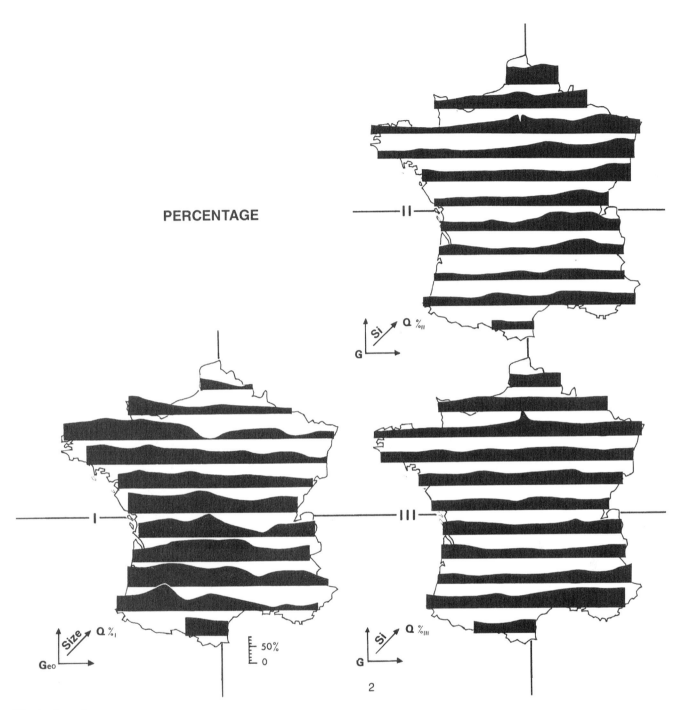

Vertical sections

Isarithms correspond to the sectioning of a volume by a series of horizontal planes. This volume can also be suggested by a series of vertical planes, drawn as sections. If the spacing is planned accordingly, these sections can be "shaded" in black, as in figure 2, and a quantitative perception results: the more black, the greater the quantity. These images are much more efficient than the preceding ones, but the information has lost much of its precision at the department level.

125

Value variation

The use of areas characterized by different value steps is a common solution. Indeed, it seems quite easy to construct such a map, while assuring excellent legibility. However, difficult problems occur in reproduction and photographic reduction (in map I, figure 1, for example, step 54 can be confused with black), and especially in the construction of the map. With what numerical classes should we associate the perceptible steps of value? There is no general method, although many have been proposed. One could use the mean, the median, an equipartition of geographic space or of the scale of quantities, various progressions, or combinations of several methods.

The interpretations in figure 1 below, based on different scales that are each appropriate to its own distribution, constitute a standard case in which the designer adapts the perceptible steps to any variation in quantity, whether it ranges from one to two or from one to one million, in order to obtain a diversified and vigorous image. As a result, white and black can represent any numbers whatsoever, and, since white can never be used as a precise unit for measuring black, or vice versa, quantitative perception is not possible with this formula. One must resort to numbers.

This solution reduces considerably the information transmitted and opens the door to unjustifiable interpretations.

In this example, value is combined with texture and shape to produce a good selection of distinct steps.

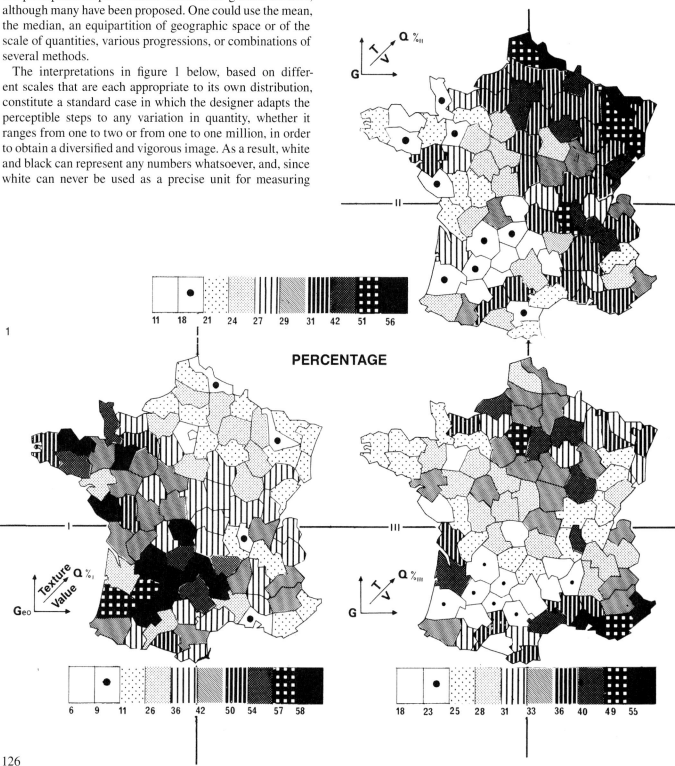

Graduated sizes in a regular pattern

A regular pattern of different-sized circles (figure 2), applied to given areas, proves easy both to construct and to copy (see also page 369). Above all, it is the most rigorous of all graphic solutions for quantitative problems. Since it is capable of espousing all the nuances of the information, it avoids the problem of a prior choice of steps, a major difficulty in the previous example. Here the reader's interpretation is not falsified by the designer's decisions.

The method of construction is universal. The information does not have to be simplified and thus "reduced." Perception is of a quantitative nature, and recourse to the legend is only necessary for rigorous discriminations (the steps given in the legend are merely reference points, since the different sizes of the circles encompass all the perceptible variations of the quantities). In short, the graphic is immediately legible, whatever the level of reading adopted (page 372).

Comparison with the preceding construction illustrates the extent to which the reader can be dependent on the designer when any quantitative information is divided into a limited number of classes, represented by steps of value.

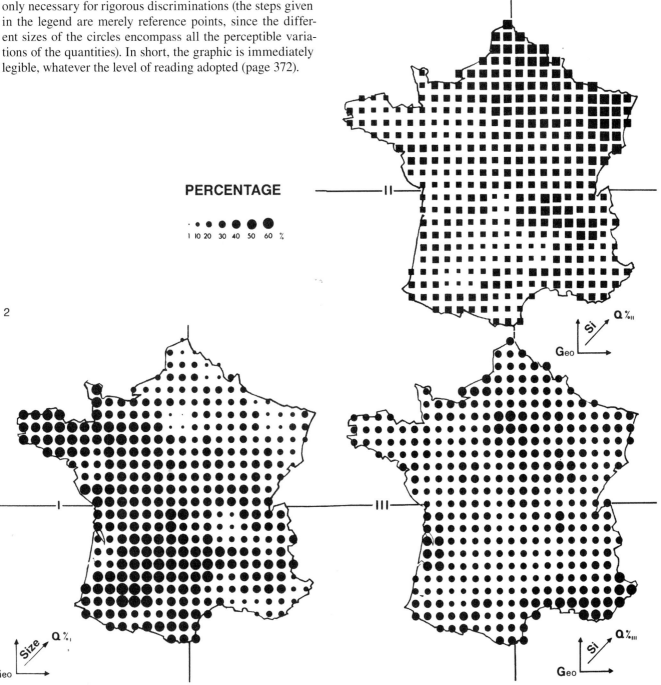

Of the same quantitative nature as the preceding construction, figure 1 has an amount of black in each department which is visually proportional to the data.

It must be noted that this solution can only be applied to the representation of percentages, since the constant width of the band corresponds to a constant total. The map is also quite difficult to construct.

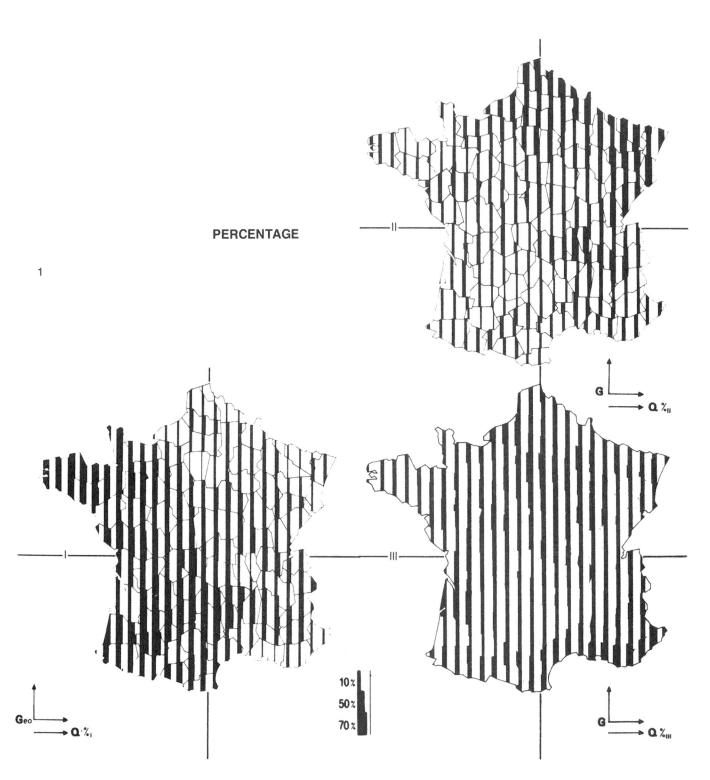

PERCENTAGE

1

Chartmaps

In the previous constructions utilizing three separate images (pages 124–128) the quantities have not yet been represented. They are perceptible in figure 2 here, as are the percentages. A rectangle corresponds to each department, although the rectangles have been aligned. Their *base is proportional to the total working population* in the department, and their *height proportional to the percentage* of each sector.

The area of the black rectangle is therefore an expression of the absolute quantities.

This method implies some displacement in relation to geographic location; the Parisian region, for example, must occupy an entire line by itself. Note again the similarity between II and III.

PERCENTAGE QUANTITY

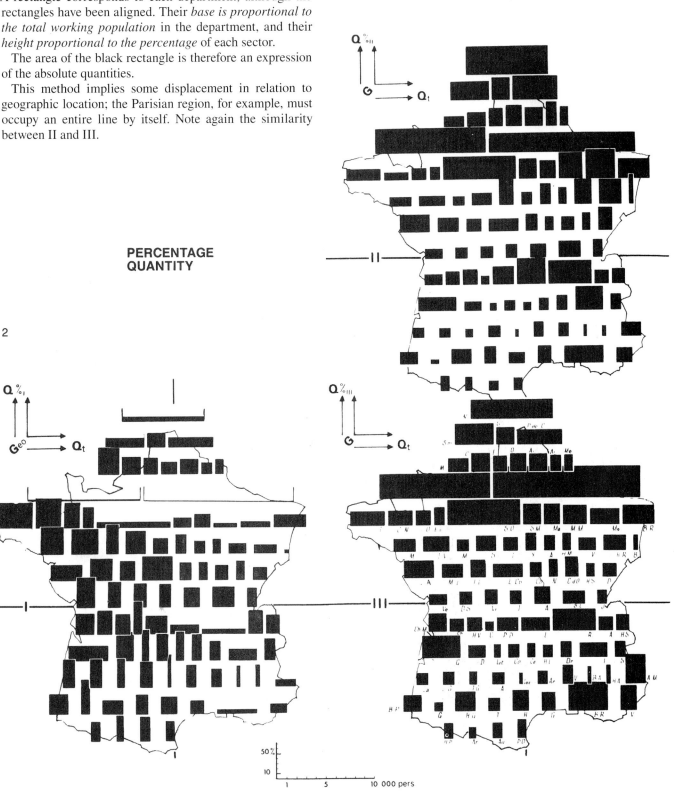

2

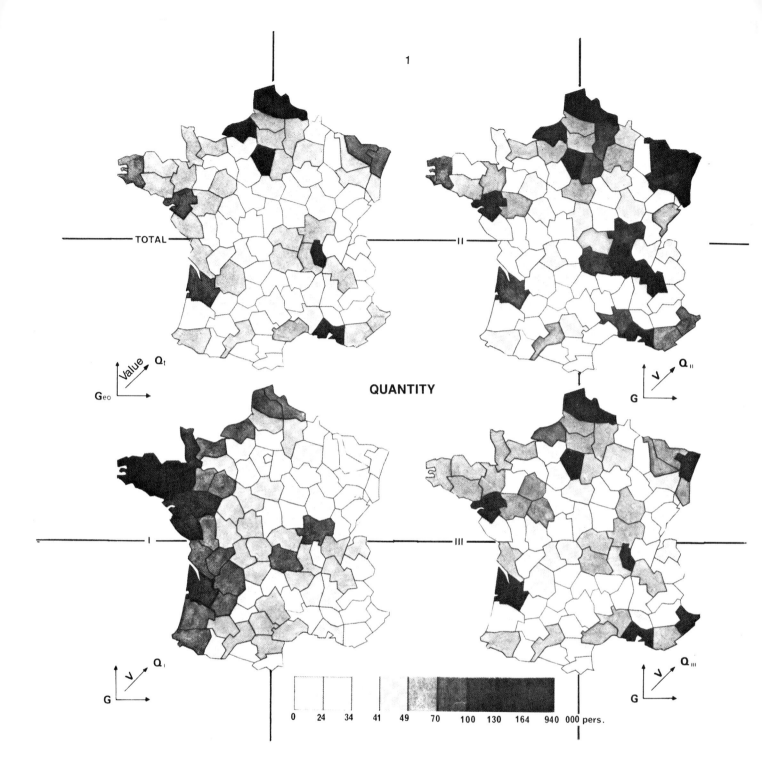

Value variation

Different value steps, applied to absolute quantities, result in completely erroneous images. It is sufficient to compare these maps to a rigorous expression of the quantities (page 137) to determine this. The values are not quantitative, and they are also applied to variable departmental areas—a variation already incorporated into the data, since they have the form QS (see page 38). With this construction, it is necessary to draw two maps, one for the quantities (figure 1), another for the densities (figure 2, opposite) in order to get something out of the information. Note that value alone is used in figure 1; there is no redundant combination with another visual variable. This creates some difficulty in accurately recognizing a given perceptible step. Note also that the values in this series have been affected by the printing process.

The map of densities in figure 2 partially corrects the preceding one (figure 1). But it amounts to an entirely different

2

DENSITY

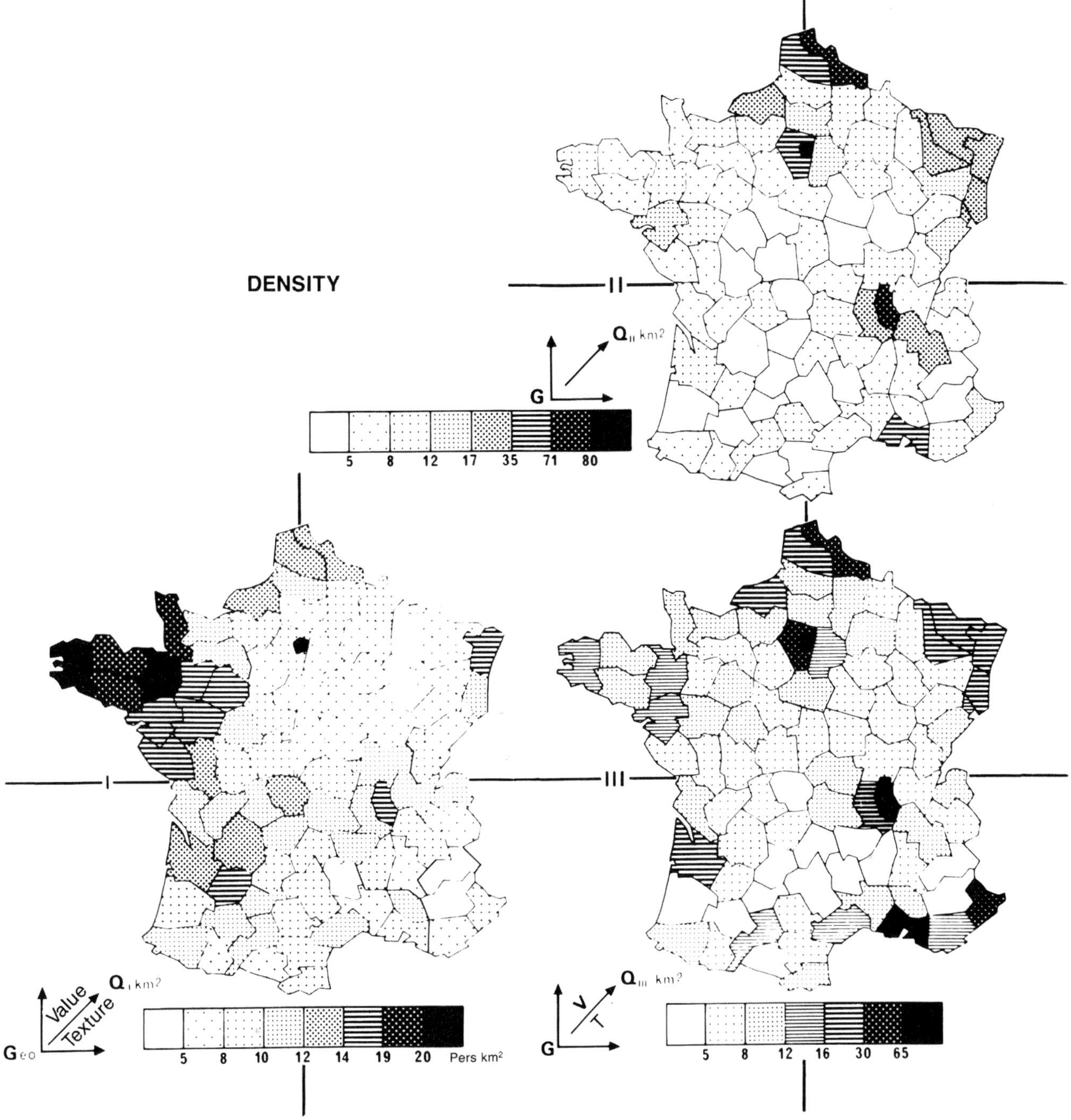

view of the information, necessitating great mental effort to integrate it with the other maps.

A comparison with the following maps fully illustrates the defects of maps based on value. They are only utilizable with a simplified and highly interpreted message, which is incompatible with a comprehensive inventory or an experimental image.

131

Perspective maps

Isarithms, applied to densities, lead to the constructions in figure 1 below. When the quantities have a wide range, as here (from 2 to 3, 104), only a perspective map allows us to make the actual differences visible.

The figure then suggests a volume proportional to the total quantity (density per km² × number of km²).

1

DENSITY QUANTITY

1 contour = 3 persons/km²

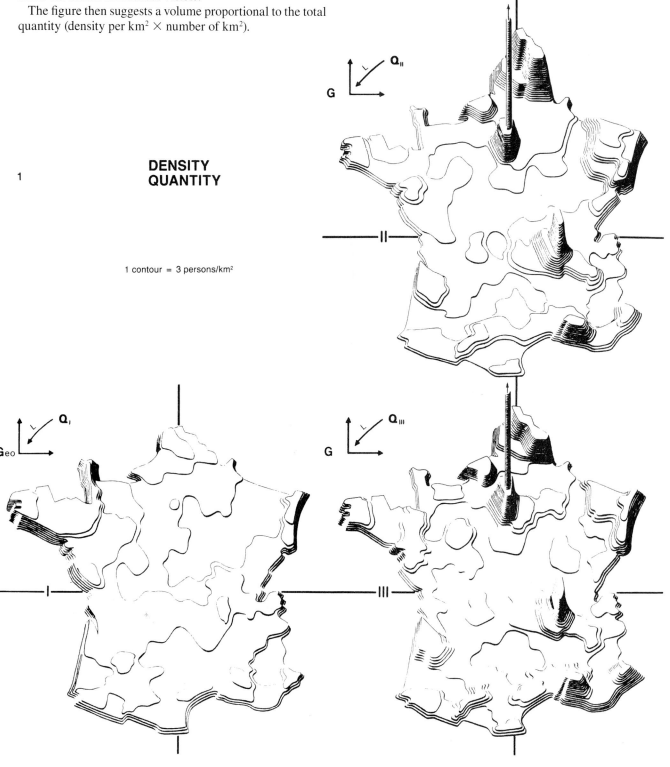

These graphics are always difficult to construct. Just as on page 125, volume can be defined by a series of vertical planes. But here (figure 2) black cannot be used over the entire visible section.

A lighting effect can render this construction quite spectacular, but it remains very difficult to draw.

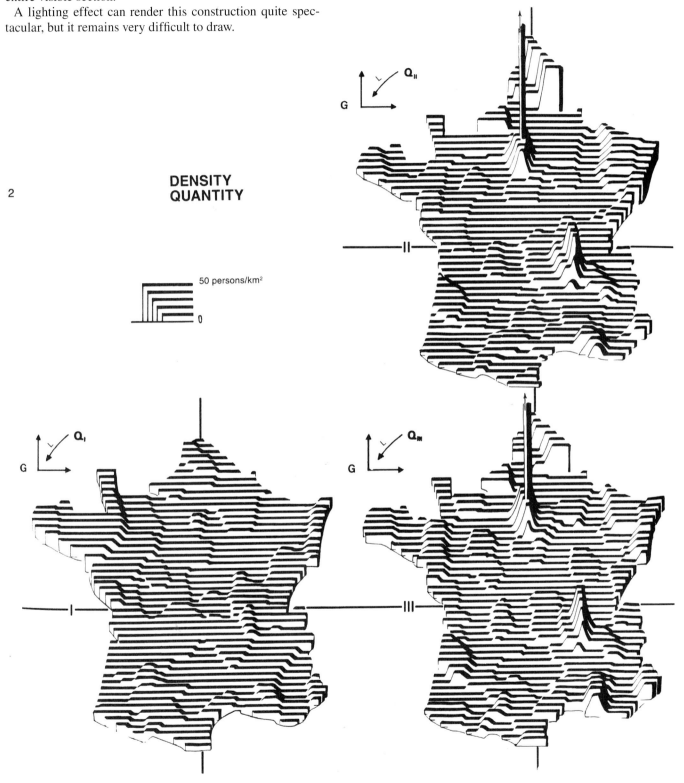

2

DENSITY
QUANTITY

50 persons/km²

133

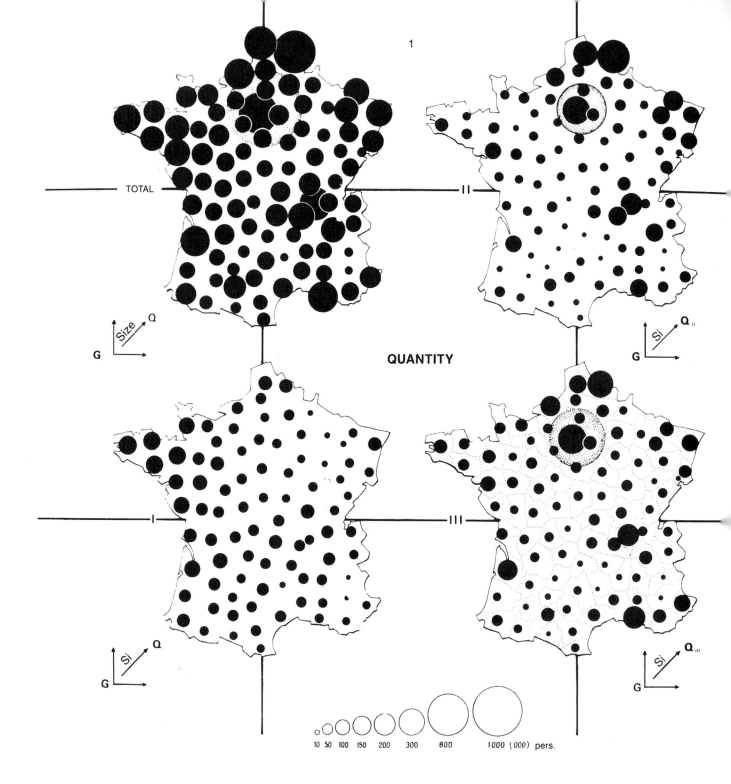

Proportional sizes

In figure 1 the quantities are represented by *different-sized circles* for each department. For the Paris area the circle must be rendered "transparent." This is an easy construction to draw, and is immediately perceptible, without a prior choice of steps. But perception of the densities is practically excluded. It would be even more difficult if the French departments were not relatively similar in size.

The total working population per department is portrayed as the total of the three other maps, which, because of the overwhelming size of the circles, makes comparison with sectors I, II, and III rather difficult.

The quantities can also be represented by *columns of different height* (figure 2). The placement of the columns denoting Paris and the department of the Seine, on the one hand, and the department of Seine-et-Oise, on the other, necessitates

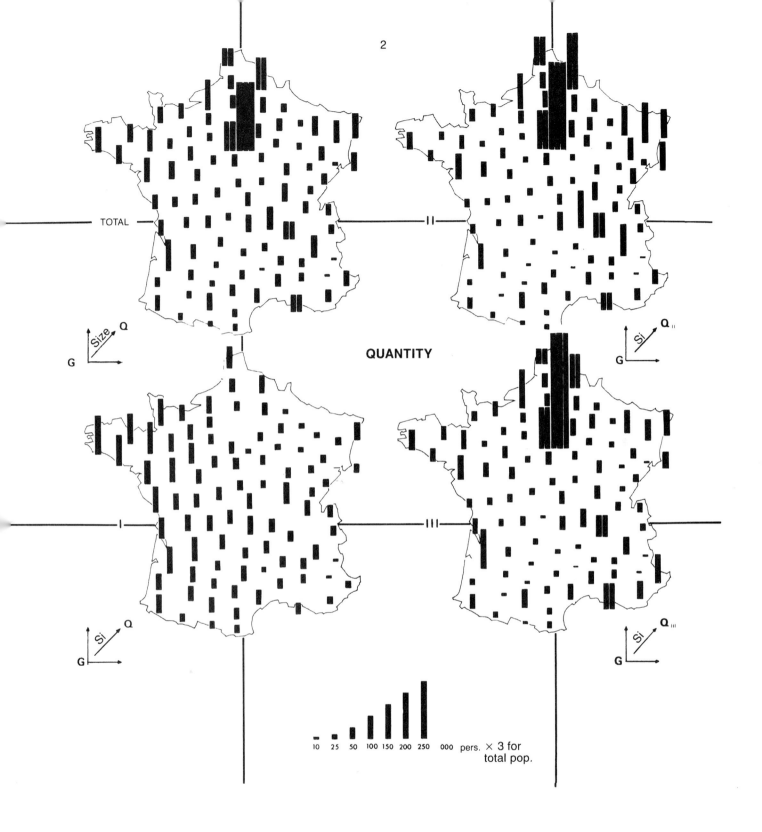

a slight displacement of the departments located north of them. For the other departments, the column is situated in the center of the department. Here the total population is represented at one-third of its actual size (in comparison with the other maps), which gives the map of total population the same visual density as the sector maps and renders their distributions comparable. The resemblance between II, III, and the total population, obscured on page 134, appears clearly here.

Equal-sized points: dot maps

The point signs are of equal size and are proportional in number to the quantities per department. There is a "dual" perception: of the density (ratio of black to white); and of the quantity (to the degree that points can be added visually). The size and number of the points, as well as their numerical meaning (here ten thousand persons), are completely interdependent.

Note that points are three times smaller for the total population, which makes this map comparable to the others. The representation of the Parisian population poses a problem, resolved in figure 1 by the use of a "horn," which suggests an elevation above the plane.

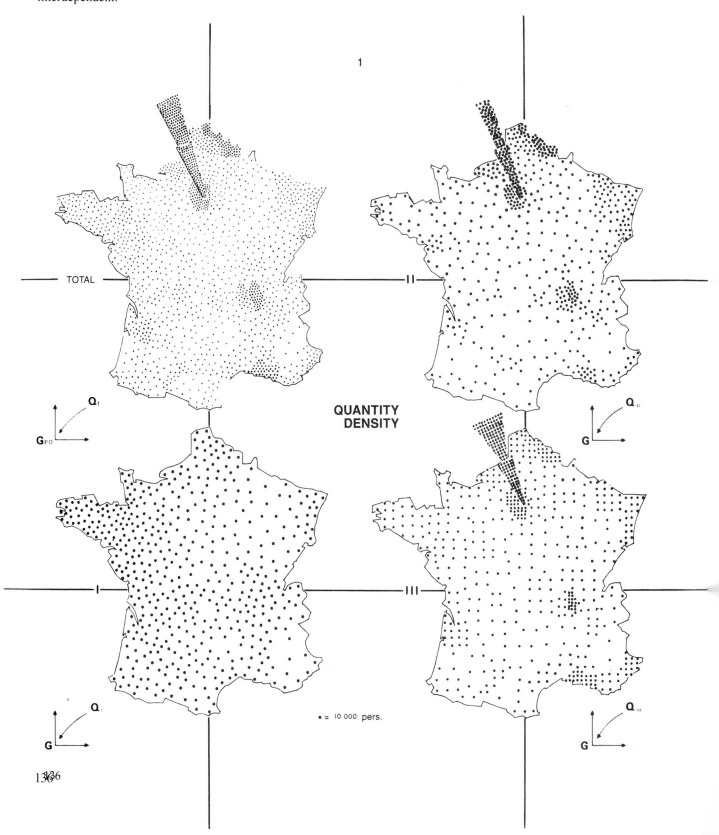

• = 10 000 pers.

Graduated sizes in a regular pattern

Figure 2 below utilizes the same formula as on page 127, but here it is applied to the absolute quantities.

Note that the legend is double. Each point simultaneously expresses a density and a quantity. Thus, whether attention is fixed on one point or on a group of points, we perceive a total quantity and an overall density, or a partial quantity and an elementary density. When the area defined by a square of four points represents 1000 km² (or 100 or 10 . . .), the same numbers express density and quantity.

This solution seems to be the best geographic representation of the information being considered. If we add to it the representation of the percentages, from page 127, the observer has, in seven images, the total content of the information without prior interpretation. Furthermore, comparisons become extremely easy, for the whole of France, as well as for all regional variations. The correlation between sectors II, III, and the total population suggested by the earlier diagrams, is particularly striking here.

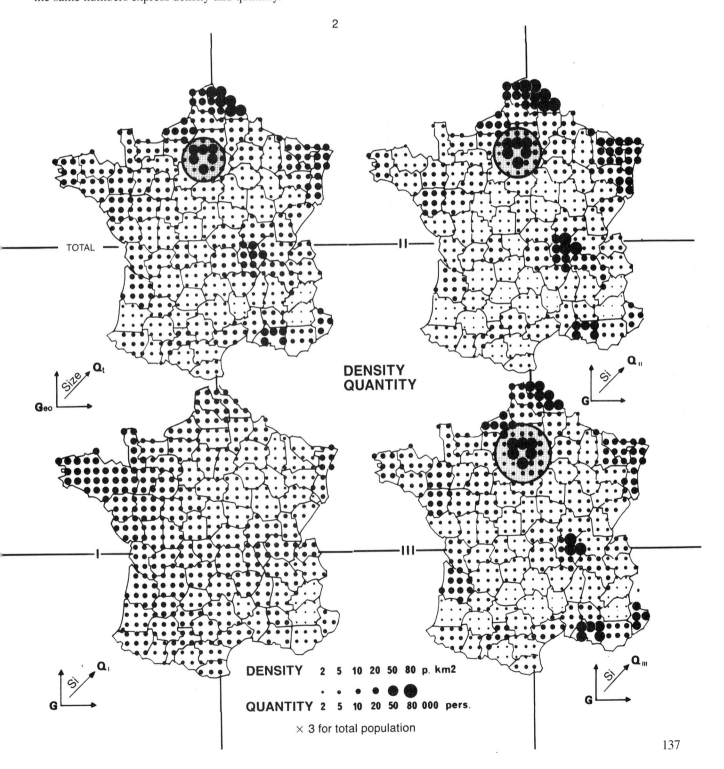

B. Image theory

EFFICIENCY

We have just examined a hundred graphics in terms of the correspondence between components and graphic variables. Some of the graphics are "good," others "worse," others simply "bad." But these opinions are purely subjective. We need only submit a dozen maps for evaluation by a group of readers in order to discover that each person will have a different opinion, based most often on considerations of an aesthetic nature.

It is important, therefore, to define a precise, measurable criterion which we can use to class constructions, determine the best one for a given case, and explain why readers prefer different constructions. We will call this criterion "efficiency."

EFFICIENCY is defined by the following proposition: If, in order to obtain a correct and complete answer to a given question, all other things being equal, one construction requires a shorter observation time than another construction, we can say that it is more efficient for this question.

This is Zipf's notion of "mental cost," applied to visual perception (George Kingsley Zipf, *The Psycho-biology of Language* [Boston: Houghton-Mifflin, 1935]).

This notion comes into play only when the drawing is utilized to semantic ends, that is, when the reader asks a precise question. In most cases the difference between an efficient construction and an inefficient one is extremely clear and can involve a considerable difference in the amount of perception time. We will consider several examples of this.

Efficiency is linked to the degree of facility characterizing each stage in the reading of a graphic. The set of observations which define reading facility enables us to formulate an "image theory." In this chapter, we shall discuss five aspects of "image theory," then show the role of this concept in defining the rules for constructing efficient graphics.

(1) Stages in the reading process

To read a graphic is to proceed more or less rapidly in the following successive operations.

EXTERNAL IDENTIFICATION

As we have seen (page 19), before all else the reader must identify, *in the mind*, the invariant and components involved in the information. Figure 2 obviously has no meaning without a prior understanding of the terms in figure 1. External identification relies on acquired habits, on the recognition of words, shapes, or colors. It permits us to isolate, from the vast realm of human knowledge, the precise domain treated by the figure. In a sense, "new" information is no more than the discovery of new relationships among concepts which are already known.

INTERNAL IDENTIFICATION

Next, the reader must recognize by what visual variables each of the components is represented *in the graphic* (page 24). In many diagrams, the arrangement of the words enables us to link the two operations. The name of each component is inscribed on the planar dimension representing it, as in figure 1. But whenever a retinal variable is involved, a legend becomes necessary to define its meaning precisely.

The two stages of identification, whose efficiency (perception time) depends on the particular case (a complete and definitive knowledge of the legend on page 156 would require several hours), always precede perception of the information itself.

PERCEPTION OF PERTINENT (NEW) CORRESPONDENCES

After identification (accomplished by the words and arrangement of the components), the reader is ready to perceive the series of pertinent correspondences (figure 2) which the drawing isolates from the vast number of possible correspondences. Obviously figures 1 and 2 are superimposed to form figure 3. To become "informed," to perceive pertinent correspondences, we must formulate a question, whether consciously or not:

"On a given date, what is the price of stock X?"

The questions can be highly diverse. However, for each data set they are of a *finite number* and can be rigorously defined.

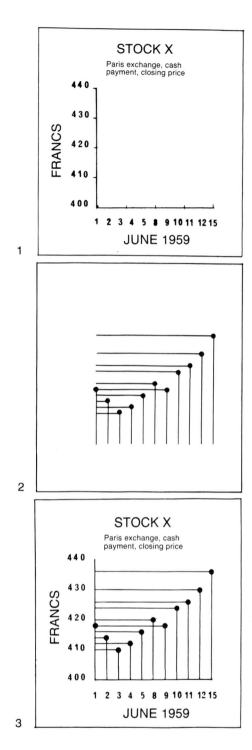

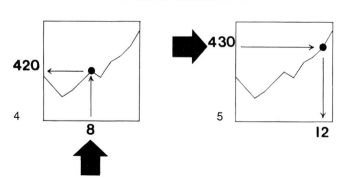

TYPES of QUESTION

(2) Possible questions

TYPES OF QUESTION

When we look at figure 3, two types of question can be formulated:
"On a given day (the eighth, for example), what is the quotation for stock X?"
Answer: 420 francs (figure 4).
"On what day(s) was a given quote (430 francs, for example) reached?"
Answer: June 12 (figure 5).

There are as many types of question as components in the information. Each component produces a type of question.

LEVEL OF THE QUESTION OR READING LEVEL

But another distinction is necessary. Indeed, for a given type of question one must distinguish the following cases.

(a) Questions introduced by a single element of the component and resulting in a single correspondence
Example: "On a given date, what is the quotation for stock X?" There is only one input date, and on that date there is only one pertinent correspondence, a single point (figure 6).
This is the elementary level of reading.

Here questions tend to lead outside the representation, to utilize the latter as a reservoir of information from which one extracts a piece of elementary information in order to transcribe it into another language or into another image.

(b) Questions introduced by a group of elements or categories and resulting in a group of correspondences
Example: "Over the first three days, what was the movement of the stock?"
Answer: The stock fell (figure 7).

Such questions are numerous, since the groups can be formed by any combination of the categories (dates) of the input component.
These questions constitute the intermediate level of reading.
Here, questions tend to "reduce" the length (number of categories) of the component in order to discover, from within the information, groups of elements or homogeneous categories which are less numerous than the original categories and consequently easier to understand and memorize. This is the internal form of "information processing."

(c) A question introduced by the whole component
Example: "During the entire period, what was the trend of the stock?"
Answer: The stock rose (figure 8).
This is the overall level of reading or "global" reading.

Such a question tends to reduce all the information to a single, ordered relationship among the components. It enables the reader to retain the whole of the information and compare it to other information. This is the external form of information processing.

We can therefore say that:
– in approaching information there are as many types of question as there are components
– within a type of question there are three levels of reading:
 the elementary level
 the intermediate level
 the overall level
– any question can be defined by its type and level.

Thus, from the moment that the components of the information are defined and their length known, and before any attempt at representation, it is possible to establish a list of all the questions which a reader can pertinently pose, which amounts to saying that:

it is possible to use the components, the variables, and their length to define all the objectives which the information can attain.

This analysis enables us to establish, if need be, a reduced list of questions which can produce an immediate answer; that is, we can construct the graphic in light of a precise objective. We can now understand one reason why opinion can be divided over the merits of a given drawing; some readers ask questions which can be answered immediately; others ask questions whose answer can only result from a series of elementary observations. But what is an immediate answer? How is it linked to the concept of an image?

LEVELS OF READING

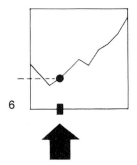
6

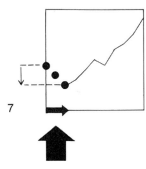
7

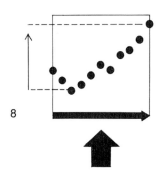
8

(3) Definition of an image

Answering a given question involves:
(a) an input identification (a given date)
(b) perception of a correspondence between the components (a point)
(c) an output identification (the answer: "so many francs").

This process implies that the eye can isolate the input date from all the other dates and DURING AN INSTANT OF PERCEPTION can see only those correspondences that are determined by this input identification, but can SEE ALL OF THESE.

During this instant the eye must be able to disregard all other correspondences. This is visual selection. We can state that in certain graphic constructions, the eye is capable of encompassing all the correspondences determined by any input identification within a single "glance," a single instant of perception. The correspondences can be seen in a single visual form.

The meaningful visual form, perceptible in the minimum instant of vision, will be called the IMAGE.

The curve of the trend of stock X (page 141, figure 8) is constructed in a single image, which means that any question will receive an immediate response.

But in some other constructions the answer to certain questions can only result from the mental accumulation of several successive images. Consider the information: volume of salaries distributed according to branches of the economy and size of enterprises.

INVARIANT —volume of salaries paid by enterprises
COMPONENTS —≠ five branches of the economy
Q percentage of salaries per branch, according to
O five categories of enterprise size

This information can give rise to the various classic constructions opposite (figure 1).

Consider the question: "For what category of enterprise size do we find the largest volume of salaries paid in the commerce branch?" All the constructions enable us to find the answer quite rapidly. Now consider the question: "In what branch do we find the highest percentage of salaries paid in enterprises of one to five workers?" which can also be phrased: "The highest percentage of salaries paid in enterprises of one to five workers, in what branch is it found?"

This question, governed by the component "size of enterprises," causes us to compare images, which must first be sought one by one, then memorized and classed in order to discover the largest. The answer to any question of this type is not immediate; it demands several successive instants of perception, at least one per branch.

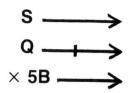

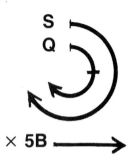

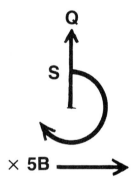

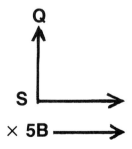

1

VOLUME OF SALARIES IN %

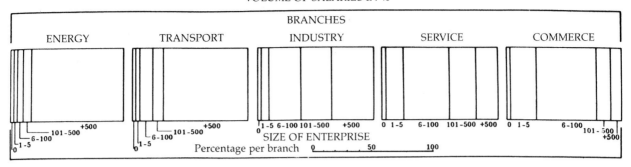

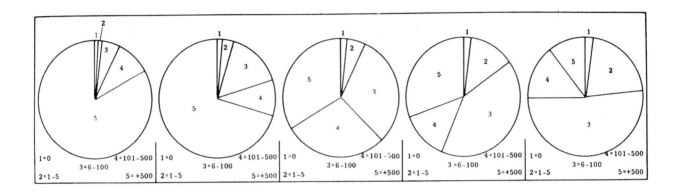

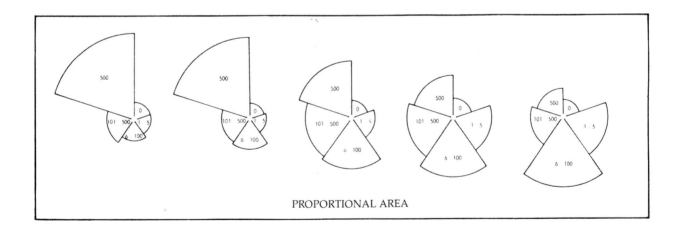

PROPORTIONAL AREA

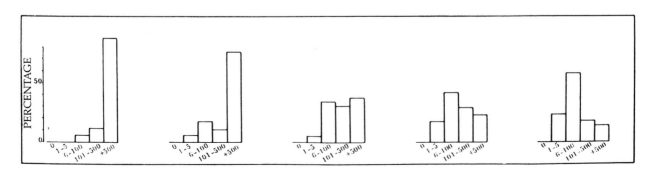

The inefficiency of figurations

Even when "blackened in," or, as we will see later, when differentiated by retinal variables, the constructions in figures 1–4 are completely useless for certain fundamental questions.

The overall level of reading translates into these terms: "In what order are the branches of the French economy classed, according to the percentage of salaries paid within a given size of enterprise?"

The answer requires a total reconstruction of the information. In order to illustrate this, it is sufficient to conceive of a question implying an external comparison of this information with other factors that define the French economy.

For example:

"According to the branches and the size of enterprise, is the volume of salaries proportional to the number of enterprises?"

If we use the same type of construction (in order to make the figures comparable), this question leads us to represent, not the volumes of salaries, but new information: the number of enterprises (in percentages), as in figures 5–8.

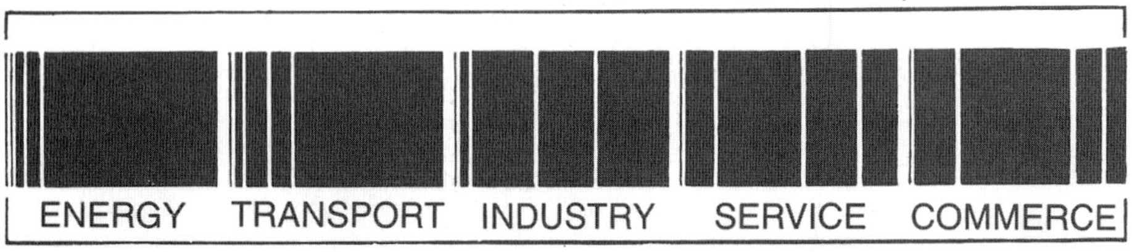

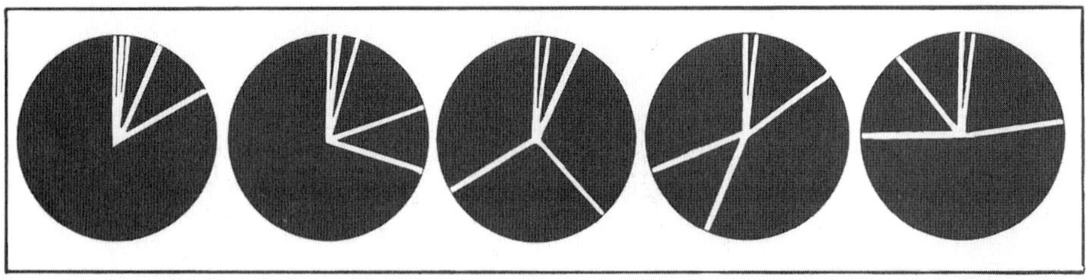

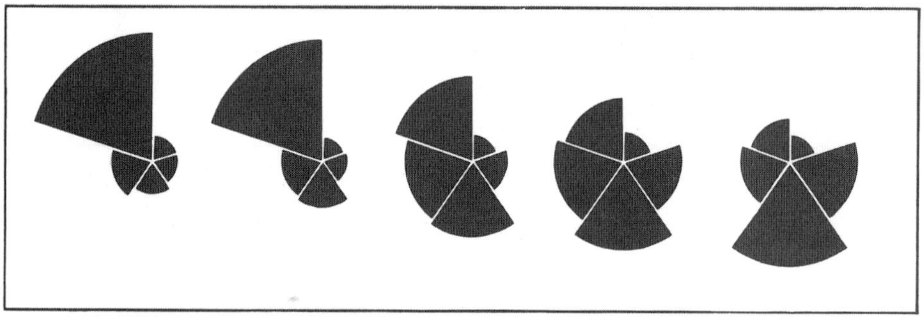

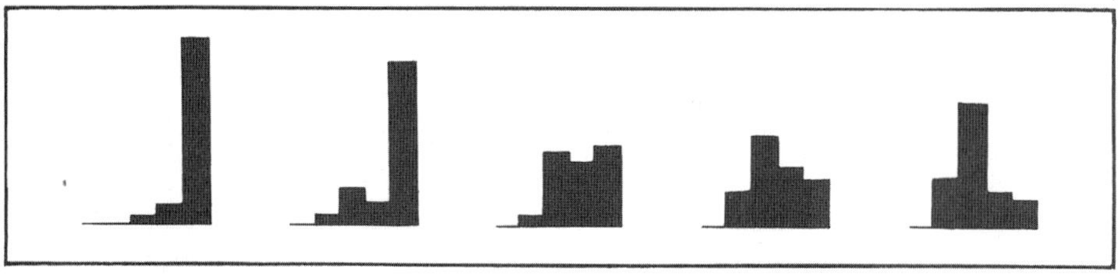

It is obvious that the comparison of, say, figures 1 and 5 or any other pair of constructions does not enable us to answer the question concerning proportionality. To do so would

VOLUME OF SALARIES (IN PERCENTAGE)

1

2

3

4

require us to select and order fifty images successively (by branches, then by categories), then class and compare them, which is visually impossible.

These constructions are inefficient for the majority of questions which the information is capable of generating. And "inefficiency" is due to the large number of images which the viewer would have to select and retain in order to obtain a correct answer. It is also obvious that the overall image produced by one of these constructions is perfectly useless. It has no meaning.

These constructions in multiple images will be termed FIGURATIONS.

Any reader who is content with elementary questions, with questions involving one number, or with questions introduced by the component "branch of the economy," will find these constructions satisfactory.

But the reader who asks questions introduced by the component "size of enterprise," or questions of an intermediate or overall level, will find these constructions hopelessly inadequate.

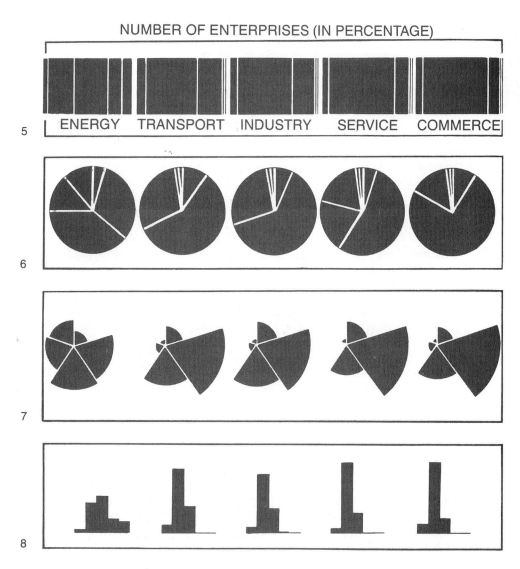

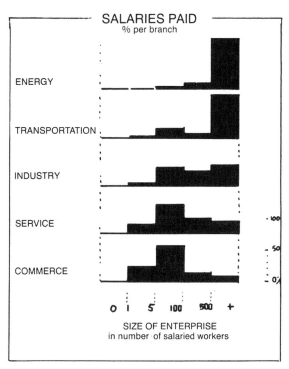
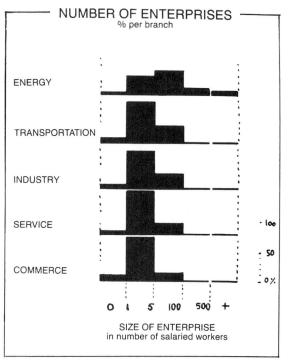

Efficiency of the image

Figure 1, on the other hand, enables us to answer all these questions effortlessly.

"In what category of enterprise is the largest volume of salaries for the commerce branch?
—In the category from six to one hundred workers!"

"In what branch is the highest percentage of salaries distributed in enterprises of one to five workers?
—In the commerce branch!"

"In what order are the branches classed in terms of size of enterprise, and according to the percentage of salaries distributed?"
—In the order: energy, transportation, industry, service, commerce." (This answer arises from the ordering of the drawing by "diagonalization," a procedure we will discuss later.)

Any question which the information could generate can in fact be answered in an instant of perception.

It is obvious that:

the most efficient constructions are those in which any question, whatever its type and level, can be answered in a single instant of perception, that is, IN A SINGLE IMAGE.

Figure 1 is a *single image*. Little effort is required to read the information. The mind is freed from the task of memorizing separate figures and is thus better able to grasp a meaningful, overall image, which can then be usefully compared to another image.

Represented by the same type of construction, the number of enterprises also produces an image (figure 2). It is clear that the volume of salaries is not proportional to the number of enterprises. Two images based on the same principles are easily comparable.

Efficiency of graphics

The negative response to the question concerning proportionality can then be complemented by the more detailed study of a particular branch or category. Note that the exact numbers, indispensable to the construction of the diagram, now lose some of their interest, in contrast to the order and proportionality of the elements constructed from them.

The properties of the image, that is, its efficiency and capacity for retention, appear here in a striking manner. For example, how could the information contained in these two images, with all their nuances, be accounted for in any other system of expression, such as verbal language, without the aid of graphic illustration?

Only the graphic sign-system, *provided it utilizes the minimum number of images*, **enables us to rapidly assimilate the complex relationships among four components.**

It is graphics which offers the best solution to the problem of computer utilization, whether it involves the programming, processing, or displaying of information.

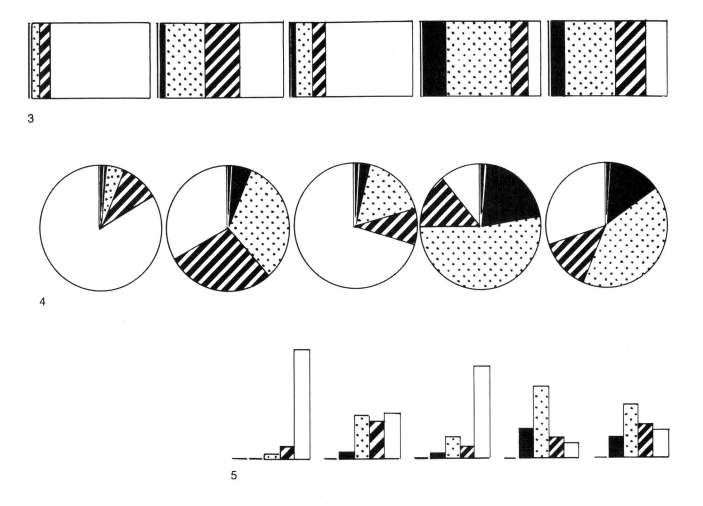

3

4

5

It is only in a semantic situation, when the reader looks to the drawing for the answer to a precise question, that the inefficiency of the constructions on pages 144 and 145 becomes evident. To find an answer we must first isolate the elements defined by the question; after that we can compare them. This is obviously the intent of the retinal signs which the designer has superimposed on the inefficient constructions here (figures 3–5). They improve the constructions in figures 1–8, pages 144–145, noticeably and enable us to identify all the elements of a given class quite rapidly.

However, these signs add nothing to the construction in figure 6 here, that is, to the IMAGE. On the contrary, they illustrate the simplicity of any visual identification when it follows the two planar dimensions.

Having examined the characteristics of an image, we shall now derive the rules of graphic construction from them.

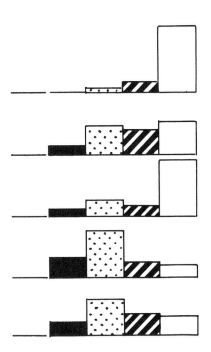

6

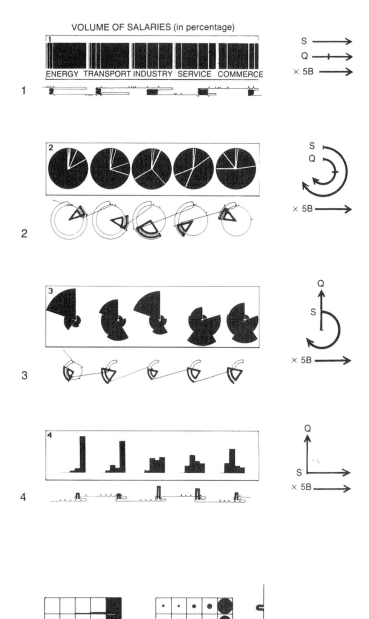

(4) Construction of an image

IMPOSITION (UTILIZATION OF THE PLANAR DIMENSIONS)

Consider the question:
In which branch is the fourth category of enterprises (100 to 500 workers) the largest?

In seeking an answer, the eye follows a specific scanning movement which has been sketched under the constructions in figures 1–4. The fine line in each sketch corresponds to an attempt at identification; the irregularities to any obstacles; the dark line to the perception of pertinent correspondences. The first four constructions (figures 1–4) compound visual difficulties: backward movements; irregularly spaced categories; overlapping of different identifications on the same line; broken scanning; etc.

In an image (figure 5), on the other hand:
– A RECTILINEAR scanning, as suggested by the drawing, groups the elements of a single category within a single component; that is, it groups HOMOGENEOUS elements.
– The two systems of identification are best differentiated by an ORTHOGONAL construction (indeed, the necessity of this differentiation leads the eye naturally toward the construction in which it risks the least confusion; in this sense the right angle seems to play a psychologically privileged role in visual discrimination).

The IMAGE is formed within a HOMOGENEOUS FIELD, in which any RECTILINEAR SCANNING, suggested by the construction, groups identical elements. The standard differentiation between two planar systems of identification is ORTHOGONAL DIFFERENTIATION.

This statement can be translated by a schema and a rule: **Any graphic problem necessitating two visual variables bases its standard construction on the following schema:**

This schema has few exceptions, which involve very limited components (those having few categories). The schema was formulated as early as 1361 by Nicholas Oresme, in his treatise "de latitudinibus formarum," although the author does not represent it as his own invention. The notion appears to have originated in the scientific circles at Oxford. (See Pierre Dedron and Jean Itard, *Mathématiques et mathématiciens* [Paris: Magnard, 1959].)

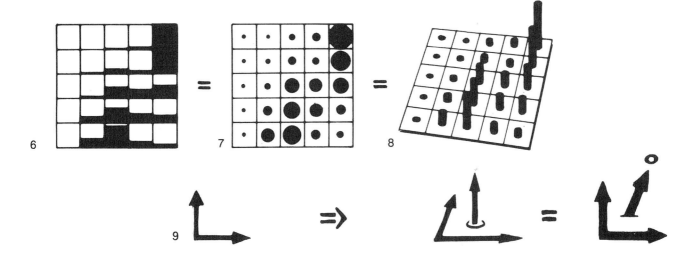

THE USE OF RETINAL VARIABLES IN FORMING AN IMAGE

The graphics on page 62 show that it is not sufficient to utilize the plane according to the standard in order to obtain an efficient construction. Indeed, any cartographic representation is by definition constructed on a homogeneous plane. However, the example on page 34 shows that a map can often be merely a figuration, necessitating the perception of numerous successive images before the information can be assimilated in a satisfactory manner. This same information can be apprehended in a single image, as on page 35, provided that the retinal variable is VISUALLY ORDERED.

Like the planar dimensions, which are naturally ordered, the variable must be VISUALLY ORDERED in order to construct an image.

Such is the case, for example, with the televised image. It permits us to understand the construction of any image. The sweeping of the screen by the electronic signal follows the two planar dimensions, while the modulation in the intensity of the signal creates an ordered variation from white to black, perceptible as a "retinal" variation, independent from the two other variables.

The image in figure 6 can be seen as a televised image (figure 7), suggesting an "elevation" above the plane (figure 8). We can translate this observation by a schema (figure 9) and a rule:

Any graphic problem necessitating three visual variables bases its standard construction on the following schema:

The letter O signifies that the retinal variable must be ordered.

Like the preceding schema, this one has only a few exceptions: problems with very limited components or with extremely simple planar arrangements which create no more than a few images that are easily distinguishable on the plane.

The necessity of using an *ordered* retinal variable entails two important consequences:
(1) Size, value, and texture are the *only variables* which will lead to the construction of a single image. The other variables can form part of an image, but only in combination with one of these three variables.
(2) A differential component (signifying only "different categories") cannot be constructed as a single image, if a higher concept cannot be used to order its different categories (see, for example, page 156).

In short:
– **THE IMAGE is formed by three HOMOGENEOUS and ORDERED variables, the two planar dimensions and an ordered retinal variable.**
– **For information to be represented as a single image, each of its components must be homogeneous and must correspond to an ordered concept.**

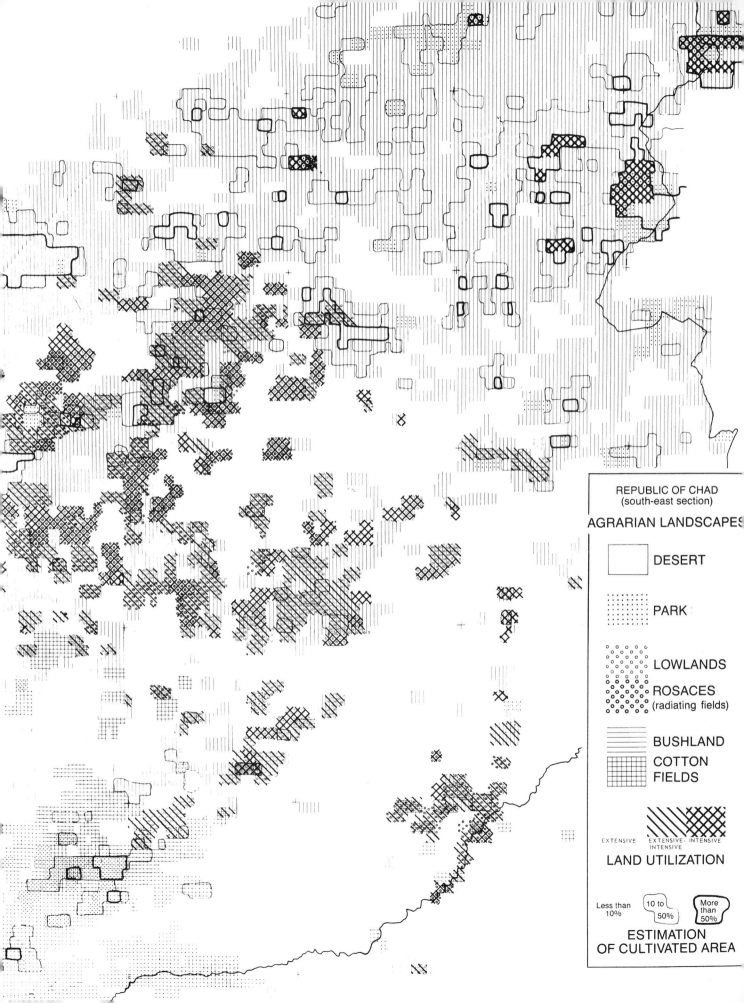

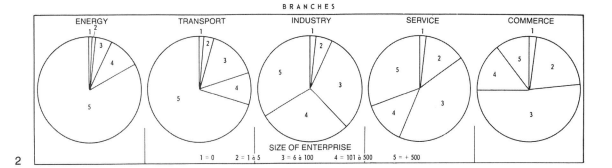

2

IMAGE AND FIGURATION

We will use the term IMAGE to describe the meaningful form immediately perceptible in the minimum instant of vision.

We should not confuse IMAGE with FIGURATION, which is the apparent and illusory unit defined by the sheet of paper, by a linear frame, or by a geographic boundary. Figure 2 does not constitute a single image, nor even five images, because the circle is not meaningful here, and memorization of the five circles "informs" us of nothing. This figure is composed of twenty-five images, one per sector, since only the sector informs us of something new. It is the only *meaningful form* in this FIGURATION.

The map in figure 1 (agrarian landscapes in the southeast part of the Republic of Chad—interpretation of the photographed surface by J. Letarte of the Laboratoire de Cartographie of the E.P.H.E.: Paris, 1964) is not an image, it is a FIGURATION. We cannot grasp the unified totality of its content without successively perceiving a multitude of images, based either on points or on small areas. The overall form is comparable to that of other maps but only gives evidence about the cultivated area, all categories combined, which is a small part of the information being represented.

To compare the "park" area, where large trees cover the land, with other factors which might explain its occurrence would require the perception of multiple partial images. Furthermore, these would have to be marked or put on tracing paper, since memory alone could not retain all these observations and construct the overall image necessary for the comparison. Incidentally, the use of color would not resolve the problem. It would reduce the time of analysis, but could not approach "immediacy." In a mixture of multicolored confetti one cannot compare the overall image constituted by the reds to that produced by the greens.

LEVELS OF READING IN AN IMAGE

Figure 4 is an IMAGE. We can read it on the *overall* level, see the image as a whole, and compare it to another (figure 3). This comparison involves the totality of the information.

But we can also settle on the *intermediate level*, isolate partial images, as in figures 5–7, and compare them with each other. Finally, we can isolate only a single element and create an image for the smallest meaningful form (figure 8). This is *elementary reading*.

With an IMAGE (figure 4), we can choose to isolate either twenty-five elementary images, or ten intermediate images, or reduce the information to a single overall image. Information constructed as an image accommodates all levels of reading.

In the map in figure 1, only *elementary* or *intermediate images* (small homogeneous areas) can be isolated. The overall image reflects a certain "sum" but not the unified totality of the information.

Figure 1 is a figuration in which the information can only be reduced to a large number of elementary or intermediate images. However, another type of construction could reduce it to six images (see page 165).

In figure 2, only *elementary images* can be created. The forms on the intermediate level (individual circles) or on the overall level (the five circles) are not meaningful, and a reading at these levels is useless.

In figure 2, the information cannot be reduced to fewer than twenty-five images. The information is constructed in twenty-five elementary images. Figure 2 is a figuration.

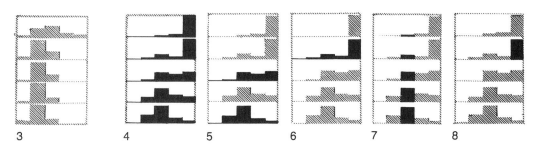

3 4 5 6 7 8

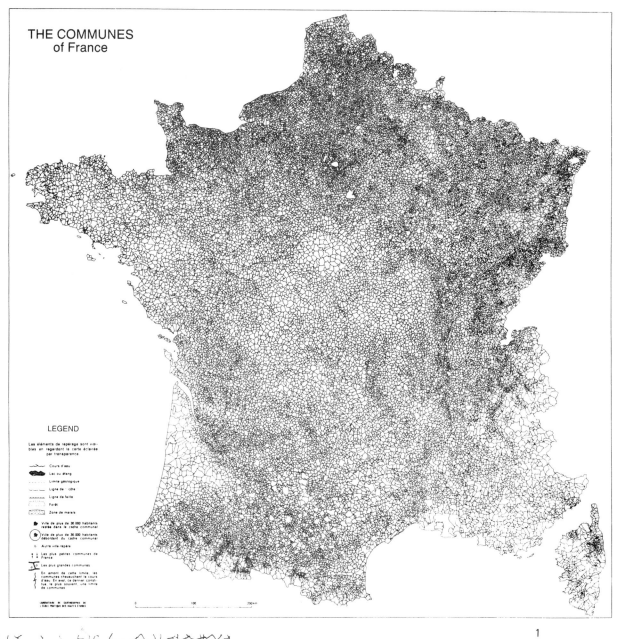

1

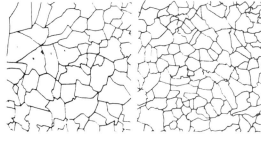

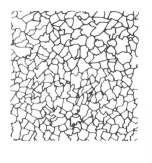

2

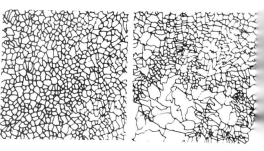

LEARNING TO READ AN IMAGE

The map in figure 1 opposite, in spite of its apparent complexity, is an image. It displays the web of the 30 000 French communes. It is readable on all levels.

On the elementary level
Although the map was not intended to be read on this level, it enables us to identify a given commune in which we have a particular interest or to spot one which stands out from the rest.

If the reader remains on this level, the map will appear as a gratuitous and inaccurate game, because "accuracy" is only relative, and never exists on the level of elementary information.

On the intermediate level
The map was meant to be read at this level, and here the reader will see the map as a document fostering different types of analyses. The most obvious characteristic, the size of the communes, lends itself to a "typology" (figure 2),* and can lead to a partition of the country. This is interesting, since the inverse of the commune "size" corresponds here to the "number of communes" in a given area.

Another characteristic, independent of commune size, is the homogeneity (figure 3) or heterogeneity (figure 4) of the web. This characteristic raises numerous problems and divides the country in a different way.

Regional bodies, such as the Vosges and Alsatian forests (figure 5) or the Cantal volcano (figure 6), foster numerous comparisons. The aim of this map is to solicit internal comparisons, and to this end a system of reference points is included on the back side of the original edition. At any time, this can be made visible by back lighting.

On the overall level
Here comparison is possible with other distributions, such as the hydrographic network of France shown in figure 7. When the map is compared to those of other countries, it raises the problem of the minimum regional partition, its origins, and its future.

Numerous studies show that the average person tends to read on the elementary level and encounters difficulties in adopting the intermediate level and, even more, the overall level. Graphic designers contribute to this habit by continuing to provide the public with figurations (haphazard curves, encyclopedic cartography, visual "puzzles"). These encourage the reader to remain on the elementary level. However, as constructions in a single image multiply, and as designers realize to what extent figurations are inefficient or anecdotal, the reader will learn to utilize better the perceptual means with which we are endowed.

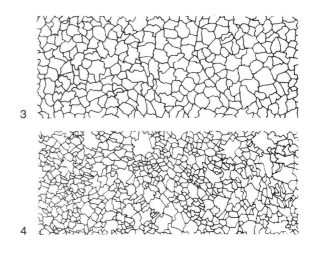

3

4

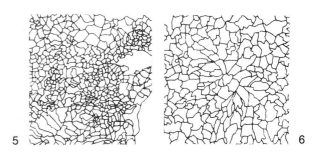

5 6

7

*The examples are at 1/2,000,000 (1 mm = 2 km).

(5) Limits of an image

(A) AN IMAGE WILL NOT ACCOMMODATE MORE THAN THREE VARIABLES

Consider the information:
INVARIANT — *anthropometric characteristics*
COMPONENTS — **Q** *quantities (or order) according to*
≠ *three characteristics (height, hair color, cephalic index)*
GEO *geographic order*

Four variables are necessary, since the geographic order alone occupies the two planar dimensions. This information can be constructed as three images (figures 1–3), or as a figuration composed of multiple images (figure 5). What are the properties of these two graphic solutions?

Construction of three images

With this solution there are three images, and the schema in figure 4 can be used to signify this. This schema indicates both that the representation has four variables (four arrows) and that it involves three images: one image per category of the component "≠ characteristics" (n = 3).

This graphic provides a response to any question introduced by the component "≠ characteristics": "Dark hair colors, where are they?" The answer is immediately obtainable from figure 1. Each characteristic is, in its totality, comparable to the others and to any other distribution: mountainous zones, density, industry, social characteristics. . . . However, this graphic does not provide an immediate response to every question introduced by the geographic component: "In a given area, what are the three characteristics?" In order to answer this, it is necessary to add information from all three maps for this area.

Construction of a figuration composed of multiple images

This graphic solution, illustrated by figure 5, utilizes:
– the two planar dimensions (for the geographic order);
– a retinal variation in size (for height of individuals);
– a retinal variation in value (for hair color);
– a combination of shape and orientation (for cephalic index).

The schema in figure 6 can be used to define such graphics. It shows that the representation involves five variables (one per characteristic and two for the geographic order), each of which could be identified by abbreviations.

This graphic provides an answer to any question introduced by the geographic component: "In a given area, what are the three characteristics?" However, there is a considerable

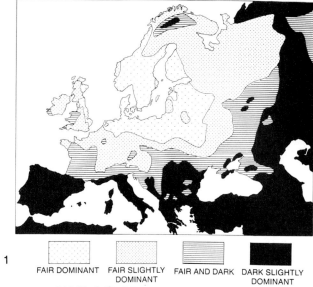

1 FAIR DOMINANT   FAIR SLIGHTLY DOMINANT   FAIR AND DARK   DARK SLIGHTLY DOMINANT
• HAIR COLOR

reduction of the information, since the geographic space is no longer represented as continuous, as it was in figures 1–3.

This graphic (figure 5) does not respond to questions introduced by the component "characteristics": "Where are the highest cephalic indices?" For this question the construction of three images is infinitely more efficient and provides an immediately retainable answer.

Therefore we can state that:

when the information necessitates more than three variables, we cannot construct a figure which will provide an immediate response to all types of question. The image will not accommodate the representation of a meaningful fourth variable.

Consequently:

in order to respond efficiently to all the types of question which can be generated by information:
– **having more than three components (in diagrams), or**
– **having more than two components (in networks and maps),**

we must construct two types of graphic, whose general schemas are given in figures 4 and 6.

Any designer who utilizes only a single construction chooses, consciously or not, to answer certain types of questions (preferred questions) efficiently, but to be inefficient for other types. The designer chooses, in effect, to fulfill only one of the three functions of graphic representation (see page 160).

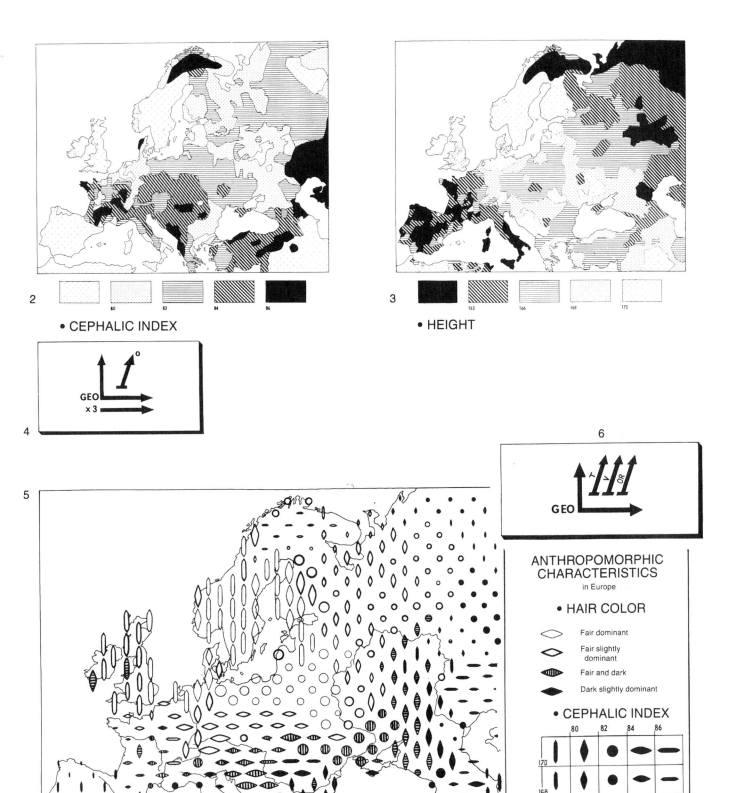

(B) AN IMAGE WILL ONLY ACCOMMODATE A REORDERABLE COMPONENT

Consider representing different industries cartographically (after a map of French industries published by the Comité National du Patronat Français. Paris, 1962).

INVARIANT — *a place where an industry is operating*
COMPONENTS — **GEO** *the geographic order*
≠ *fifty-nine categories of industry*

The information has two components. Its graphic representation thus necessitates three variables: the two planar dimensions for the geographic order, and shape variation to identify the fifty-nine categories (figure 2).

But no concept can order these fifty-nine categories (except for a quantity involving persons or production; but in this case the information would be transformed by the addition of a new component, thus modifying the graphic problem). The schema for this construction is given in figure 3.

The length of the differential component (fifty-nine categories) naturally leads to the variable which offers the highest number of distinct signs: shape variation (figure 1). The construction is thus logical; but is it efficient?

Any elementary question introduced by the geographic order can receive a quite rapid response. "At a given place, what is there?" The selection of a place on the plane is always easy since the plane itself is highly selective.* The eye isolates the place in question and easily disregards all other places during an instant of perception. If one recognizes a shirt, and provided that the legend has been memorized (which nonetheless represents a great preparatory effort), one has the answer: a knitted goods factory.

Efficiency is not great; however, a reader asking any question of this type will find an answer in this construction.

But any question of an intermediate level: "In a given region, what is there?" and especially **any question introduced by the component "different industries"** will only be answered after lengthy study. For instance, to answer the question—"The knitted goods factories, where are they?"— one must isolate the "shirts," see *only* them, and *all* of them, that is, be able to visually disregard all the other signs.

This map (figure 2) does not permit such an operation. To select all the signs for each industry would take several hours. A reader asking any question of this type will not find an answer. The representation is inefficient for this type of question.

A nonordered retinal variable cannot be perceived in a single image.

Electric household appliances	Mineral water
Automobiles	Foundry
Hydroelectric plant (Dam)	Solar power plant
Jewelry	Gloves
Cookies and cakes	Natural gas
Lumber	Foods
Knitted goods	Chemicals
Buttons	Timepieces
Brewery	Dairy products
Rubber products	Musical instruments
Atomic energy	Toys
Heating plant	Eyeglasses
Ceramics	Agricultural equipment
Champagne	Heavy equipment
Hats	Mechanics
Shoes	Mine
Cement	Paper goods, printing
Garments	Umbrellas
Candies	Oil refinery
Canned goods	Pipes
Fruit preserves	Bridgeworks
Aeronautics	Port
Electronics	Ironworks
Ship building	Sugar refinery
Cutlery	Tobacco
Leather goods	Textiles
Bicycles	Metalworks
Distillery	Tubing
Lace	Glassware
Electrometallurgy and electrochemistry	

---

*The original document identifies all the places involved.

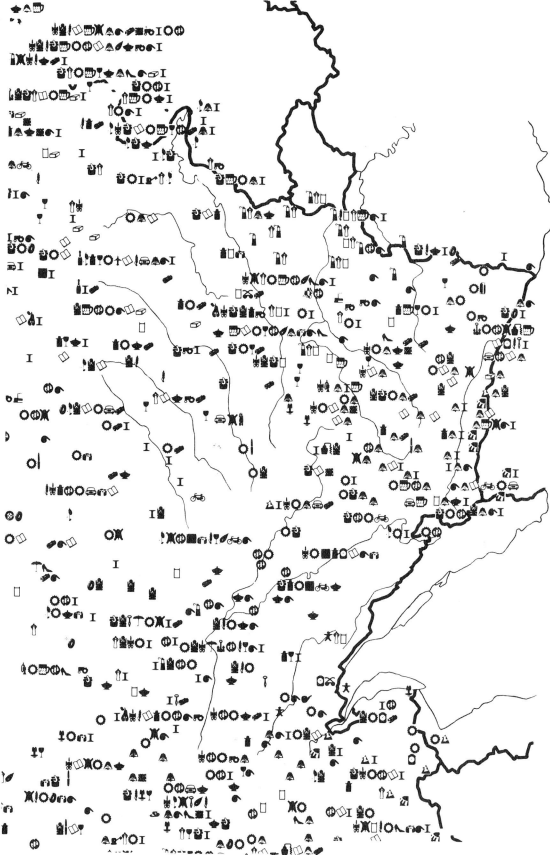

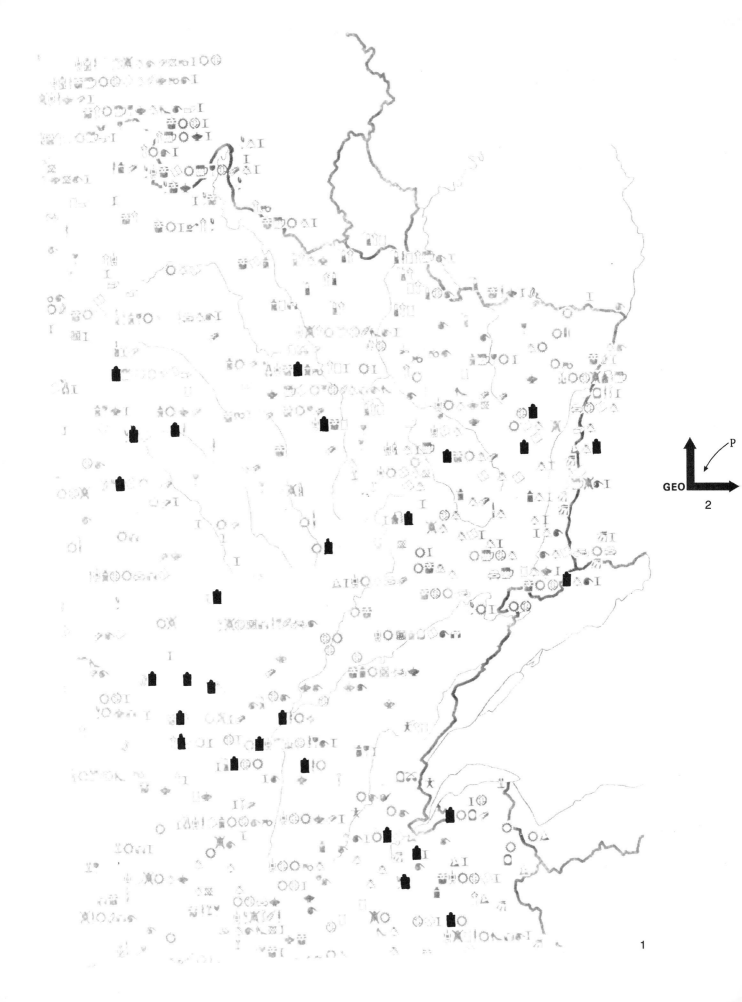

To obtain an answer, it would be necessary to blacken in the shirts and leave all the other shapes in a lighter value.

One could then isolate the black signs from the light signs (value variation) and perceive the pertinent information corresponding to the question.

In figure 1opposite, the black signs form an image which is easily retainable. Thus, we encounter again, but for a representation involving three variables, the problem illustrated on page 154. And the conclusion is the same:

In order to efficiently answer all types of questions in a representation involving three variables, where one component is not reorderable, two graphic constructions are necessary.

(1) One map, based on the schema in figure 2, for each industry. This schema indicates that each industry constitutes partial information, having one component, which can be expressed as follows:

INVARIANT —*a place where a given industry is operating*
COMPONENTS —**GEO**: *the geographic order*

The geographic component occupies the two planar dimensions.

The correspondences on the plane are points (P), which are not differentiated from each other (there is no component, such as industry size, which differentiates them). Undifferentiated correspondences are schematized by a curved arrow pointing downwards.

The whole of the information, i.e., fifty-nine images of this type, is expressed by the schema in figure 3 (n = 59). In practice, one can always group several industries on these partial maps—either those having very different distributions or even those having similar distributions—and thus reduce the number of these figures to about fifteen.

(2) A map representing all the industries, such as the one on page 157, whose schema is given in figure 4. (A redundant combination with color, grouping the industries by types, would improve the efficiency of this construction somewhat.)

Another example is found in the map of crop distribution, in the *Atlas de France du Comité National de Géographie* ([Paris: Editions Géographiques de France, 1951–59], plates 41, 42, and 43). Since the component (crops) is not in itself reorderable, the various crops are not ordered in the graphic representation (diversification is achieved by color and shape). It is impossible to obtain a complete and immediate answer to the question: "A given crop, where is it?" It would take eight small maps in black and white—that is, photographic reductions of each crop distribution—to provide an instantaneous answer.

Again it is obvious that any designer who uses only a single construction is limited to answering only one preferred type of question.

In order to choose this type with full knowledge of the consequences, it is necessary to analyze the functions of graphic representation (see p. 160).

The idea of simplification

Certain figurations can lead one to believe that order and homogeneity are not requisite attributes of an image. The reading of figure 5 is easy. However, note that the component "different categories of voters" is very limited (three categories), the superimposed forms are simple, and their graphic selection is very deliberate. In this case, the visual discrimination of the three superimposed images does not increase the mental cost of perception appreciably.

We could thus superimpose several images in a figuration, and it would still remain efficient, provided that the images were *not very numerous, that they were very simple*, and were differentiated in the most efficient graphic manner.

Passing from information which produces a simple image to the simplification of complex information is a short step, but it involves reflection and a sound knowledge of the functions of graphic representation.

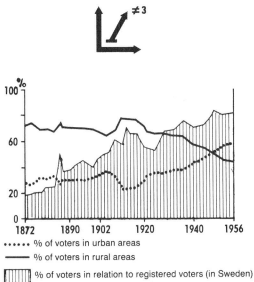

5 ••••• % of voters in urban areas
   ——— % of voters in rural areas
   ▥▥▥ % of voters in relation to registered voters (in Sweden)

C. Three functions of graphic representation

When information having three components is represented according to the standard schema depicted in figure 1, it forms an image and offers an immediate answer to all potential questions, whatever their type and level. It is not necessary to choose "preferred" questions, and information portrayed in this way meets all three of the functions of graphic representation.

However, the image is limited to three visual variables, and modern research involves a continuing effort of comparison which, in fact, tends to increase the number of components introduced into a study. Communicating information with more than three components is the major challenge we face. The choice of "preferred" questions is thus a permanent problem, which can only be defined if we go beyond special cases and analyze, within the framework of graphic representation, how we can take advantage of human perceptual means to extend our knowledge—in short, *how we utilize our memory.*

To understand graphic information is to visually memorize one or several images; it is obvious that for given information, memorization will be more difficult as the number of images increases. In the final analysis, it is the conditions of memorization, linked to the amount and conceptual level of the information, which permit us to distinguish and define three functions in graphic representation, and probably in any system of communication.

(1) Recording information (inventory drawings)

A graphic can function as convenient and comprehensive inventory of information. The plane and its visual signs are utilized to record all the correspondences in a given informational set, in order to:
– create a storage mechanism
– which avoids the effort of memorization.

The subway diagram which can be put into one's pocket, the highway map, the data table enable us to avoid the task of memorizing all the lines, all the correspondences, all the numbers. They are available in one document which assembles them and renders them conveniently accessible.

What matters is that all the information "announced," that is, conceivable within the domain defined by the title, is recorded there. This is COMPREHENSIVITY.

What matters less is the time necessary for extracting the sought correspondences, that is, the number of images required for a correct answer. Taken as a whole, the document can be NONMEMORIZABLE.

With this function the graphic is an INVENTORY, which favors reading at the elementary level. This function thus authorizes the construction of complex FIGURATIONS, with multiple images, limited only by the rules of legibility.

The researcher, investigator, or explorer who gathers information and records it in the form of numbers, letters, shapes, or colored symbols, in a table or on a map, is employing the best system of graphic recording. It is the easiest, as well as the least ambiguous, since it eliminates confusion in rereading. It constitutes an inventory, often vast, which is the essential starting point for any further utilization of the information.

Such is the case with the map of agrarian landscapes (figure 2) at 1/1 000 000 (here reduced more than three times), whose legend is given on page 150. At its original scale it displays agrarian landscapes point by point. It does not provide immediate answers for questions of the type: "A given category, where is it?" It thus prohibits us from entering such answers into the interplay of external comparisons. But it does contain these elements; we only need to bring them out. Such is the case with the diagram on page 259, which contains the elements of numerous studies. This is also true for the map in figure 3 and the diagram in figure 4. They both display the same information with two components—working population employed in the tertiary sector, according to:
GEO–*departments*
Q –*working persons in sector III (% of work force)*

Information represented in this way involves only the inventory stage of graphics. All these documents must be reread point by point. Their overall visual memorization is practically useless (except for figure 2, where the overlapping of landscapes is meaningful). They are formed of multiple images, which in fact makes these documents comprehensive. They contain the totality of known information, within the framework announced by the title.

This is the first state in communication. Imagine the amount of lost knowledge if the expert were to stop at the inventory stage!

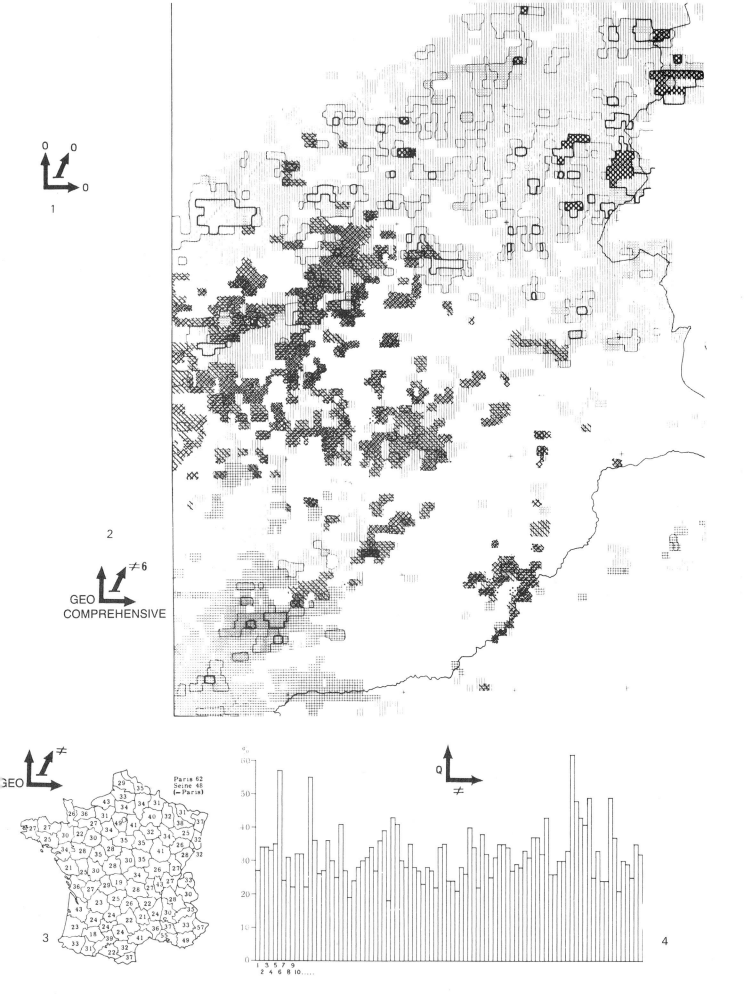

(2) Communicating information
(simplified drawings or "messages")

A graphic can furnish the means of retaining information with the help of visual memory. The plane and its visual signs can be utilized for communicating information, in order to:
- **create a memorizable image**
- **which inscribes THE OVERALL INFORMATION within the field of assimilated knowledge.**

School maps, blackboard sketches, and all representations of a pedagogic nature tend to inscribe information in the viewer's memory, to make it become assimilated knowledge, capable of being *recalled* at the time of an exam, a conversation, a research project, or a decision.

What matters is that the information is MEMORIZABLE.

What matters less is the number of correspondences retained, provided that these are essential ones. The image can be NONCOMPREHENSIVE.

With this function, the graphic is a "MESSAGE." Its efficiency increases as the number of images (superimposed or separate) **and their complexity diminish and as reading approaches the OVERALL LEVEL.**

Such is the case with the map in figure 2, which, by means of a considerable simplification, permits us to retain in several instants the essential features of the information contained in figure 2 on page 161.

This is also the case with the simplified map in figure 3, which offers in one easily memorizable image the essential points of the information in figure 3 from page 161.

The graphic in figure 4 shows the extent of this simplification and illustrates the reduction in the number of correspondences depicted by figure 3 in relation to the comprehensive information.

In all three cases, the "essential" features result from a choice. But is this choice a good one? The maps in figure 1, which represent the same information displayed in figure 3, show that numerous choices are possible, that what is "essential" remains to be defined. How can the comprehensive information be reduced, the smallest number of simplified images be determined, so that the information becomes communicable and memorizable, *without being falsified or destroyed*? How can this choice be justified? Is a choice always necessary?

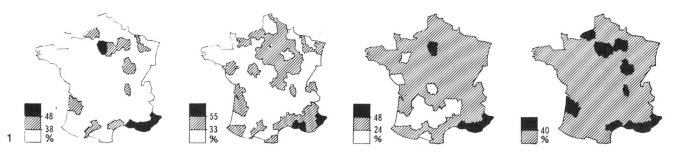

1

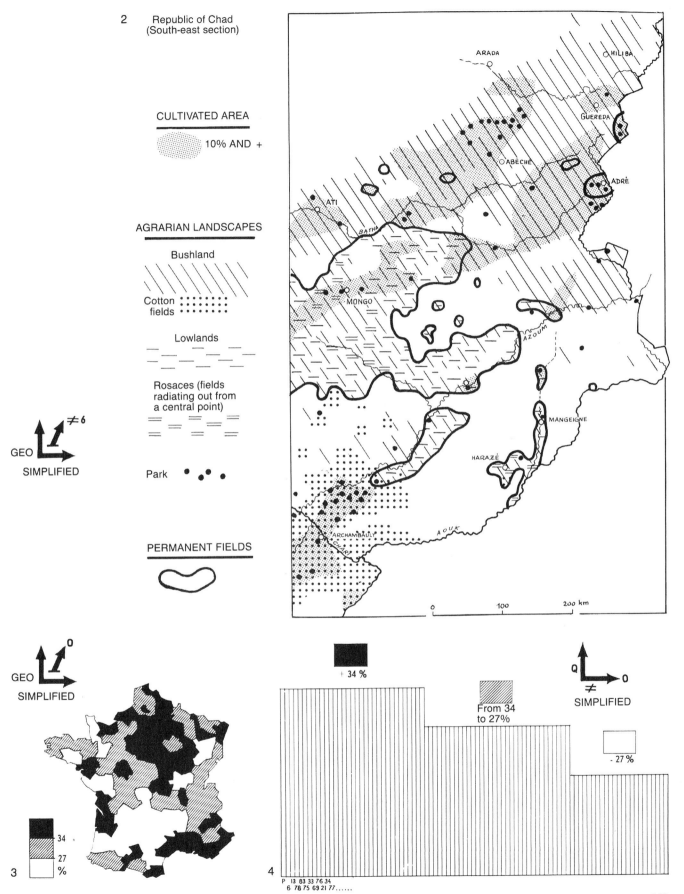

(3) Processing information
(graphics used for processing)

Graphic representation can be used to reduce the comprehensive, nonmemorizable inventory to a simplified, memorizable message. For the message to be useful, simplification must not involve eliminating part of the information, but only "processing" it, that is:
– utilizing the mechanisms of ordering and classing, for the purposes of discovering
– the groupings contained in the information being processed, and
– deriving from it new components or categories, reduced in number and consequently easier to memorize than the comprehensive information.

The combination of these new elements must enable us to recall and understand the whole of the initial information.

Collections of comparable diagrams, networks, or maps are found in working atlases and graphic files, and foster all types of comparisons and classings. Such collections allow researchers to discover the correlations contained in a finite set and to derive from them "lines of force," subsets, and new components. These elements broaden the scope of the message, increase the rapidity of comprehension, and thus lead to better "communication."

What matters is to avoid any prior reduction of the information, to use the complete information, which alone provides all the givens for pertinent correlations and choices. The representation must be COMPREHENSIVE.

But it also matters that all types of comparisons and classings are possible and easy. The most useful questions will obviously involve the overall level of reading, where their answer will be found in a limited number of comparable images. The representation must be reduced to the smallest number of MEMORIZABLE IMAGES.

With this function, the graphic is an experimental instrument leading to the construction of collections of comparable images with which the researcher "plays." We class and order these images in different ways, grouping similar ones, constructing ordered images to discover the synthetic schema which is at once the simplest and most meaningful.

The images used for processing and analysis must be free to enter into any interplay of comparisons or superimpositions. They must not be, as with inventories, tied to each under the pretext of a dependency which, in fact, can always be put into question and which reduces freedom of analysis and manipulation.

The information in figure 2 on page 161, for example, is merely the superimposition (obviously confusing) of the images in figure 1 opposite. We discover here that the characteristics displayed earlier present different distributions, that they are not related geographically, and that it would be desirable to compare each one to other phenomena. This would enable us to discover characteristics of distribution which are similar and therefore geographically related. The coincidence of similar distributions (U and L + R) emphasizes important groupings, which justify the simplified contours of the message in figure 2 on page 163.

But graphic simplification is not always necessary. Information having one, two, or three components, *represented as a comprehensive image*, permits the eye itself to simplify to the necessary level, without being dependent on the graphic designer's choice. Represented as an image with three variables (figure 2) as an ordered diagram (figure 3) the information offers the reader the means of *regionalizing* the image, of *categorizing* the diagram, *while remaining informed of the level of the steps retained and of those eliminated*. The reader can judge this better still with the images on page 373 and following.

The reader can even retain only the simplest image, generally perceived as composed of three steps: black, gray, and white or, in diagrams, top, middle, and bottom. Such images correspond to the three functions of graphic representation. Comprehensive, they are inventories; memorizable, they are messages;—and these two properties together make them instruments for the graphic processing of information.

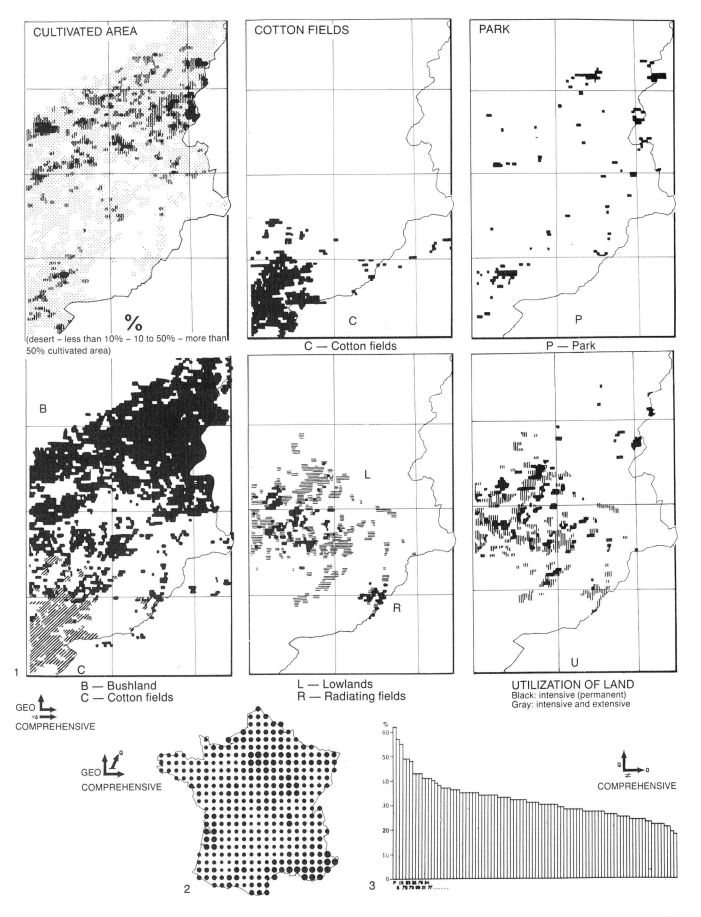

GRAPHIC PROCESSING OF INFORMATION

The functions of graphic representation form a sequence, defined by the role of memory and tending toward simplification (see the table on this page).

Simplification is an obligation of the communication process

Whether communication involves a verbal statement or a graphic representation, it starts from complex information and aims at making it understood, that is, at discovering combinational elements which are less numerous than the initial elements yet capable of describing all the information in a simpler form. When it is logical, the simplification is creative. By revealing concepts of a higher order, it enables us to know more than our predecessors, and more still with each generation.

Logical simplification, that is, information-processing

This can operate either verbally, mathematically, or graphically. Current studies are defining the conditions which will determine a choice of the most efficient of these three operative systems (the most economic one, all other things being equal) in terms of the information and the level of the intended result. What are the modalities proper to graphic information-processing?

Graphic information-processing operates by simplification of the image

The graphic transcription of information can arrive at a complex image (column A opposite) or a simple image (column B). Visual complexity varies between two extremes: In figure 1 everything is scattered everywhere; in figure 2 division of the plane is minimal (binary). These two extremes exclude graphic representation; verbal language would suffice to express them. A graphic is therefore justified only for transcribing intermediate complexities, that is, problems relating to statistical studies (figures 3–12).

Complex images, the map of communes on page 152, for example, generally offer numerous levels of reading, among which the reader may hesitate. On the other hand, simple images offer few levels and reduce this hesitation. They are more easily memorized and can be used in superimpositions of images (see, for example, figure 5, page 159 and figure 2, page 163).

Georges Th. Guilbaud characterizes a simple visual form by two qualities: connectivity—not having gaps, that is, being homogeneous, or in a network avoiding meaningless intersections; and convexity—being delimited according to convex angles and thus forming a uniform area inside of which any straight line will cross the figure only once. Any visual simplification must tend toward these characteristics.

We achieve this in two ways: (a) by ordering a qualitative component; and (b) by eliminating certain correspondences in ordered components.

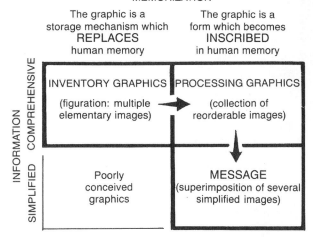

(A) ORDERING A QUALITATIVE COMPONENT: TRANSFORMATION, DIAGONALIZATION

The perceptual characteristic of a *qualitative component* is instability. Each person can arrange products, objects, or individuals in a different way. To understand a qualitative series amounts to identifying all the elements of the components one by one, and the instability of the concept leads to renewing this effort for each new message.

On the other hand, what characterizes an *ordered component* is stability. Whenever an ordered concept—time, size, geography—is introduced into the message, all the elements assume a fixed place. This order will grow progressively stronger, since, all other things being equal, it will be the same at any moment and for any individual. A *single effort* will suffice to assimilate it definitively.

The discovery of an ordered concept thus appears as the ultimate point in logical simplification, since it permits reducing to a single instant the assimilation of series which previously required many instants of study.

The ordering of a qualitative component is the basis of information-processing involving computers. It is also the basis of graphic processing.

Simplification of the image by ordering does not eliminate any correspondence and preserves the integrated totality of the information. It involves the DIAGONALIZATION of diagrams, and the TRANSFORMATION of networks.

The totality of the information in figures 7 and 9 is preserved in the diagonalizations displayed in figures 8 and 10. The same is true for the network in figure 11 and its transformation in figure 12. The diagonalization of diagrams forms a part of the fundamental rules of construction. Its principles are outlined on pages 168 and 169 and will be developed throughout the diagram section of this book (pages 193–268). The transformation of networks is sketched out at the beginning of the section on networks (pages 269–283).

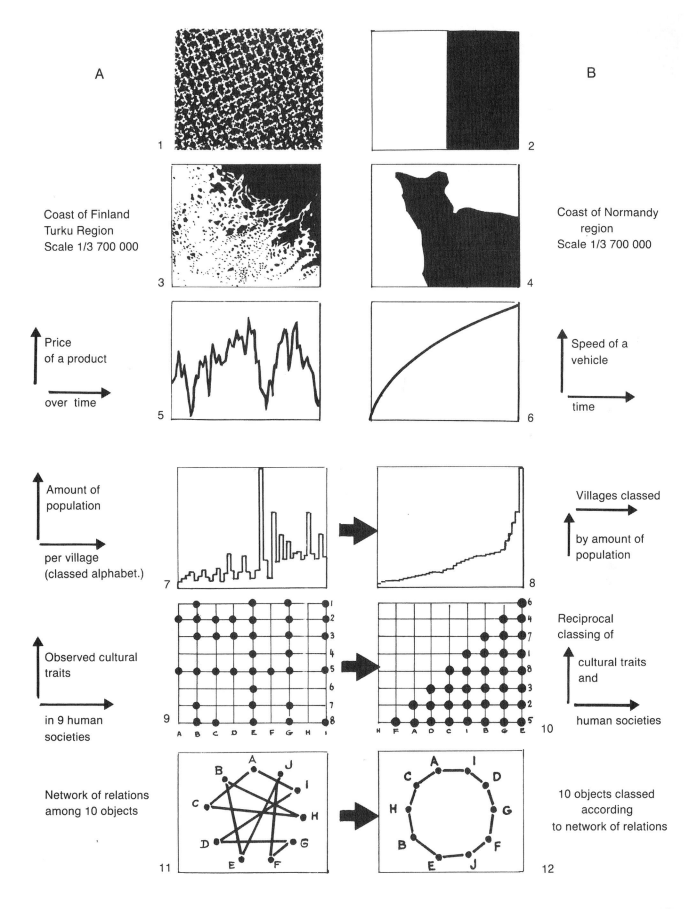

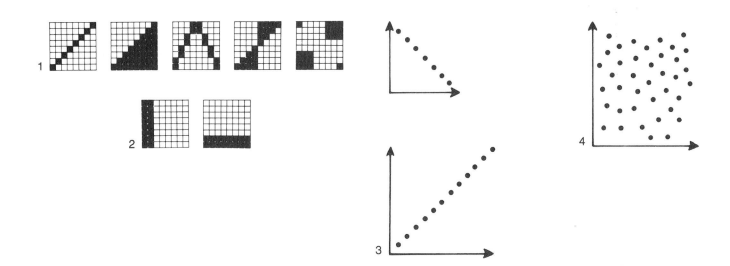

The diagonalization of diagrams

The term "diagonalization" is based on the observation that any operation of classing from orthogonal coordinates tends toward the diagonal of the field or toward simple forms derived from it (figure 1). Any classing which arrives at a form parallel to one of the coordinates simply furnishes proof of the nonvariation of one of the components and thus the triviality of the problem posed (figure 2).

"Diagonalization" encompasses the operations known as permutations, triangulations, scalograms, hierarchic analyses, and Guttman scales, which all result from the same process of visual simplification, namely:

– the perfect ordered correspondence between the two planar dimensions will take the form displayed in figure 3 (see page 249);

– lack of order or noncorrespondence will produce a very different image, as shown in figure 4.

It is therefore desirable to aim for the form in figure 3 whenever there is a reorderable component (≠).

Any deviation from the form in figure 3 reveals that the correspondence between the two orders is situated at a certain "distance" between perfect correspondence (figure 3) and null correspondence (figure 4). This distance can be numerically defined (page 249).

Diagonalization of two components

In order to express these rules, the standard schema is completed by signs signifying that the ≠ components must be ordered on each of the planar dimensions.

When the information involves only one ≠ component, it is ordered according to the component O (or Q).
The schema is therefore:

When the information involves two ≠ components, they are ordered reciprocally.
The schema is:

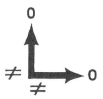

Diagonalization most often necessitates several successive attempts, whose number depends on the length of the components. In order to avoid numerous drawings, R. Pagès and A. Lévy-Schoen have developed a magnetic table (figure 9), the "Permutator" (patent C.N.R.S.), which admits components up to a length of eighty categories and eliminates successive drawings by making any permutation of rows or columns possible. Diagonalization is an operation frequently required of computers.

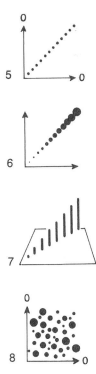

Diagonalization of three components

The perfect ordered correspondence between two components, as we have seen, takes the form of figure 5. A third component, if perfectly ordered in relation to figure 5, would take the form of figure 6 and be perceived as in figure 7. A lack of order among the three components would produce a very different image (figure 8).

It is desirable, in a problem having three components, to aim for the form of figure 6 whenever there are qualitative components.

The standard schema for a construction having three components must therefore be:

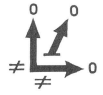

Numerous examples are given in the chapter on diagrams. The Permutator can be used in problems involving three components when two are ≠ components. We represent the component **O** or **Q** by ordered signs (size or value variation) glued onto small blocks, and the two ≠ components are represented by the two planar dimensions.

A simpler permutation instrument has been developed by the Laboratoire de Cartographie at École Pratique des Hautes Études (E.P.H.E.). This device, called "Domino," involves eleven visual steps.

It allows us to represent (figure 11) and diagonalize (figures 10 and 12) any information involving a third component **Q**, even when a positive-negative series is involved (see examples, pages 231 and 397).

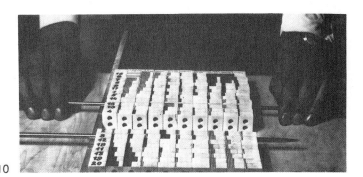

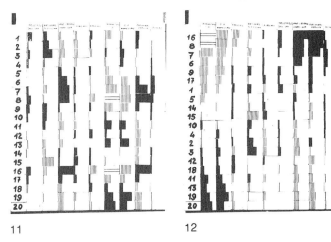

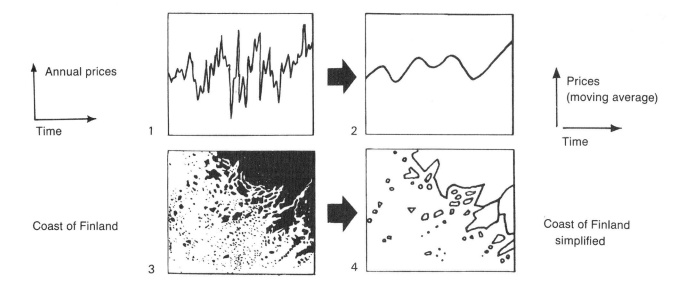

(B) ELIMINATING CORRESPONDENCES: SMOOTHING, REGIONALIZATION

An ordered component is by definition not reorderable. When its graphic transcription leads to a complex image, the latter can be simplified only by the elimination of a certain number of details.

The simplification of an ordered image can be accomplished only by the elimination of certain correspondences, by a diminution of the information. This is called "smoothing of curves" in diagrams, "regionalization" and "generalization" in maps.

The totality of the information in figures 1 and 3 is not preserved in the moving (running) average (figure 2) or the generalization (figure 4). Only what is essential is retained.

But what is essential? How can the simplified form which we will offer to the reader be determined? Smoothing can proceed by two different methods, depending on whether or not external information is introduced into the problem.

The internal method

This method encompasses all the processes which rely uniquely on the correspondences contained in the information being processed. This includes, for example, the calculation of moving (running) averages (page 216) and the numerous internal mathematical operations which can be undertaken when the information has two ordered or quantitative components (e.g., when the information is a time series). These operations are detailed by M. Barbut and C. Fourgeaud in *Eléments d'analyse mathématique des chroniques* (Paris: E.P.H.E., 1965).

The internal method also includes certain visual operations shown on pages 239, 260, and elsewhere, the structural generalization of geographic forms (described on pages 300 and 307), and the reduction of geographic information. The drafting of the "message" in figure 2, page 163 is based solely on the information contained in figure 2, page 161. No call has been made on external information: population density, geology, pedology, relief, etc. The same is true for pages 397, 401, and 405.

The external method

This method encompasses all the processes which call upon external information in order to justify a given choice in relation to another. But in fact this means introducing the information being processed into a *higher set*, involving an additional differential component (different concepts) so that we can class and order these concepts once again, determine "tendencies" and "lines of force," and define their simplified form. We can thus simplify a given curve by introducing a collection of comparable curves, from which we can derive the general tendencies, which we then apply to the initial curve. Likewise, we can simplify a map by utilizing the lines of force in a series of comparable maps. This is the principle of conceptual generalization.

The scope of external concepts is unlimited. It is also subject to current concerns. *Therefore, the external method always involves a choice*, which depends on the delimitation of a finite domain of research.

It is a hypothesis, or if one prefers, a mental orientation which can always be put into question. In the final analysis, the expert's skill depends less on the *internal processing* of a given informational set, *which becomes automatic*, than on the timeliness of the external concepts introduced into the problem, that is, on the extent of the expert's knowledge and "inspiration" in the delimitation of the research domain.

This delimitation is the main attribute which we will never be able to ask of a computer. A general discussion of graphic information-processing is given on page 254.*

---

*See also *GIP*, pp. 20–21 (translator's note).

D. General rules of construction

A rigorous definition of the components of the information, specifying their number, level, and length, must precede any graphic construction.

"Knowledge" involves a continuing effort of comparison, which means that one objective of any new information is to enter into the most extensive possible interplay of comparisons.

To achieve this the graphic must favor external comparisons and thus foster an efficient reading at the highest level, the overall level. Remember when this level is attained, all lesser levels of reading, and consequently all internal comparisons, are still possible; whereas the opposite is not true.

The preceding observations determine the general rules of graphic construction:

To represent the information in a single image, or in the minimum number of images necessary (to render it perceptible in its entirety, in the number of instants of perception), is the first rule of graphic construction.

To simplify the image without reducing the number of correspondences is the general rule which applies to any information having one or several reorderable components.

To simplify the image by reduction and thus create a clear and efficient message is the general rule which applies to any information having several ordered components.

Depending on the imposition (diagrams, networks, or maps) and the number of components, these rules are represented by the **standard schemas** which follow.

Diagrams

FOR PROBLEMS INVOLVING TWO OR THREE COMPONENTS

The schemas in figures 1 and 2 indicate the necessity of:
– utilizing the two planar dimensions in a homogeneous, rectilinear, and orthogonal manner;
– employing an ordered retinal variable (size, value or texture) for representing the remaining component at the appropriate level of organization;
– ordering any reorderable component (by diagonalization).
Information represented according to these rules will be perceptible as a single image.

FOR PROBLEMS INVOLVING MORE THAN THREE COMPONENTS

The circumstances particular to the gathering of the information (door-to-door surveys, library research, etc.) or to its utilization on the elementary level (looking for a site, a street, an itinerary) will lead the designer to adopt:

An inventory drawing (figure 5, comprehensive)
This is easy to draft but must be reread, point by point, for comparisons. It displays the comprehensive information in a single figuration, which can provide a visual response only on the elementary level and will exclude any question on a higher level.

These inventories are no more than the raw materials for constructing:

Processing graphics
These are capable of responding immediately to any overall question. The designer then follows the schemas in figures 3 or 4 and represents the information in n separate images, each involving three components:
– n is generally the length of the shortest component or the product of the lengths of the shortest components;
– it is understood that all the images are comparable, and that the planar dimensions are affiliated with the two *most ordered* components, those most capable of producing a useful domain of comparison;
– the n separate images can then be classed linearly (figure 3), distributed over one or several double-entry tables (figure 4), and thus become the object of new groupings.

The drawing of a "message" (figure 5, simplified)
Here the drawing superimposes the result of the processing operations in several simplified images. For diagrams, this generally involves an *image-figuration superimposition*. It is more efficient when the superimposed images are simple and few. In any case, SPECIAL CONSTRUCTIONS (linear, polar, triangular, . . .), which differ from the standard are justified only by the presence of very limited components. These reduce the necessary images to a number small enough so that it does not appreciably increase the mental cost of perception.

STANDARD SCHEMAS

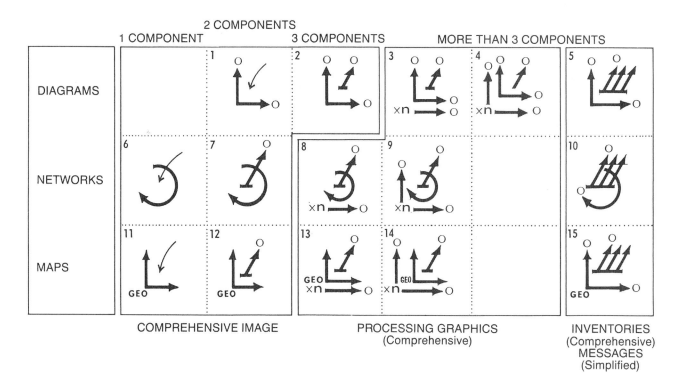

Networks

The circular schema (figure 6) is the construction which permits the most rapid transcription of a verbal analysis of the information into a two-dimensional drawing.

This construction immediately displays the principal characteristics of the information; it allows an initial ordering; it permits observing the structure of the system of correspondences and, if appropriate, discovering a simpler construction (by transformation) based on that structure (see page 271). Additional components will utilize the retinal variables (figure 7), then produce supplementary images which can be juxtaposed (figures 8 and 9) or superimposed (figure 10).

Maps

DIAGRAM, NETWORK, OR MAP

Information which includes a geographic component can be constructed according to any of the three impositions.

A diagram
Constructed in a line, the geographic component occupies only one dimension of the plane. It is reorderable; it can be treated in a diagonalization procedure involving three variables and simplified, as with any information having three components, without a reduction in the number of correspondences. A diagram thus enables us to discover the internal categorization which characterizes the information being processed in a much shorter time than does a map. In the information represented from page 100 on (working population according to the three sectors I, II, and III, by department), the diagram shows that the component ≠ sectors can be grouped into two categories—the agricultural sector and the two other sectors—and that the geographic component ≠ departments can be reduced to a certain number of groups, derived from the triangular diagram. But these groups are independent of the geographic order, and the latter cannot be exploited as a basis for memorization and external comparison.

A diagram permits the rapid and precise internal processing of information having three components, but it does not permit introducing the information into a universal system of visual memorization and geographic comparison. It is a closed graphic system, limited solely to the information being processed.*

A network
When constructed as a reorderable network, the geographic component occupies both planar dimensions. Thus we can only construct a network as a single image with information which does not exceed two components.

But we can attempt to find the order which achieves the simplest structure of the network of correspondences and derive characteristic groups from it (see figure 9, page 51).

A map
Like the reorderable network, the network ordered according to geographic order occupies both planar dimensions (figure 11).

Thus we can only construct a map as a single image with information having at most two components. A map invariably takes longer to construct than the corresponding diagram. In any problem involving more than two components, it leads to a greater number of images. But mapping introduces geographic order and thereby inserts the information into a universal system of visual memorization and unlimited external comparisons.

STANDARD SCHEMAS FOR MAPS

In problems involving two components, the schema in figure 12 indicates the necessity:
– of utilizing an ordered retinal variable for representing an additional component
– at the appropriate level of organization.

Information constructed according to this rule will be perceptible as a single image.

In cartographic problems with more than two components, the geographic component plays the same role as the two ordered components in diagrams with more than three components and obeys the same general rules. These are represented by the same schemas: figures 13, 14, and 15 correspond to figures 3, 4, and 5 for diagrams.

---

*In order to introduce a diagram into a universal system of comparison, it is necessary to call upon the mathematical notions of:
location—mean, median, mode
dispersion—variance, standard deviation
shape—skewness, kurtosis
all of which have a universal definition and are thus comparable in any circumstances.

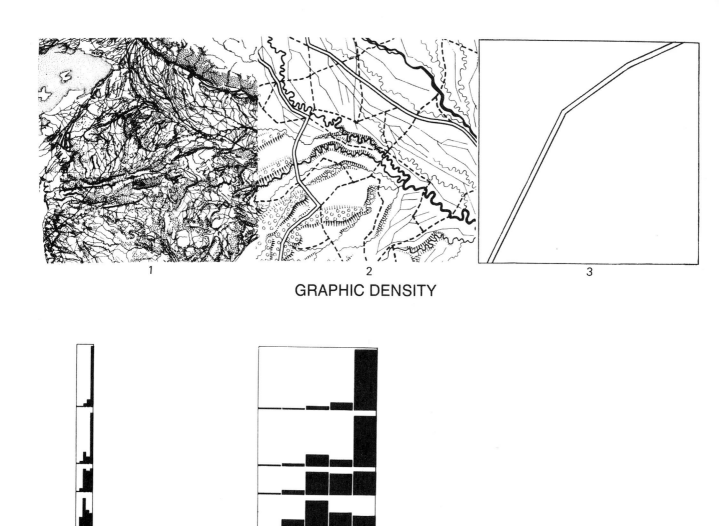

GRAPHIC DENSITY

ANGULAR SEPARATION

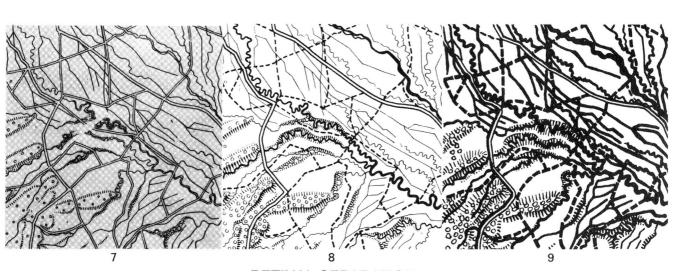

RETINAL SEPARATION

E. General rules of legibility (or rules of separation)

GRAPHIC DENSITY
ANGULAR SEPARATION
RETINAL SEPARATION

The rules of construction that we have just examined govern the choice of visual variables. With some constructions, a reading may only be possible on the elementary level; with standard constructions, it will be possible on all levels. But this still implies that a reading is possible, that the image is legible.

If the rules of construction can be compared to the principles that apply to the DRAFTING of a speech, then the rules of legibility are comparable to those governing the AUDIBILITY of the speech (pronunciation, sound quality). *They assure a separation among the variables and among the steps of each variable.* A speech which is excellent on paper can be practically inaudible if the pronunciation or any audibility factor is defective. Conversely, a very bad text can be delivered by an excellent speaker under perfect conditions of audibility.

The same is true for graphics. A good construction can be done under conditions of legibility which are such that the image will be difficult to read; and conversely, inefficient constructions can be reproduced by excellent draftsmen who will make them appear as good drawings to the inexperienced viewer.

Let us extend this comparison with verbal language in order to uncover the factors of legibility. The speech can be inaudible because the speaker speaks too rapidly. The amount of sound per second goes beyond human capacities of attention and comprehension. If the orator speaks too slowly, the listener has the impression of wasting time. Thus there exists an optimum temporal cadence for auditory delivery.

Likewise there exists an optimum number of marks per cm^2 (see figure 2) between a density which is too great (figure 1) and one which is too sparse (figure 3). This is GRAPHIC DENSITY.

The speech can also be inaudible because the orator mispronounces the words or speaks too low. The sounds become incomprehensible; they mix with each other and with the "background noise" of the room. The sounds must be differentiated from one another and separated from meaningless noises. The orator who speaks too loudly loses the benefit of vocal inflections. Thus there exists a range of differences and powers which the speaker must utilize in order to benefit from all the auditory differentiation available.

Likewise, a graphic must utilize the range of perceptible differentiation afforded by the visual variables, in such a way that the eye can:
– Separate the TWO PLANAR DIMENSIONS (figure 5) and avoid a "squashing" of the plane which limits angular differentiation, as in figures 4 and 6. This is ANGULAR SEPARATION.
– Separate the meaningful marks from the meaningless ones (separate "figure" from "ground," "content" from "form") and, within each RETINAL VARIABLE, separate the steps in order to avoid an image which is too weak (figure 7) or too bold (figure 9). This is RETINAL SEPARATION (figure 8).

Both figures 7 and 9 utilize only a small portion of the perceptual differentiation available.

(1) Graphic density

There is an optimum number of marks per cm^2 in figurations, i.e., graphic representations which superimpose different images. For example, the map of India in figure 1 makes sense only if the reader can separate, on the elementary and intermediate levels, the image of the highways from the image of the relief, from the image of the rivers, from the image of the forests, etc.

A density which is too great (figure 2) provides no more than the sum of the images, all signs combined (or nearly so), and this sum is rarely meaningful. It would seem that an average density of some ten signs per cm^2 is an upper limit. However, this optimum varies according to the number of different images (length of the component $\neq$), the utilization of differences in implantation, the retinal variables employed, and the reading habits of the individual. A precise study, objectively measuring these differences, remains to be done.

The legibility of a FIGURATION is drastically altered by too great a density of signs. Ten signs per cm^2 represents a maximum limit.

However, there is practically no maximum density in an *image*, i.e., a homogeneous graphic representation.

An image can accommodate very great densities and consequently substantial photographic reductions. Such is the case with the maps on pages 319 and 373 or with the map of buildings in Poland (figures 3 and 4) authored by Professor Uhorczak (Warsaw, 1954), in which the eye encounters meaningful information whatever the density of signs.

What disappears with increasing density is the elementary precision. However, the eye, unencumbered by variations in detail, discovers regional characteristics more easily and defines several levels of regionalization without difficulty (figure 3). Finally, an overall reading, based on density, is meaningful and easy to accomplish (figure 4).

An IMAGE remains legible while accommodating great graphic densities and thus substantial photographic reductions. Consequently, reading on the intermediate and overall levels is generally found to be easier.

However, for both figuration and images, it is obvious that *too weak a density* corresponds to underutilization of graphic capacity. The only exception is a representation aiming at high precision measurements (plans, surveys, triangulation), since dimensional accuracy is proportional to the size of the drawing (see page 298).

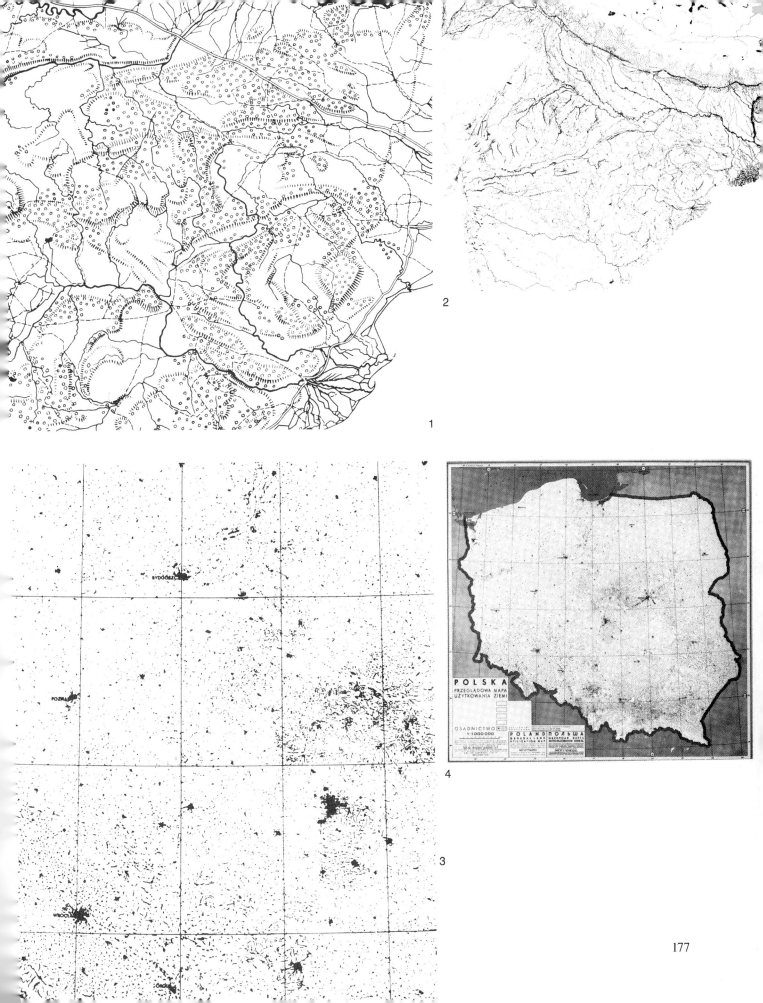

(2) Angular legibility

ANGLES

A visual form is delimited by a series of more or less clear lines which determine angles.

Angular legibility diminishes:
– as the clearness diminishes
– as the angle approaches 0 degrees or 180 degrees
– as the lines forming the angle shorten.

The choice of scale in a diagram is based on angular legibility.

The time-yield curve in figure 1 presents two types of angle:
– clear and legible angles, which are commonly found and readily perceptible on the level of elementary reading;
– an unclear and practically invisible angle (the one that the whole of the curve forms with the horizontal axis of the plane), on the overall level of reading. This angle is meant to express the ordered relationship between the two planar dimensions, that is, the correlation between time and yield.

This angle is perfectly visible in figure 2, but at the expense of a reading on the elementary level, since the eye encounters angles approaching 0 degrees. On the elementary level optimum legibility is located near the right angle. But angular differentiation is perceived through the intermediary of the lines forming the angle, which are more comparable when brought closer together. We therefore conclude that:

– **On the elementary level optimum angular legibility is located near 70 degrees.**
– **On the overall level the image tends toward the form of a square where optimum angular legibility is provided by the diagonal.**
– Since these two conditions can be contradictory, angular legibility results from a compromise between the conditions of legibility for the two extreme levels of reading.

This compromise is shown in figure 3. The designer should thus avoid elementary angles which are too sharp (or too flat) as well as diagrams which are too wide (or too high).

SHAPES

The perception of angles governs the perception of shapes and consequently of signs differentiated by shape variation.

However, as the length of the lines diminishes, approaching a certain threshold, the angle is no longer legible. Therefore it is not surprising that with reduction in size, the most varied shapes tend to become confused and end up as a point or a dash (figure 4). Consequently:

– **On the elementary level of reading a meaningful shape must have a minimal size of about 2 mm in order to be legible as such.**
– **With smaller sizes there are only three distinct shapes:**
  (a) the point
  (b) the dash
  (c) the intersection of two dashes (i.e., the cross).
– **On the overall level of reading, provided they are of sufficient size, the dash, the point, and the cross produce three steps which permit visual selection. Any other shape, being only intermediate, will diminish perceptible differentiation and eliminate selectivity.**

The dash must still be sufficiently long (at least four times its width), since we are sensitive to the difference between a point and a line. With the cross, we are affected by the difference between a "vibrating" sign and a "nonvibrating" sign (see the "vibratory effect of texture," pages 80–83).

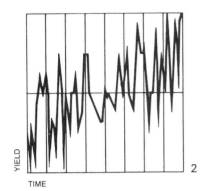

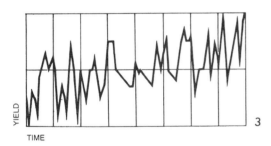

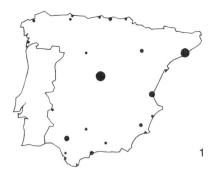

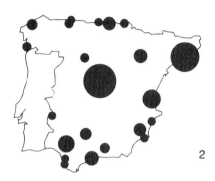

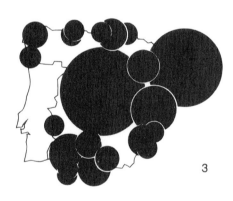

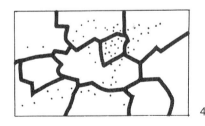

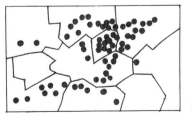

(3) Retinal legibility

If angular legibility concerns meaningful perceptible differentiation on the plane, then retinal legibility applies to differences in "elevation" *above the plane*, to differentiation among the steps of a retinal variable. In order to achieve maximum differentiation, it is necessary:

(1) *to have a total amount of "black"*:
- sufficient so that the smallest signs are visible, stand out from the background, and cover the "visual noise," such as blotches, errors, printing irregularities, etc. (in figure 1 the total amount of black is insufficient);
- but limited so that the largest signs are separate (in figure 3 the amount of black is too great).

(2) *to obtain the greatest amount of differentiation*, that is, to utilize the entire perceptible range of a given variable. But here, it is necessary to distinguish between:
- information which involves only quantitative perception;
- and that which involves selective perception, combined most often with ordered or associative perception.

THE AMOUNT OF BLACK AND THE BACKGROUND

Experience shows that when the total amount of black varies between 5% and 10% of the meaningful area of the plane, legibility is optimum (figure 2).

Consequently, this percentage enables us to calculate the scale of proportional signs. The hypothetical sign which would correspond to the total of the quantities to be represented should have an area less than a tenth of the meaningful area of the plane. The total amount of black can necessitate a *density adjustment* (see page 374) for series of signs of graduated size and for automated cartography. But this amount of black concerns only the representation of the pertinent (new) correspondences which constitute the information itself, and which must *contrast* with the background, that is, with the "reference" components represented by the two dimensions of the plane.

First perceptible step: the contrast between subject matter and background (figure and ground)

What is visible in figure 4, utilizing the major portion of "black," is the geographic component, the "predictable" information which anyone can reconstruct by opening an atlas. The representation is illegible.

The representation in figure 5 is legible. What dominates are the pertinent (new) correspondences, the "unpredictable" elements which, theoretically, no other document could furnish. *The first perceptible step must separate the subject matter from the background.*

The recipe is simple, and its use is a graphic necessity. Take the diagram in figure 6, in some ways well constructed, but nonetheless illegible. It represents the evolution of the number of ships linking Spain to the different areas of Spanish America.*

It illustrates the most common errors of legibility. The background (figure 7) is more visible than the subject matter (figure 8).

A REDUCTION IN THE VISIBILITY OF THE BACKGROUND is obtained by:
- eliminating known signs or ones which can be reconstructed if an elementary reading of the details is called for (figure 9);
- decreasing the visibility of the remaining signs (use of fine dashes, dots, light values);
- increasing the predictability of the identification signs, by accentuating the presumption of continuity.

A single vertical scale accentuates the presumption of homogeneity for all the curves. Several characteristic dates categorize the time, make its perception immediate, and accentuate the presumption of its continuity.

Conjointly the effort will involve:

AN INCREASE IN THE VISIBILITY OF THE SUBJECT MATTER.

For example, by "blackening in" the proportional columns (figure 10) we obtain a double perception involving the heights of the columns on the plane (which alone was visible in figure 6), and the amount of "black."

The superimposition of figure 10 on figure 9 produces figure 11, which is legible, and inclusion of the subtitles makes identification easier.

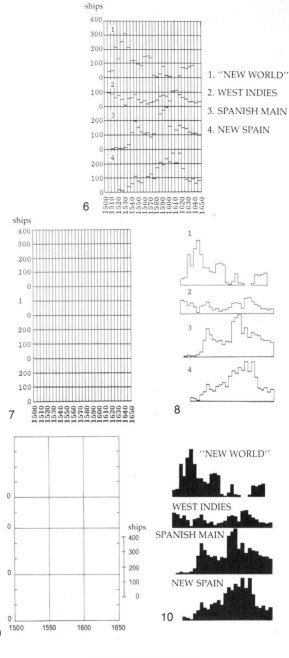

*Data from P. and H. Chaunu, *Séville et l'Atlantique. Atlas* (Paris: S.E.V.P.E.N., 1956). It is a question of the number of ships leaving Seville, apportioned according to their destination and the decade.

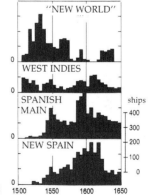

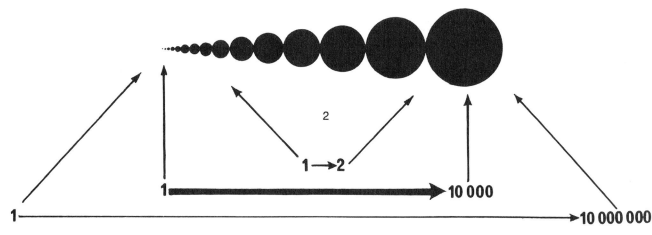

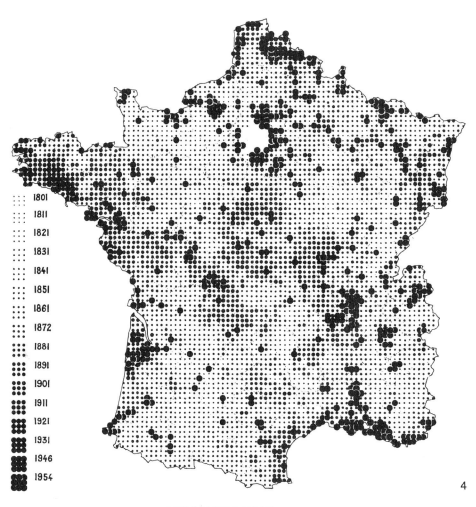

DATE OF MAXIMUM POPULATION

PERCEPTIBLE DIFFERENTIATION IN QUANTITATIVE PERCEPTION

Quantitative components (absolute quantities, measurements, ratios, etc.) do not call for *selective perception*. The question "A given precise quantity, where is it?" is generally useless and results from the bad habit of elementary reading. Note, however, that such a question could be answered by using a redundant combination of size (quantitative) with shape (selective), as on pages 357 and 377.

A quantitative variation accommodates a large number of steps (see example on page 373) and is all the more accurate for it.

Quantitative legibility essentially depends on the utilization of the maximum range (ratio of smallest to largest signs) in a series based on size differences.

It is easy to construct (in point representation) two circles whose area ratio is 1 to 10 000 (figure 1).

A ratio of 1 to 20 can still produce a legible image, but **legibility is practically nil at a ratio under 1 to 10.**

Conversely, the ranges of quantitative series can vary greatly (see page 357), and **legibility leads us to extend (figure 2) or to reduce (figure 3) the quantitative series in order to adjust it proportionately to the perceptible series. This is called a "range adjustment."**

The different modalities of this adjustment are described, according to the three implantations, from page 357 on.

PERCEPTIBLE DIFFERENTIATION IN ORDERED PERCEPTION AND SELECTIVE PERCEPTION

The perception of order does not in itself require the selection of particular steps. Consequently, these can remain numerous so long as selective perception is not called for. Ordered legibility only depends on the utilization of *the maximum range of the ordered variables: size and value.* Texture can also create a perceptible order, on the intermediate level of reading, without destroying the associativity of the signs (see figure 4, page 357).

However, ordered information generally calls for selective perception as well.

One experiences difficulty in reading the map in figure 4 (dates of maximum population per canton). The time component is represented by a variation in size. This component is relatively long, which makes it difficult to isolate the distribution of a precise date, that is, to answer the question "A given date (1850, for example), where is it?"

With the map in figure 5 (year of worst harvest), where ordered information is represented by a value variation, the

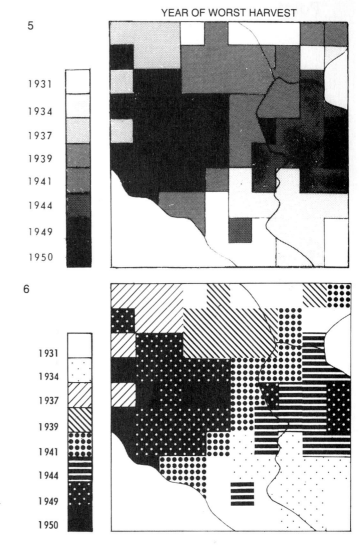

reader can also attempt to isolate one precise year: 1941 or 1944, for example. This can be accomplished better than in the preceding map, since the variation is shorter and the area signs are larger than the point signs. However, this is achieved better by the map in figure 6, which combines value, texture, orientation, and shape.

The combination of several variables reinforces selective legibility.

We will now define a combination of variables and examine its properties.

183

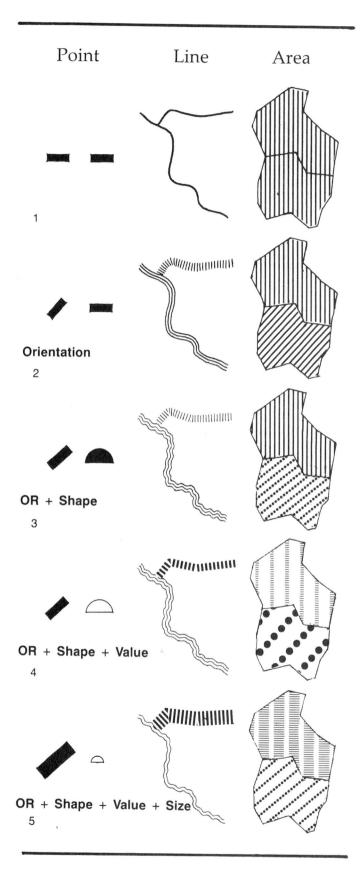

COMBINATIONS OF VARIABLES

The pairs of point, line, and area signs, which are similar in figure 1, are differentiated in figure 2 by the use of orientation. In figure 3 they are differentiated by both orientation and shape. This is an orientation-shape combination. The differentiation in figure 4 combines orientation, shape, and value. In figure 5 we utilize orientation, shape, value, and size.*

All the combinations of retinal variables are possible.

The figure on page 185 displays all the combinations which can differentiate two point signs. In group 1 (first row) a single variation separates the two signs of each pair. In the other groups, each pair is differentiated respectively by two, three, four, five, and finally six variations.

When the thirty-two combinations with size are aligned in the same column, one can see that color, for example, or any other variable is represented thirty-two times. Excluding repetitions, there remain sixty-three possible combinations between two signs.

*When value variation, normally obtained by a difference in inks (black, dark gray, light gray inks), is obtained, as here, by a visible line pattern or dot pattern, this will add to the perception of value a perception of texture, which is obviously different from black, which has no texture.

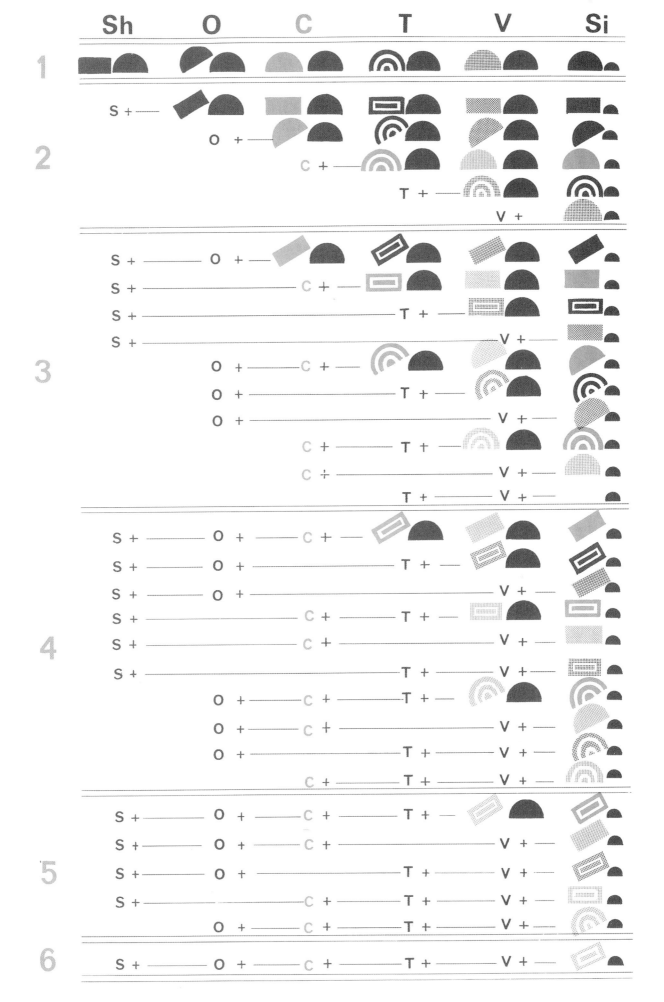

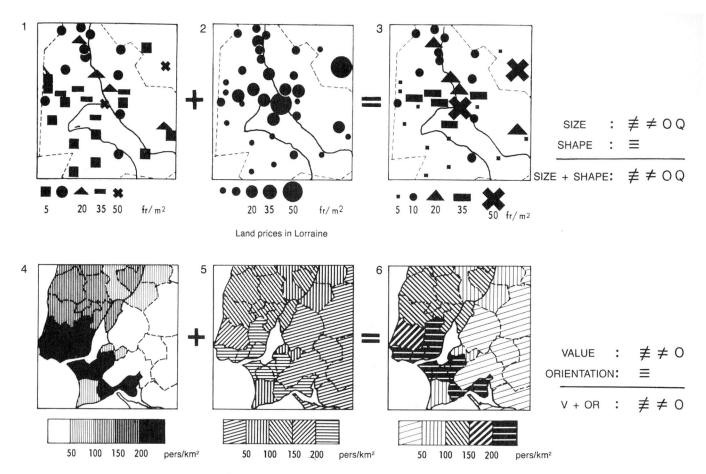

Land prices in Lorraine

Population density in Portugal

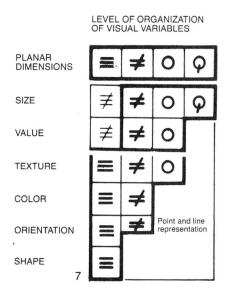

Properties of a combination of variables

The level of organization, i.e., the perceptual properties of each variable are known (see the table in figure 7). Note that the associative variation of shape (figure 1) and the dissociative variation of size (figure 2) produce a dissociative combination (figure 3). The density of the sample points, which is perceptible in figure 1, is dissociated in figure 3.

Likewise a value variation (figure 4) and an orientation variation (figure 5) produce a combination having the properties of value (figure 6).

A combination of variables will retain the properties of the variable having the highest level of organization, as determined by the table on page 96 (with the exception of the size-value combination).

Size and value, both being dissociative, that is, both having variable visibility, can be combined:
– either in the same direction (the concurrent combination in figure 9);
– or in opposite directions (the compensating combination in figure 10).

186

REDUNDANT COMBINATIONS

In figure 3 the combination of shape and size represents a single component: quantities of francs per m². Likewise, in figures 6 and 8, each combination represents only a single component.

A combination of several variables, utilized to represent only a single component, is a redundant combination.

Selective differentiation is always greater between two steps of a combination of variables than between two steps of a single variable, all other things being equal (with the exception of compensating combinations).

Consequently:

Redundant combinations increase the separation between the steps of the retinal variables. They are the basis for selective legibility.

The concurrent combination of size plus value (figure 9) enables us to reinforce the legibility of a quantitative variation (figure 11) whose range does not exceed 1 to 10.

The use of redundant combinations

The graphic designer encounters two main types of problem in the use of retinal variables.

(1) The information *will admit variable visibility* (see page 323), that is, a spread from weak to strong or from light to dark. In this case, the designer can create an *image which is ordered and selective*, and base the combination on size or value, that is:

size plus texture, color, orientation, shape or

value plus texture, color, orientation, shape.

(2) The information *will not admit variable visibility*; all the signs must have the same visual power. In this case, the designer can create a *selective-association* figuration, basing the combination on texture, and *preserve an order* in the visual variation.

texture plus color, orientation, shape.

If order is not involved, one of several possible combinations can be based on the last three variables: color, orientation, shape.

The modalities for using redundant combinations and determining their length for each of the three implantations can be found from page 323 on.

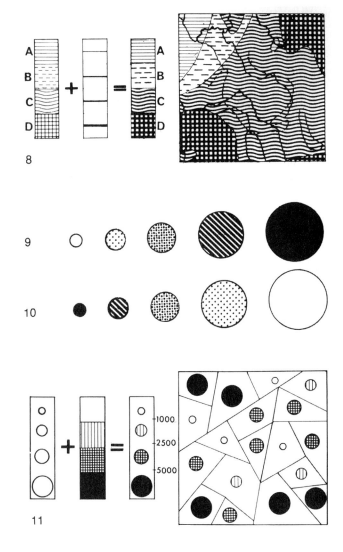

187

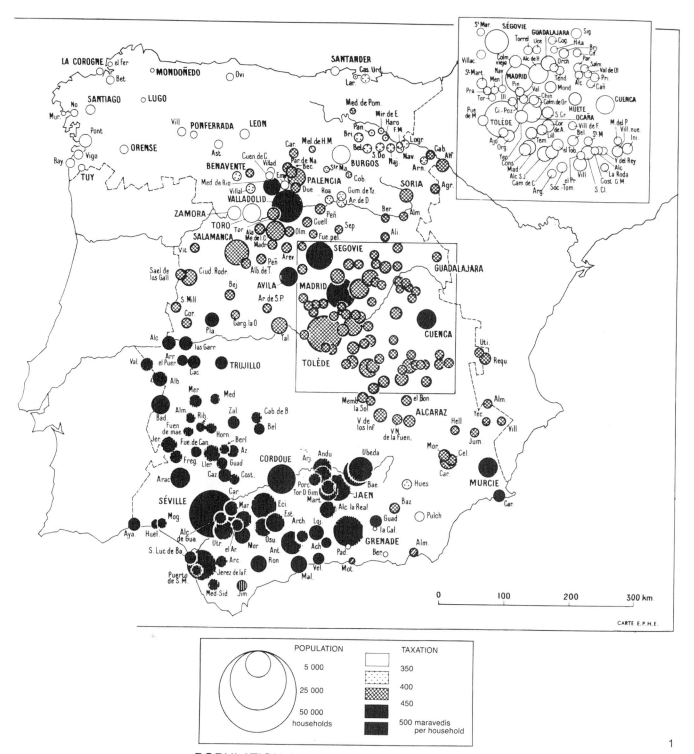

POPULATION AND TAXATION IN CASTILLE
TOWARD THE END OF
THE SIXTEENTH CENTURY
after Alvaro CASTILLO

"MEANINGFUL" COMBINATIONS

In the same implantation, when two variables are each associated with a different component, the combination is no longer redundant, it is "meaningful." The map in figure 1 is an excellent example of this. Each retinal variable represents one component, but the visual properties of the combination of variables are obviously applicable to meaningful correlations within the information. Therefore note that:

in a meaningful combination, the image is based on size or value, or on the visual sum of the two.

We should thus affiliate size and value with components which are particularly meaningful or whose combination is meaningful. For example, in the *map of Spain* in figure 1: population × tax rate = total amount of taxes.

In the *wind map* on page 353: force × length = air turbulence.

In a meaningful point combination, size and value cannot benefit from the entire range of perceptible variation. The lowest step must be sufficient to ensure the legibility of the other components.

SELECTIVITY WITH DIFFERENCES IN IMPLANTATION

Superimpositions involving two or three implantations—points, lines, areas—ensure an excellent visual separation (see figure 5, page 159) in graphic representations.

In mapping, for example, it permits superimposing figure 3 (a quantitative variation of inhabitants) on figure 4 (an ordered variation whose steps represent the number of doctors per 10 000 inhabitants), thus producing the map in figure 5. Easy to use and efficient, this is the recommended selective formula for sketches and simplified pedagogic figurations (figure 2).

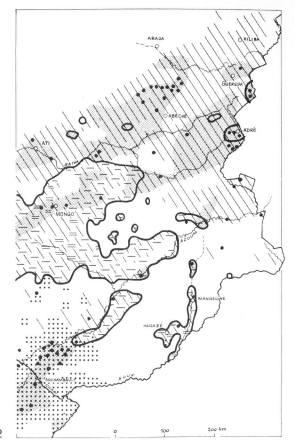

See page 163.

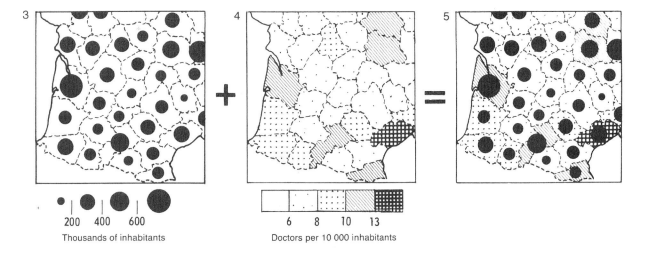

Summary of the rules of legibility

In any graphic

The overall figure should tend toward the form of a square (or a rectangle).
 Generally the ratio of the sides should not exceed 1/2.
 The entire range of a visual variable should be utilized.
 It is thus necessary to use a total amount of "black" which will occupy 5% to 10% of the area of the figure. This will eliminate graphics which are too light and become invisible as the eye moves away.
 The first perceptible step should separate the "subject matter" from the background, thus emphasizing the pertinent (new) correspondences and differentiating them from the signs which merely identify the reference elements of the plane.

In an image

Graphic density can be very great.

In a figuration

A figuration, by definition, poses problems of selection. Remember that: (a) the best visual selection is produced by the plane, that is, by the construction of the information in as many separate images as there are additional components (beyond the number represented by the planar dimensions); (b) the number of selective steps in the retinal variables does not generally exceed five; and (c) in order to benefit from these:
graphic density should not exceed ten signs per cm^2, whether these signs are points, lines, or areas;
separable shapes should be at least 2 mm in size; in very small sizes, there are only three separable shapes: the point, the dash, and the cross;
redundant combinations increase the differentiation among the categories of a component;
a difference in implantation creates the best differentiation among two or three superimposed components.

Part Two

Utilization of the graphic sign-system

Classification of graphic problems

A first order according to:
THE GROUP OF REPRESENTATION (IMPOSITION)

    diagrams
    networks
    maps (cartography)
    problems involving symbolism (*see shape and color*)

A second order according to:
THE NUMBER OF NECESSARY VISUAL VARIABLES

    1
    2 } representation as a single image is possible
    3

    3+   function of graphic (inventory, processing, message) must be considered

A third order according to:
THE LEVEL OF ORGANIZATION OF THE COMPONENTS

    ≠   simplification by ordering
    O
    Q } simplification by smoothing and regionalization

A fourth order according to:
THE LENGTH OF THE COMPONENTS (diagrams)

    limited   special cases
    extensive   standard constructions

THE TYPE OF REPRESENTATION (IMPLANTATION)
(networks and maps)

    point
    line
    area

I. Diagrams

Definition
The graphic is a diagram when the correspondences on the plane can be established among all the elements of one component and all the elements of *another* component.

Process of construction
In order to *construct* a diagram, it is necessary:
(a) to determine a form of representation for the components;
(b) to record the correspondences.

The standard construction is expressed by the schema in figure 1, which implies: (1) the orthogonal utilization of the planar dimensions; (2) the utilization of an ordered retinal variable for the third component; and (3) the ordering of qualitative components by diagonalization (see page 168).

Unity of the image
Any diagram involving two or three components can be constructed as a *single image*. Special constructions are justified only by the presence of very limited components.

A. Diagrams involving two components

Such diagrams exclude systematic comparisons, which would necessarily lead to introducing a third component into the problem. Consequently, the purpose of these diagrams is the internal reduction of each component by the ordering and grouping of categories. The display below (figure 1) represents the principal forms which can be used, according to the level of the components.

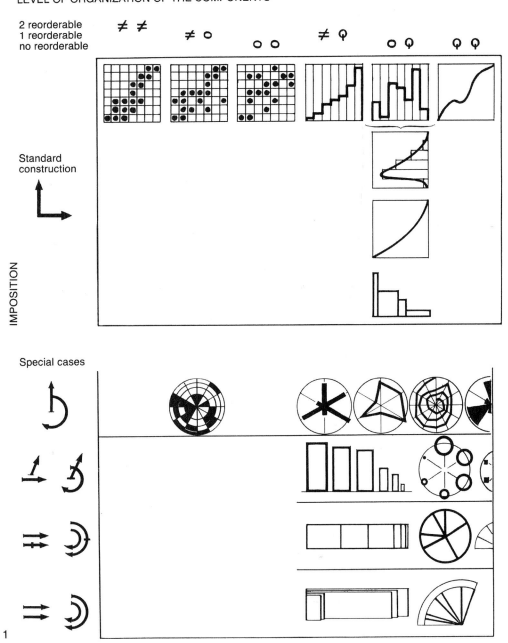

1

1. Nonquantitative problems

≠ ≠
Two reorderable components

First example: Sales organization
Consider a list of clerks A, B, C, ... and a list of articles sold l, m, n, ... in a store.
- ≠ eight clerks
- ≠ twelve articles

For various purposes, the management compiles lists, which are generally arranged alphabetically, and then sets up a table of correspondences between the two lists (figure 1) in order to know the articles sold by A, B, ...

But these lists also contain information relative to the organization of the sales force. In the example given it is possible to arrive at a simpler image than that in figure 1, without changing any of the correspondences among the elements, by adopting a new order within each list. This new order is discovered by permutating several columns A, B, C, then several rows l, m, n, up to maximum simplification. One thus obtains figure 2. The process of diagonalization, utilized here, reveals the structure of the sales organization. Here it is simple and apparently logical: two clerks are assigned to the same group of articles, and each article is sold by two clerks. With a poorly organized store, the process of diagonalization would have produced a complex image, which only changes in sales assignment could simplify. Thus the image can be a means of reorganizing the work in a more logical manner.

Second example: Significance of cultural traits in nine human societies
(After Robert L. Carneiro, "Scale Analysis as an Instrument for Study of Cultural Evolution," *Southwestern Journal of Anthropology* 18, no. 2 [1962]: 149–169.)

Take nine societies, A, B, C, ... in South America, in which the presence or absence of cultural traits 1, 2, 3, ... has been observed.
- ≠ nine human societies
- ≠ eight cultural traits

During the survey stage, traits and societies were correlated in a double-entry table (figure 3) immediately after their identification.

The search for a simpler image, by permutation of rows and columns, leads to the triangulation shown in figure 4. It is sufficient to class the societies by number of traits: Inca 8, Chibcha 7 ... and the traits by number of societies: agriculture 8, pottery 7 ....

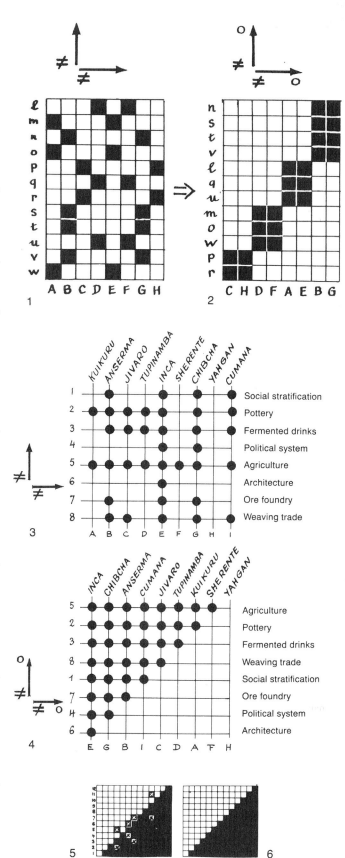

Thus, societies and traits are found to be reciprocally classed: societies in the decreasing order of number of traits; traits in the increasing order of rareness. This classing also displays the inclusive or cumulative nature of the cultural traits: the presence of the weaving trade, for example, implies the presence of all the preceding traits (fermented drinks, pottery, agriculture), and its identification will affirm their existence also. But weaving bears no information relative to traits such as social stratification and ore foundry, which can be present (Inca) or absent (Jivaro).

Scalograms rarely lead to a perfect image.

Louis Guttman, who was the first to demonstrate the importance of this method ("The Basis for Scalogram Analysis," in *Studies in Social Psychology in World War I*, vol. 4: Samuel A. Stouffer, et al., *Measurement and Prediction* [Princeton: Princeton University Press, 1950], pp. 60–90), proposes a numerical evaluation of the difference between a given scalogram (figure 5) and its ideal model (figure 6). This is the "reproducibility coefficient" calculated according to the formula:

$$C = 1 - \frac{\text{number of differences}}{\text{length of the first component} \times \text{length of the second}}$$

Here, a "difference" is either an empty box ("cell") that ought to be filled, or a filled cell that ought to be empty, relative to the ideal figure. Thus in the example from figure 6:

$$C = 1 - \frac{8}{12 \times 13} = 0.949$$

In the social sciences this coefficient can be used to class several diagrams in a logical order, as a function of the degree of correlation between pairs of concepts.

Third example: The district of Paris for various administrative bodies

(After *Etude statistique de la région parisienne*, vol. 1, *Délimitation de l'agglomération* [Paris: I.N.S.E.E., 1947], p. 85.)

Both departments and administrative bodies are reorderable components. Ordering each according to the number of correspondences (the numbers from the table in figure 7) enables us to construct figure 8, which immediately reveals principal anomalies (circled on the figure) concerning either departments (Oise, Aube, Yonne) or administrative bodies (Appeals Court). The drafting of such an image is within the reach of any typist, and a map can be derived from it.

List of departments (other than Seine) comprising the Paris district for certain administrative bodies

Administrative bodies	Seine & Oise	Seine & Marne	Oise	Eure & Loir	Loiret	Loir & Cher	Eure	Aube	Marne	Yonne	Cher
Academy of Paris	+	+	+		+		+		+		+
Archdiocese of Paris	+	+			+	+					+
Appeals Court	+	+		+				+	+	+	
Social work	+	+									
Public Insurance	+	+	+	+							
Automobile registration	+	+									
Water, forestry	+	+									
Veterans Administration	+	+									
Electricity	+	+	+								
Gas	+	+	+								
Business Inspection	+	+		+	+	++	+				
Police force	+	+	+								
Civil engineering	+	+									
Mail, telephone,	+	+	+								
Military district including Paris subdivision	+	+		+	+	+	+				
Planning and zoning	+	+	+								
Public Health	+	+	+								
Social Security	+	+	+							.	
Employment	+	+									

(1) The Act of 28 August 1941 establishes the planning area as extending to the departments of Seine, Seine-et-Oise, Seine-et-Marne, as well as to the communes in the five cantons of Creil...Senlis in the department of Oise.

7

```
Veterans Administration .......  X X                      2
Social work ..................   X X                      2
Water, forestry, mining ......   X X                      2
Civil engineering ............   X X                      2
Employment ...................   X X                      2
Gas and electricity ..........   X X X                    3
Police force .................   X X X                    3
Mail, telephone, telegraph ...   X X X                    3
Planning and zoning ..........   X X X                    3
Public Health ................   X X X                    3
Social Security ..............   X X X                    3
Public Insurance .............   X X X X                  4
Catholic Diocese .............   X X     X X    X         5
Business Inspection ..........   X X     X X X X          6
Military district ............   X X     X X X X          6
Appeals Court ................   X X  X            X X X  6
Academy ......................   X X   X      X X X       7
                                 17 17 8 4 4 3 3 2 2 1 1
                                 S  S  O E E L L C M A Y C
                                 e  e  i u u o o h a u o h
                                 i  i  s r r i i e r b n e
                                 n  n  e e e r r r n e n r
                                 e  e    -  e  e  t  e    n
                                 -  -    e  t  t           e
                                 e  M    t     L
                                 t  a    -     o
                                    r    L     i
                                 O  n    o     r
                                 i  e    i     
                                 s       r     
                                 e       
```

8

≠ O
A reorderable component
An ordered component

First example: Sales organization
Take:
≠   eight clerks A, B, C, . . .
O   twelve articles ordered according to their monetary value.

By placing the classed list of articles on one coordinate, it is possible to class, on the other coordinate, the list of clerks in a corresponding order D, F, A, . . . (figure 1). We form the image which most closely approaches the diagonal. This table permits us to answer any question based on the order of value of articles sold: for example, "Who sells the least expensive articles?" or "What is the performance (in value of articles sold) of a given clerk?"

Second example: Historical classing
For a series of silver mines in Yugoslavia, the dates of initial activity and of closing are known (these dates can be more or less precise).

This historical information, although brief, assumes its full importance when we represent it as in figure 2, which enables us to class the mines and discover potential groupings. It also constitutes the time basis for a quantitative evaluation of production, but this involves a new component and will be discussed later (see page 222).

OO
Two ordered components

First example: Sales organization
Consider:
O   eight clerks, classed according to their age
O   twelve articles, classed according to their monetary value.

Unlike figure 1, figure 3 cannot be reordered. We can only evaluate the meaning of the correspondence between the two given orders. In this example the most expensive articles are assigned to middle-aged clerks, whereas the least expensive articles are assigned to a young-elderly tandem. The manager's organizational principle appears clearly.

Second example: The "day" of a mountain peasant
Take:
(1) places of work or activity, ordered in space, according to altitude
(2) a sequence of days.
This means two ordered components, through which one can follow the activity of a family or individual.

Graphically transcribed, according to the standard construction, the information forms the classic image in figure 4, which would lead to numerous discoveries if extended over a longer time period.

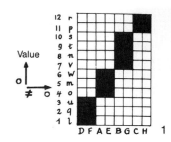

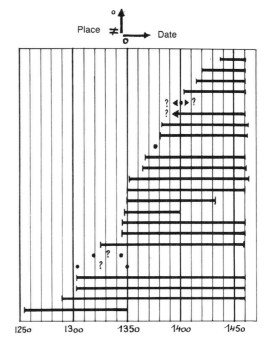

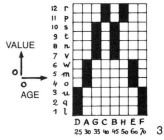

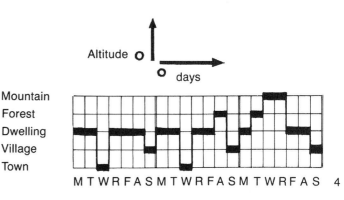

2. Quantitative problems

Q ≠
A limited reorderable component
A quantitative series

A quantitative series enables us to order a ≠ component.

A limited or "short" ≠ component (up to about six categories) authorizes "special constructions" (differing from the standard schema), since the number of necessary images will be at most equal to the length of the ≠ component.

The appropriate construction will depend on the eventual use of the graphic, as outlined in the display below (figure 6). All these figures apply to the same set of information (figure 5).

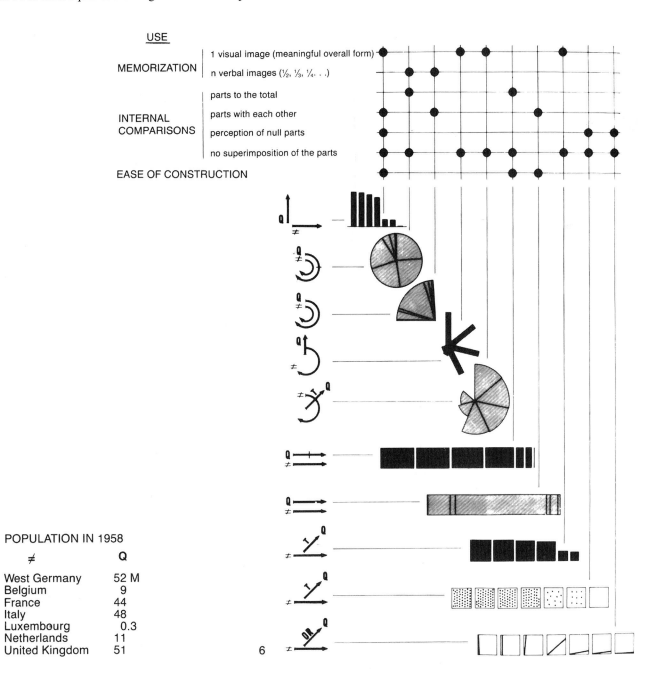

POPULATION IN 1958

≠	Q
West Germany	52 M
Belgium	9
France	44
Italy	48
Luxembourg	0.3
Netherlands	11
United Kingdom	51

5

6

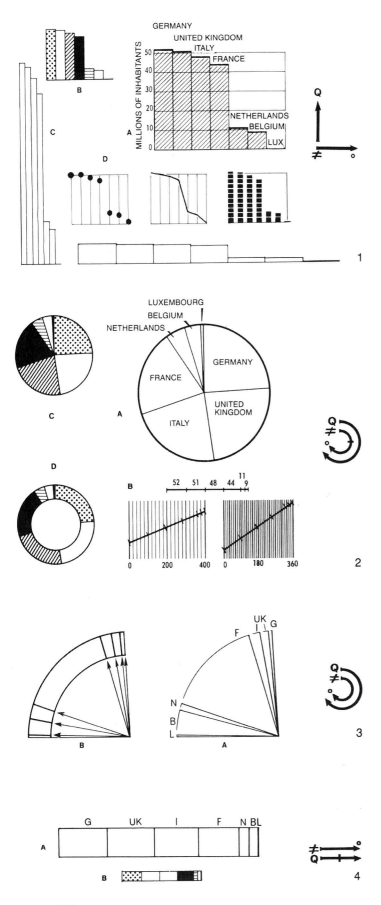

Take the information from figure 5 on page 199.

Standard construction

Figure 1: The image resulting from the standard construction favors comparisons among the parts, but it does not allow comparison of the parts with the total. The overall form is meaningful. The use of an additional retinal variation (B) is not necessary. The rules of legibility exclude drawings such as C. Of the three representations in D, the one on the right is preferable, since it produces a countable figure. This is particularly useful for portraying very small quantities (number of daily entries into a port, for example).

Special constructions

Figure 2: Circles and proportional sectors. If external visual comparisons are not envisaged, a problem ≠ (limited) Q can be represented in a construction where the overall form is not meaningful. The internal forms (the sectors), provided they are few in number, are readily (though roughly) perceived as fractions of the total: 1/4, 1/3, 1/2. Thus the essential point of information—the relationship of the parts to the whole—is depicted. Each part is defined in relation to the total, which is considered as the basic unit.

The angle about the center plays an important role in evaluating this relationship (compare C with D).

In order to construct this figure, we must have the total and transform it into "grads" or degrees. This operation can be carried out graphically (B). The parts are added along a straight line, which is then placed onto millimetered paper, between two parallel lines separated by 400 grads or 360 degrees. We can then read the corresponding angles and place them on the circle. We recommend that the parts (here the names of the countries) be identified on the figure itself. The redundant addition of a retinal variation in C is not necessary.

Figure 3: Superimposition of sectors. In order to facilitate a comparison of the parts with each other, which is difficult to achieve in figure 2, the sectors share a common radius in figure 3. It is generally important that the entire circle, which is easily imaginable, correspond with the total of the parts, whose dimensions are then meaningful.

This construction is often utilized as in B to represent various speeds.

Figure 4: Linear construction. Although simple to draw and easy to read, this construction has little visual efficiency in spite of portraying the total. It is difficult to estimate the numerical value of the parts without taking precise measurements. This construction also poses a problem in positioning the names on the graphic, though this can be avoided when the line is represented vertically.

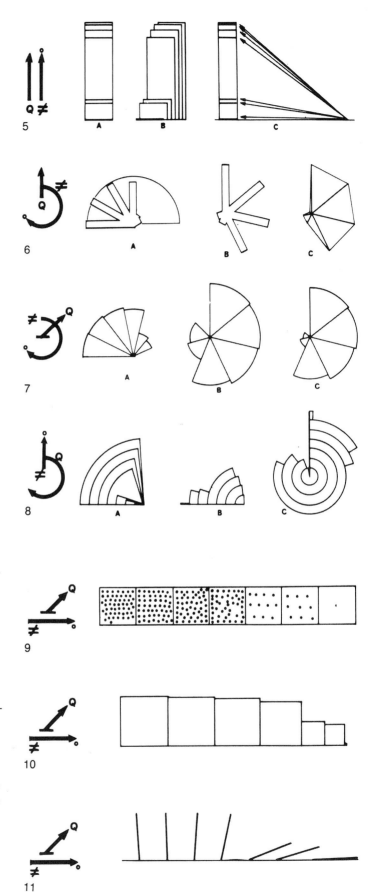

Figure 5: The same construction as figure 4, but with **quantities which are not cumulated**, favors comparisons among the parts. Again it is difficult to position the names. The variants given in B and C illustrate the necessity of showing the common origin of the quantities: zero.

Figure 6: Polar variations of figure 1 are striking, though not very efficient. The variation in C has the advantage of constructing a retainable "form." However, the quantities are represented in an ambiguous manner, since they are not proportional to the areas depicted.

Figure 7: This is another polar variation of figure 1, but here the **quantities are portrayed by the areas of the sectors**. The angles about the center are equal, as in figure 6. The length of the radii must be proportional to $\sqrt{Q}$. The error of constructing radii proportional to Q, as in C, should be avoided, since the areas would be proportional to Q^2. These constructions represent a retainable form, as in figure 6, C.

Figure 8: The constructions resulting from the **inversion of the components** in relation to the schema in figure 6 lead to irregular images, which should be familiar so that they can be avoided.

Figure 9: A variation in size can take the form of a **number of points proportional to the quantities**. This graphic solution, which leads to a density perception (ratio of the quantities to the area), is appropriate for problems such as comparison of regional densities, densities per product, etc. However, this solution only works for very great differences in density.

Figure 10: A variation in size (conveyed here by the **areas of the squares**) allows us to evaluate the quantities, since the side of the square is proportional to $\sqrt{Q}$. But the reader can better evaluate differences in height, which makes construction 5 preferable.

Figure 11: In a **nonpolar variation using orientation**, the order of the parts is perceptible, but the lack of quantitative precision is readily apparent.

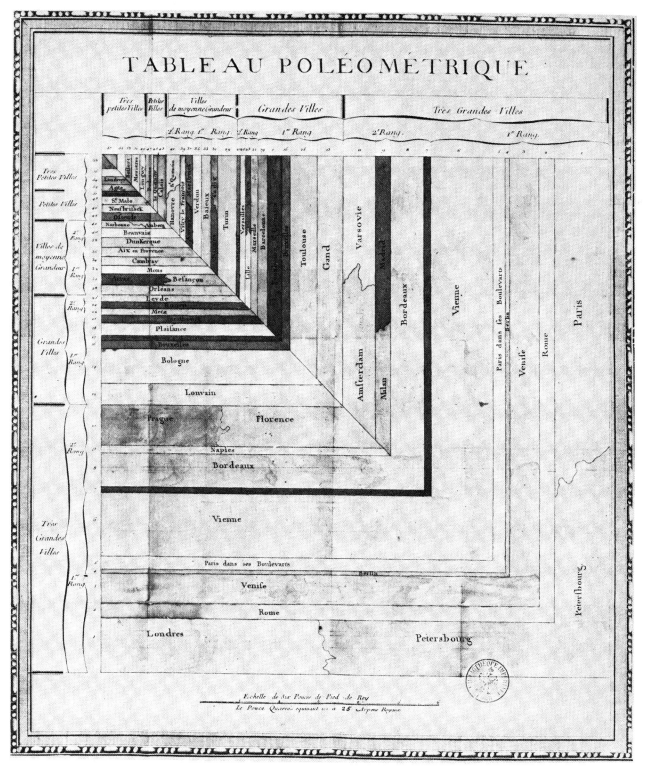

Q ≠ extensive
Repartitions
(information known for each element*)

The "Poleometric Table" (figure 1) published in 1782 by Dupain-Triel, is one of the oldest proportional representations of human phenomena which is currently known. François de Dainville has demonstrated ("Grandeur et population des au XVIIIe siècle," *Population* 13, no. 3 [July–Sept. 1958]: 459–480) that its author was Charles De Fourcroy, a director of fortifications.

Each city is represented by a square whose area is proportional to the geographic area occupied by the city (and for the smallest cities, by a half square only, divided by the diagonal line).

When superimposed, the squares are classed automatically. This results in visual groupings which lead the author to propose an "urban classification." This example allows us to appreciate the evolution of graphic representation and the efficiency of more recent solutions, based on the standard construction.

A repartition

Consider the information given in figure 2, that is, a list of villages 1, 2, 3, . . . 40 (≠ 40) and their population (number of households).

To begin with, we should note that special constructions, such as those in figures 5 and 6, lose all efficiency here. The length of the ≠ component is too extensive for them.

In place of the diagram in figure 3, which directly represents the initial classing, the standard construction substitutes the diagram in figure 4, where the villages are classed according to their quantities of population.

This creates at least two visual plateaus or "levels," that is, villages which have about 70 or 200 households. We must consider these levels cautiously, since statistical evaluations have a well-known attraction for rounded numbers. This also shows the villages in the extremes of the distribution (i.e., very small and very large villages). The reordering process enables the reader to pass from the notion of quantity per village to a general typology of villages, independent of special cases and exact population figures. The diagram in figure 4 is called a **repartition**.

---

* As opposed to information which is known only by classes of quantities (page 206).

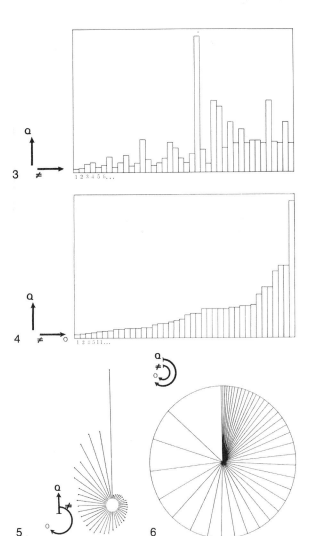

POPULATION OF VICINITY OF MADRID in the 16th century
per village (≠), in number of households (Q)

≠	Q	≠	Q	≠	Q	≠	Q
1	25	11	45	21	68	31	300
2	30	12	65	22	130	32	200
3	60	13	230	23	950	33	220
4	70	14	90	24	160	34	220
5	37	15	47	25	60	35	200
6	50	16	65	26	500	36	500
7	110	17	100	27	460	37	210
8	35	18	210	28	170	38	200
9	70	19	170	29	350	39	350
10	120	20	100	30	200	40	200

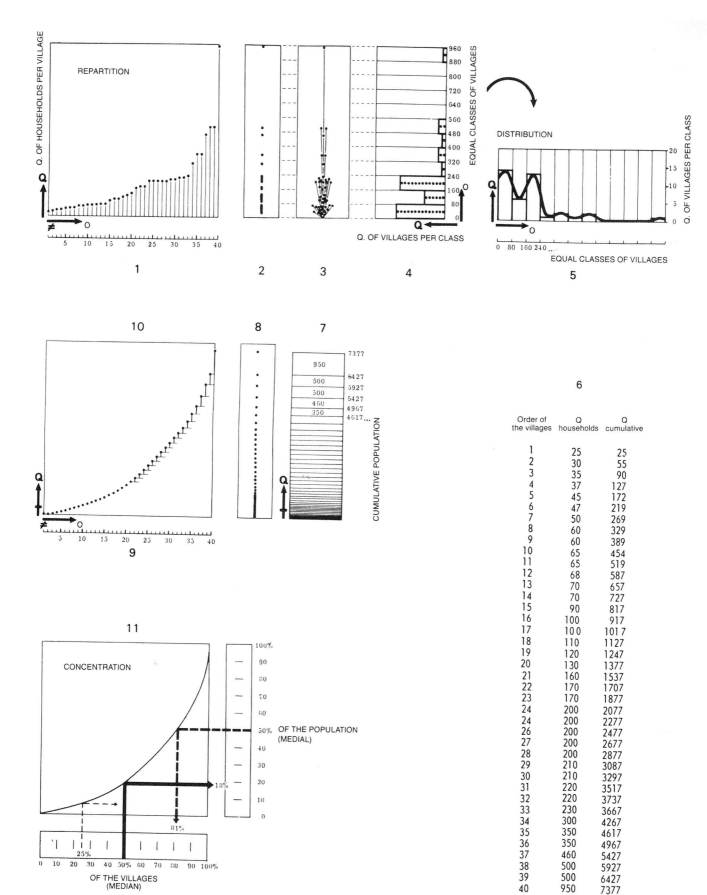

Repartitions, distributions, concentrations

The diagram obtained by ordering the villages is a **repartition** (figure 1). It causes levels, i.e., villages of the same type (size), to appear. How can these levels be defined?

Suppose that instead of forty villages, we are faced with forty pupils in a class, each one having a ball of twine whose length (25, 30, 35 cm . . .) is expressed by numbers similar to the number of households per village.

One can join all the twine at the same point and ask the students to line up, each one holding the other end. They will not be able to line up as in figure 2, since, at certain distances from the common point, they will be so numerous that it will be necessary for them to be distributed in larger *groups*, as in figure 3.

These groups are the levels in question. For reasons which remain to be determined, the balls of twine are more numerous for certain lengths, the villages more numerous for certain quantities of population.

To define these groups we divide the alignment in figure 2 into equal parts (here classes spaced at intervals of eighty households), chosen in such a way that the groups are not split up (figure 4). Next we count the number of villages (or pupils) in each division. The diagram thus obtained is a **distribution**, more often represented as in figure 5. It emphasizes the levels of the repartition (figure 1).

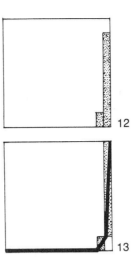

But it is also interesting to know how much population half of the villages (the twenty smallest, for example) contain, or the length of twine in the twenty largest balls. Nothing is easier if we join, in order of length, all the balls of twine, end to end, or if we arrange the numbers of households per village, one after another.

This is what we obtain arithmetically, by calculating the cumulative quantities (figure 6), or graphically, as in figures 7 and 8. The villages or the pupils are identified and numbered on one of the coordinates (figure 9), in order to be easily referred to. This leads to figure 10, which is generally replaced by a continuous curve (figure 11). This is a **concentration** curve, which provides an answer to the question posed earlier (i.e., the amount of population in half of the villages). This answer is expressed as a percentage and bears evidence of the more or less large "concentration" of the population in one or several villages (in which case the "repartition" would have the form of figure 12 and the "concentration" the form of figure 13). Conversely it can provide evidence of an equal dispersion of the population across all the villages (in which case the curves would look like those in figures 14 and 15).

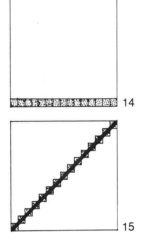

While a concentration lets us numerically define (as in figure 11) several striking aspects of a quantitative series,* it is especially useful in problems involving three components. It enables us to compare and class several phenomena, independent of their potential internal levels, which practically disappear in the wake of the cumulative calculation (see figure 14, page 111).

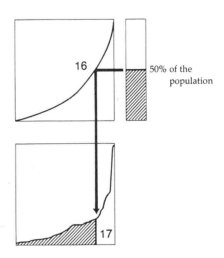

---

*This is due to the property of the definite integral. Note (figure 16) that a point on the concentration curve corresponding to half of the total population divides the repartition diagram into two equal areas (figure 17). Any point on figure 16 rigorously determines the corresponding area, represented as in figure 17, and we say that the concentration curve is the "integral" of the repartition.

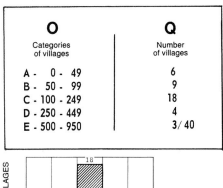

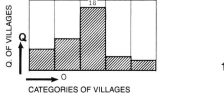

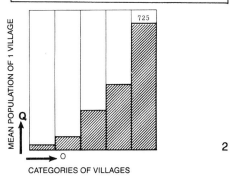

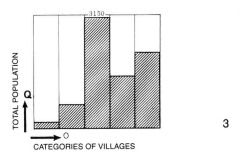

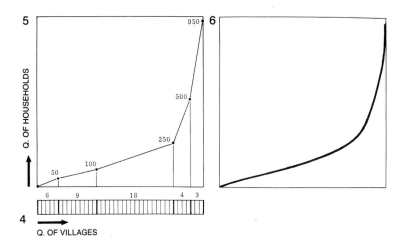

Q O (Q ≠)
Repartitions
Distributions
Concentrations
(information known by classes of quantities)

Consider the information in figure 1, where the previous statistical series is given in classes of quantities.

It can be considered as a problem Q O. The different classes A, B, C, . . . are considered as similar and ordered categories: categories small, medium, and large villages . . . , and their graphic representation leads to the diagram in figure 1.

However, the statistical series contains other information: the mean population of a village, an approximation which is situated between the smallest and the largest village in each category and which we can calculate and portray as in figure 2. From these data, it is easy to derive the total population in each category and to depict it (figure 3). But these graphics, linked as they are to the eminently variable definitions of each category, cannot be compared to any other series. They inform us neither about the types of village (the levels of the repartition); nor about the number of villages which would have to be grouped in order to obtain a half or a quarter of the population; nor about the amount of population contained in a half or a quarter of the villages. . . . These are, however, problems which must be resolved, since *the greater part of statistical information is provided by classes*, and necessarily so. However, to answer these questions, to resolve these problems, *we must know the population of each village*, the length of twine held by each pupil, just as on the preceding page. In general terms, it is necessary to know the statistical series by individual element and not just by class or group of elements.

It is one of the objectives of graphic representation to attempt to reconstruct a statistical series by element from information given by class. The rigor of this reconstruction depends solely on the pertinency of the enumeration units (classes).

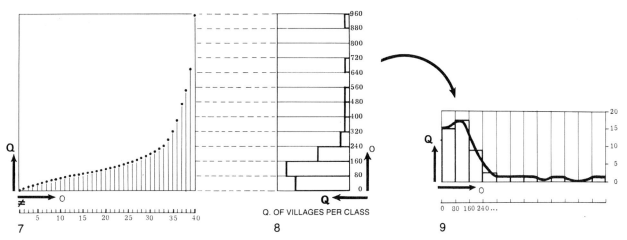

Q. OF VILLAGES PER CLASS

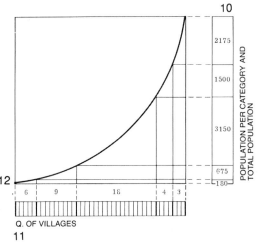

Q. OF VILLAGES

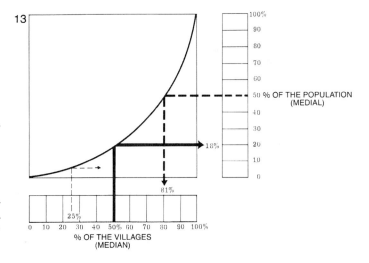

% OF THE VILLAGES (MEDIAN)

To reconstruct a series by element:
the information in figure 1 must be considered as a problem $Q \neq$, that is, as an alignment of all the villages, ordered according to their population. We can then calculate the total population of all the villages: amount (**Q**) of population per village × number (**Q**) of villages. Indeed, we know:

(1) the total number of villages: forty. They can be aligned (cumulatively) then divided according to categories (figure 4). These are the **Q** of villages per class.

(2) the population of the smallest and the largest village in each class: 0 and 49, 50 and 99, 100 and 249. ... The straight line joining these points constitutes an approximation of the repartition (figure 5). A curve is generally more appropriate (figure 6). It has a greater chance of approaching the real repartition. This curve depicts the population of each village (figure 7), much as it was known in figure 1, page 204, but with an accuracy linked to the degree to which the class limits approach the actual village populations.

From this construction (figure 7), all the operations described on page 205 are possible.

We can construct a distribution (figures 8 and 9), or a concentration (figures 10, 11, and 12); we can calculate summary values (figure 13).

From the ten numbers of the information in figure 1, we can derive the same information as in a series known by the element, although admittedly with less precision and certainty. This series is now defined by its proper characteristics and can thus be introduced into a general typology of repartitions, distributions, and concentrations.

A **repartition** (R) or a "cumulative curve" tends to define each individual observation (villages, pupils) as it is classed.

If we know each observation, the curve is accurate. When the information is known by class, the curve is an approximate reconstruction.

A **distribution** (D) emphasizes the levels of a repartition.
(1) *It tends to verify the homogeneity of the observations* and to uncover unknown variables. The villages are grouped around two classes (see figure 4, page 204); there are, therefore, a certain number of conditions (to be determined) common to the first group and other conditions common to the second group.

(2) *It is a means of prediction and calculation.* A distribution involving a large number of observations is stable, so long as the conditions of observation are constant. For example, the distribution of the daily rate of response to a circular letter enables us to predict the total number of responses which will be obtained, based on the number of responses during the first days. By the same token, knowing the distribution of missized pieces coming out of a machine permits us to calculate the cost of a piece, according to various tolerances of accuracy.

The distribution of deaths according to age allows us to calculate the cost of life insurance, based on the age of the insured.

(3) *The comparison of two distributions is a means of control.* When a distribution remains fixed, two distributions observed in identical conditions will be different only if unforeseen factors are introduced into the observation. Consequently, we could uncover, in the preceding examples:
– a source of error in the writing of addresses . . .
– a mechanical deterioration or a lack of quality control for the machine . . .
– a modification in living conditions affecting life span . . . .

A **concentration** (C) enables us to calculate numbers (median, medial, fractile) which are summary values of the series being studied and allow comparisons with other series.

The comparison of concentrations, by superimposition on the same drawing (see figure 14, page 111), permits us to class several series reciprocally and rearrange them into groups of similar series.

A **frequency calculation** also makes series comparable. The comparison of several distributions is possible only if they involve the same number of observations (which is rarely the case) or if they are reduced to a common total (see page 247). We can choose as a total 1, 100, or 1000. The absolute numbers of observations per class are then expressed as *frequencies* per 1, per 100, per 1000, and are all comparable. The term "absolute frequency" is sometimes used to denote the absolute number of observations per class, in a distribution. In this case, the frequency per 100 is called "relative frequency."

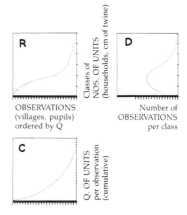

THE GRAPHIC CALCULATION OF SEVERAL SUMMARY VALUES

Summary values are the only means of introducing a diagram into a universal system of comparison (see page 173).

Mean

$$= \frac{\text{Total number of units (no. of households, cm of twine)}}{\text{Total number of observations (villages, pupils)}}$$

$$= \frac{\text{Total number of households}}{\text{Total number of villages}}$$

$$= \frac{\text{Total length of twine}}{\text{Total number of pupils}} = \frac{7{,}680}{40} = 192 \text{ (cm/pupil)}$$

In the distribution (D), the mean is in the fourth class (150–200). Mean class = 4.

Median

= number of units (no. of households, cm of twine) at the point where half of the observations (villages, pupils) have been classed
= number of households at the halfway point (20) of the villages classed
= amount of twine held by the first half of pupils classed
= 1460, that is, 18%

In the distribution (D), the median (18%) is in the third class (100–150). Median class = 3.

First quartile. Number of units contained in the first quarter of the observations classed.
First decile. Number of units contained in the first tenth of the observations classed.

Medial

= number of observations (villages, pupils as classed) grouping half of the units (households, twine)
= number of villages containing half of the households
= number of pupils holding half of the twine
= 33, that is, 81%.

In the distribution (D), the medial is in the fifth class (200–250). Medial class = 5.

Mode

Class of maximum frequency. Modal class = 2.

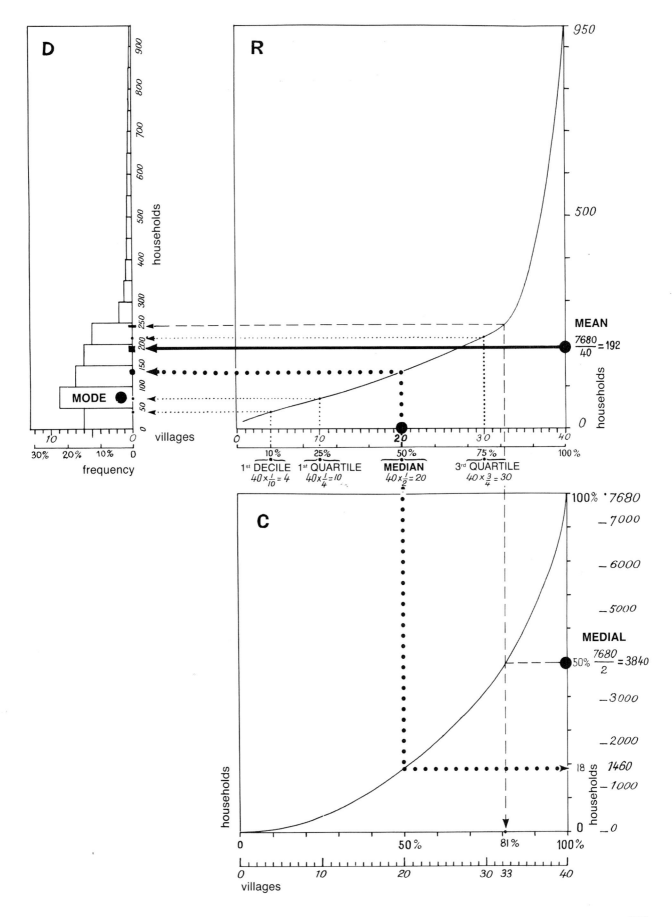

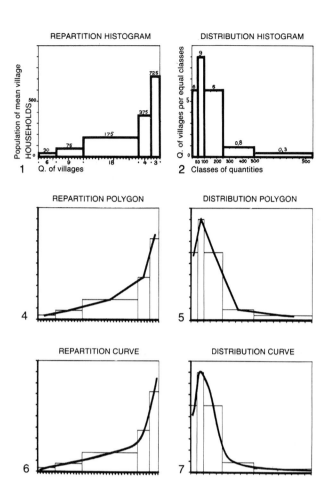
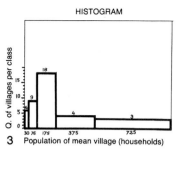
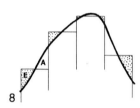

HISTOGRAM, POLYGON, CURVE

These are three forms of representation which are applicable to repartitions, distributions, concentrations, and various other constructions.

The term **histogram** applies to a graphic composed of rectangles (columns), which are juxtaposed in such a way that *the area of each rectangle and the total area of all the rectangles are meaningful*. In other words, the product of $Q1 \times Q2$ is meaningful. It can represent either the actual total (number of households) or the number of calculated units (number of meters of twine) in repartitions and other constructions (figure 3). The product can also represent the number of observations (villages, pupils) in distributions (figure 2).

The histogram is the safest type of graphic and the easiest to apply when the classes are represented on a coordinate axis by segments of a straight line.

The straight lines linking the midpoints of each successive category in a diagram form a **polygon**. It can either be constructed directly (see figure 5, page 206), or from a histogram, as in figures 4 and 5 here. Its form is simpler than that of the histogram, thus *favoring perception* and fostering the superimposition of several constructions which remain legible and separable.

A line linking the midpoints of each successive category in a diagram with minimum inflection is a **curve**. It is the goal of any graphic representation by class: *reconstruction of a series by element* (figures 6 and 7). It can be drawn from the midpoints of a polygon (see figure 6, page 206), or directly from a histogram.

The drawing operation is relatively delicate and must fulfill two conditions (figure 8):
(1) have a maximum radius of curvature;
(2) eliminate from each rectangle of the histogram an exterior area (E) equal to the interior area which is added (A); that is, the curve must "subtend" the same area as the histogram.

The histogram, the polygon, and the curve represent three levels of a bidimensional relationship. The histogram most closely conforms to the objective conditions of a particular experience. At the other extreme, the curve eliminates classes, which vary from one experience to another, and tends to generalize the relationship to include all possible experiences. These three terms cannot designate specific constructions. Rather, we will speak of repartition histograms, distribution histograms, etc., reserving the term histogram on its own for constructions such as figure 3, which are neither one nor the other. The same will hold true for polygons and curves.

CONSTRUCTION OF A DISTRIBUTION

A "distribution" (that is, a distribution of the observations along the scale of the quantities) is possible only when it is based on equal classes. In order to construct it, several precautions should be taken. Thus:

Figure 10 (classes and quantities per class) is poor. The area of the rectangle is meaningless (**Q** of villages × difference between largest and smallest village in each class).

Figure 11, which aligns all the villages (by accumulating the quantities of villages per class), forms a repartition polygon.

A distribution can be constructed in three ways:

By calculation of the number of observations in classes equal to a unit class, which means setting up the table below.

The smallest class of the information (here fifty) will generally be chosen as the unit class:

Classes	Difference	Number of intervals	Q of villages per class	Q of villages per equal classes
0- 49	50	1	6	6 : 1 = 6
50- 99	50	1	9	9 : 1 = 9
100-249	150	3	18	18 : 3 = 6
250-499	250	5	4	4 : 5 = 0,8
500-950	450	9	3	3 : 9 = 0,3

This table permits us to draw a distribution histogram (figure 2 or figure 9 on another scale). This is the surest method.

By graphic use of the repartition. Here, the repartition histogram (figure 1) is constructed from the tables in figures 1 and 2, page 206, and the repartition curve (figure 6) is derived from it. This curve is then divided into equal classes (figure 12), and the distribution histogram is drawn (figure 13). This is the most precise method, but it requires a precision drawing (the distribution in figure 13 is more precise than that in figure 14).

By direct drawing. We construct figure 10 and graphically divide each rectangle by the number of unit intervals (by drawing the diagonal of the rectangle, then a horizontal line from intersection S, as shown in figure 14). This is the most rapid method.

CONSTRUCTION OF A REPARTITION FROM A DISTRIBUTION ("CUMULATIVE" CURVE)

We can obtain a "CUMULATIVE" diagram from a distribution by operating as in figure 15, where we use the equal classes established in figure 14. The curve is generally adopted. It is similar to the "repartition" (figures 6, 11, and 12), but the order of the villages is reversed.

It can be constructed in two ways:

Figure 15: Increasing cumulative or "number of observations smaller than . . ." or "cumulative less than. . . ." (How many villages are there with fewer than 200 households? Answer 27.)

Figure 16: Decreasing cumulative or "number of observations greater than . . ." or "cumulative more than . . . (How many villages are there with more than 250 households? Answer 7.) Here, one reconstructs the repartition (figure 12).

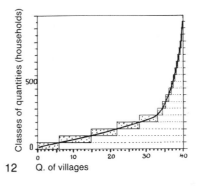

12

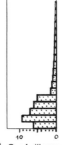

13

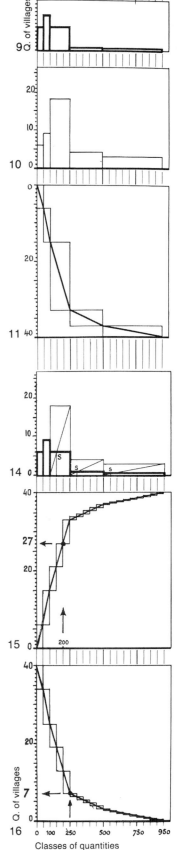

Q O Q Q
Time series
(two nonreorderable components)

With this type of problem, obviously simplification cannot be achieved by reordering either of the components. The type of graphic adopted will depend upon the **intended function**.

INVENTORY

The graphic transcription of two ordered series enables the reader to read a value on a curve (figure 1), that is, to retranscribe one of the correspondences of the relationship into verbal language. The curve is an inventory and, as such, a (poor) replacement for the data table. This kind of utilization involves elementary reading, and thus the scales must be detailed (millimetered paper).

PROCESSING

The essential function of diagrams involving two components is the reduction of the length of the components by internal processing (the information will involve three components when external comparisons are envisaged, and the graphic problem will be different). The reader can look for:

Types of structure in the relationship
On the intermediate level of reading, the curve displays parts which are similar to each other and which can be defined by shape and "duration."
 For example:
In the price curve (figure 2), parts A (stability), B (crisis periods), and C (upward movement) can be encountered several times along the ordered component.
In the yield curve (figure 3): wheat yield in France (1840–1939; statistics in E. Morice and F. Chartier, *Méthode statistique* [Paris: I.N.S.E.E., 1954]), there is a homogeneous structure.
In the age pyramid (figure 4), "low" classes disrupt the continuity.
In the seismographic curve (figure 5), phases A, B, and C can be detected.
In the land section (figure 6), we discern types of structure: A (monocline), B (synclinal ridge), C (hills), D (depression), E (mountain).
In seeking types of structure, the graphic designer can link all the points together or draw columns.
 Comprehensivity is imperative. The acute angles must remain legible.

Categories in the component O
The reader can attempt to reduce the number of categories in the component O; for example:
Figure 2: reduce the time component to four periods—m, n, o, p;

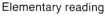

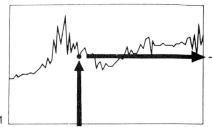

Elementary reading

Figure 3: discover three periods m, n, o separated by an indentation defined by the moving (running) average;
Figure 4: see the age pyramid as three categories of generations separated by the "low" area;
Figure 5: posit four crisis periods;
Figure 6: define six geographic regions.
 The level of the reduction is free, and the number of groupings retained (simplification by smoothing) will depend on the intended audience. In figure 6, for example, the teacher may want to define only two types of relief: mountains (B + E) and hills (A + C + D); and five regions (m, n, o + p, q, r). The number of groupings results from a choice. It is here that we can apply the mathematical processes (see page 170) of simplification (moving [running] averages, least squares, periodic functions . . .), which also imply a choice but have the advantage of a definition which is rigorous and thus transposable.

A correlation between the two concepts
With overall reading, a curve will highlight the general tendency, which can be meaningful.
Figure 2: prices, all other things being equal (which obviously remains highly problematical) increase over time.
Figure 3: yield (and here "all other things being equal" can be meaningful) increases over time, clearly showing the "progress" of human productivity.
Figure 4: an overall reading produces an evaluation of population vitality.
Figure 5: the curve suggests a general tendency toward calm.
Figure 6: here the curve shows the general slope of the region being considered.
 "Clusters" of points favor overall reading, and one should try to manipulate the scales so that the data fit into a square with a diagonal of 45 degrees.

MESSAGE
The simplified message, resulting from processing, can be transcribed by a curve which is generally superimposed on the comprehensive information and distinguished by greater visibility (a thicker line).

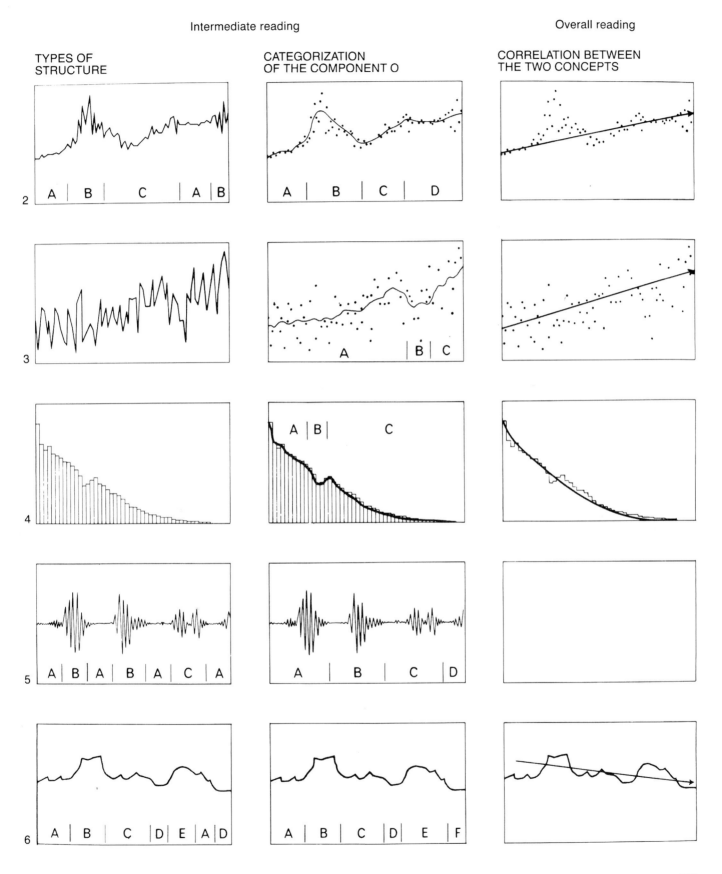

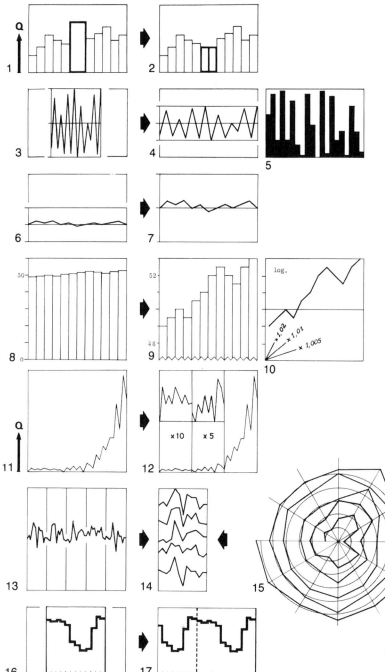

GRAPHIC PROBLEMS POSED BY TIME SERIES

Scale in years
With a scale in years, a two-year total (figure 1) should be divided by 2 (figure 2). A total for six months should be multiplied by 2.

Pointed curves
For overly pointed curves (figure 3), the scale of the **Q** should be reduced; optimum angular perceptibility occurs at around 70 degrees (figure 4).

If the curve is not reducible (large and small variations), filled columns can be used (figure 5).

Flat curves
For overly flat curves (figure 6), the scale of the **Q** should be increased (figure 7).

Small variations
For small variations in relation to the total (figure 8), the total loses its importance, and the zero point can be eliminated, provided the reader is made aware of this elimination (figure 9). The graphic can be interpreted as an acceleration if a precise study of the variations is necessary; here, we use a logarithmic scale (figure 10). (See also page 240.)

Large range
For a very large range between the extreme numbers (figure 11), we must either:
(1) leave out the smallest variations;
(2) be concerned only with relative differences (logarithmic scale), without knowing the absolute quantities;
(3) select different parts (periods) within the ordered component and treat them on different scales above the common scale (figure 12).

Obvious periodicity
If there is obvious periodicity (figure 13), and the study involves a comparison of the phases of each cycle, it is preferable to break up the cycles in order to superimpose them (figure 14). A polar construction can be used, preferably in a spiral shape (figure 15), but we should not begin with too small a circle. As striking as it seems, it is less efficient than an orthogonal construction.

Annual curves
For annual curves of rainfall or temperature, if a cycle has two phases (figure 17), why depict only one (figure 16)?

A contrast
Unlike what we see in figure 18, the pertinent or "new" information must be separated from the background or "reference" information. The background involves: (a) the invariant, highlighted by a heading (Port St. Michel); (b) the highly visible identification of each component (tonnage and dates). The new information (the curve) must stand out from the background (figure 19).

Reference points
It is impossible to utilize a graphic such as figure 20, except in a general manner. There is confusion concerning the position of the points, and no potential comparison is possible, as it is in figure 21.

Precision reading
A precision reading (utilization on the elementary level, as in figure 24) is difficult in figure 22, which results in a poor reading of the order of the points, and in figure 23, where there is ambiguity concerning the position of the points. On the other hand, figure 22 does favor overall vision (correlation).

Null boxes
Curves accommodate null boxes poorly (figure 25). Columns (figure 26) are preferable.

Unknown boxes
The drawing must indicate the unknowns of the information in an unambiguous way (figures 28 and 30). The reader might interpret figure 27 as a change in the structure of the curve and figure 29 as involving null values.

Very small quantities
Except in seeking a correlation (quite improbable here) the number of ships entering into a port is represented better by figure 33 than by figures 31 or 32. The reader can perceive the numerical values at first glance.

Positive-negative variation
This is in fact a problem involving three components O, Q, ≠ (+ −), and it must be visually treated as such. Figure 34 can be improved by utilizing a retinal variable (in figure 35 a value difference: black–white) to differentiate the ≠ component and thus highlight positive-negative variation.

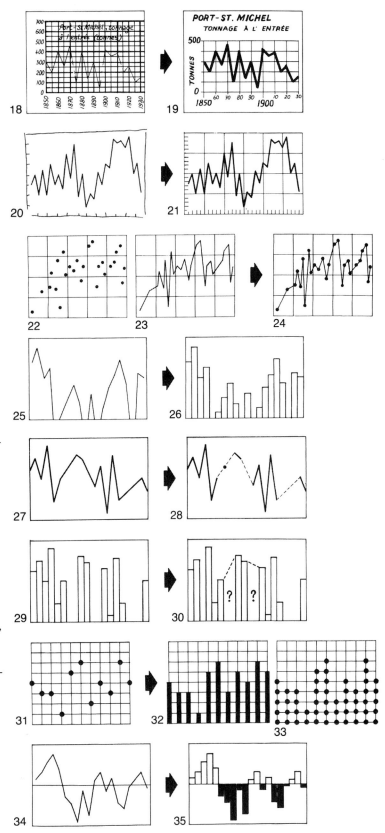

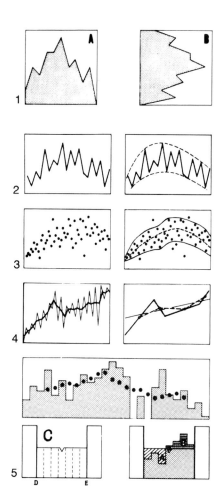

Figure 1: Should the graphic appear as in A or B? With a time curve, A is preferable; with an IQ test, B would be preferable. Given that lateral discrimination is more developed than vertical discrimination (due to the muscular development of the eye, the habit of panoramic vision), we should affiliate the more familiar component (time, for example) with the easiest perception, and conversely the component which is less familiar (socioprofessional groups, the test elements) with the type of perception which requires more attention, more concentration. This favors assimilation of the figurative "profiles," instead of already known reference points (such as dates over time).

Figure 2: "Envelopes" permit us to discover and display the general trend of the curve—a sight drawing is most often sufficient here, as it is in the following cases.

Figure 3: A cluster of points can be "enveloped," in which case the axis of the envelope and the axis of the line of points (which can be different) suggest the overall trend.

Figure 4: The simplification obtained by taking the center of each element on the curve shows the sequence of possible levels. This always aims at approximating a straight line.

Figure 5: The graphic calculation of a centered moving (running) average, over seven years. With transparent material, one cuts out the form (C), whose width (DE) is equal to that of seven columns (years); then one moves it onto the curve, and, column by column, one seeks to equalize areas A and B. Finally, one plots the average level in the central slot.

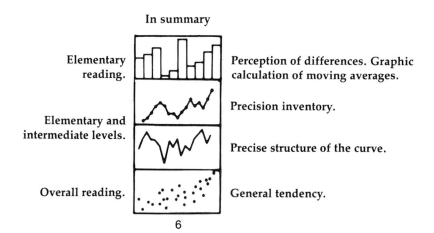

B. Diagrams involving three components

Information having three components excludes systematic comparison with any external information entailing the introduction of additional components into the problem. An image will accommodate only three components. All the following problems can be constructed as a single image, and, as such, are capable of visual retention.

The display below (figure 7) recapitulates the principal constructions which are possible. The *standard construction* generally suits all problems involving three extensive (long) components. In this case, the principle of three-component diagonalization (page 169) can be applied, whenever one or several reorderable components ($\neq$) are encountered.

The special cases result from the notion of a continuum, or from the presence of limited (short) $\neq$ components.

The construction of a collection of two-dimensional profiles leads to the perception of the information across several images.

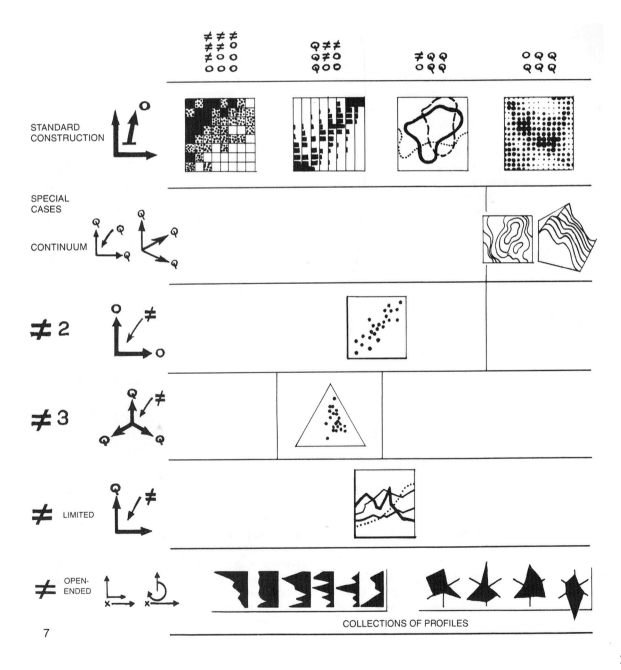

7

COLLECTIONS OF PROFILES

Nonquantitative problems

≠ ≠ ≠
Example: Cards and files

The composition of a file is an application of the laws of visual perception to problems of classification.

The simple inscription of a series of data is more logically accomplished by using two dimensions (figure 2) than a linear order (figure 1). For extensive and sometimes unlimited series, the double-entry table must give way to a FILE (figure 3).

THE FILE IS A MEMORY BANK

The composition of a file is governed by the notion that a file is first of all a memory bank, which uses known specifications to arrive at ones which are unknown. This means, for example, using a known place, period, and subject matter to discover an author's name. An author's name is merely a position, in the alphabetic order, discovered by the *coincidence of positions which are known* from other orders.

In composing a card, *we must consider the collection of cards*, since, at the time they are used, the researcher must explore at least two and more often three or four different series of specifications before being able to concentrate on a single card. From a practical point of view, a card is not a set of information, but *the convergence of several sets at a single point*.

Research proceeds by using successive sets

This means consulting different series of cards, within a type of specification (time, for example). Within the temporal order, one needs to refer to a precise date. It is therefore important that all the dates be easily identifiable during this research, that attention can be concentrated on them alone, and that they can be *visually selected*. The same is true for each series of specifications (figure 5).

Since spatial selection is by far the best and easiest to achieve, we arrive at the following fundamental principle:
A type of specification must always be in the same place from one card to another (figure 4).

The spatial organization of the card, particularly refined for computer usage, varies according to the number of specifications, the number and nature of combinations, the length of the series, and the type of processing envisaged.

At the outset, a FILE is a graphic problem involving three components:
≠ different objects
≠ different types of specifications (date, place, subject matter, …)
≠ different specifications

When we have determined the arrangement of the types of specifications on the card, they become ordered. Later, when the *collection of cards* is ordered according to a type of specification (time, for example) the file becomes a

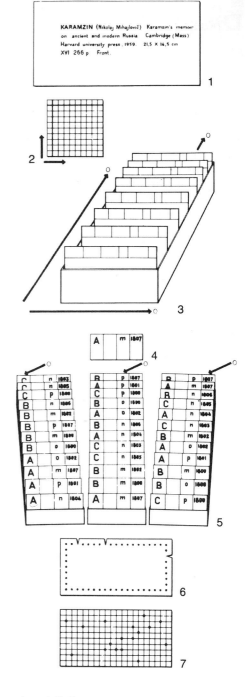

construction ≠ O O.

On the vertical dimension (collection of cards), any specification is capable of ordering or classing the entire collection. This is the first order of classing (figure 5). Classings of second, third and fourth order, or subclassings, are borrowed from various other specifications.

Examples of use are given in figures 8 and 9. Note that files which are organized spatially need no supplementary inscription (or heading) in order to be introduced into any classing whatsoever. It is sufficient to compose the record once and to reproduce it mechanically. The card (figure 4) can be introduced into all three file drawers in figure 5.

THE CARD IS A STATISTICAL ELEMENT

Any card can be considered as a statistical element (an object, an individual) possessing certain characteristics. We can thus *count the cards* by category of characteristics (by category of socioprofessional specification, for example; or by classes of an ordered sequence: age group) and can proceed with all the correlative studies afforded by the characteristics being analyzed (figure 10).

Such a counting is facilitated by the *marginal perforation card* (figure 6). A series of perforated places are reserved for given specifications (geographic specification, for example), and each perforation corresponds to a category (a department). When a given department is involved, we punch the corresponding hole on each card to make it into a slot. A rod, slipped into this hole for the entire collection of cards, frees the cards corresponding to that department and retains the other cards.

However, the number of possible perforations is limited. A substantial increase in the number of holes is obtained by using the entire area of the card, which leads to a *perforated card*, as in figure 7. It is readable, sortable by electronic systems, and can accommodate very extensive, precisely categorized series.

DESCRIPTIVE CARDS, ANALYTIC CARDS

All these cards are *descriptive*, that is, *one card exists per object* or individual; the collection of data relative to a given specification (to one characteristic) necessitates the collection of cards.

Because of the precision of spatial identification, such that 1000 to 10 000 numbered holes can be entered on each card, we can invert the problem and establish *a card per specification (per characteristic)*, each hole being reserved for a given object or individual. This is an *analytic* card, which furnishes immediately the number of holes (that is, of objects or individuals) corresponding to a given specification. The superimposition of four cards (for example, the department of Nord–price–wheat–1960) furnishes *all* the documents pertaining directly to this problem.

These documents correspond to the spaces which are punched on the four cards and which are immediately visible by transparency ("Selecto" system).

The decentralized documentation of tomorrow will be based either on series of descriptive cards, recorded on magnetic storage, or on series of analytic cards, which we can now foresee as capable of containing more than 100 thousand objects each.

The great problem with this documentation is not the number of documents, but the constitution of a *code* of analysis, which must reflect a fundamental logic, applicable by anyone (researchers, authors, analysts, . . .) whatever the research subject and whatever the sign-system (verbal, written, graphic, photographic, cinematic, . . .) in which information is recorded.

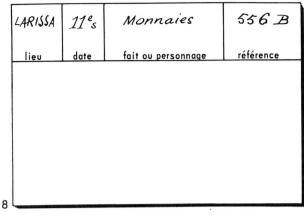

8

Bibliographical card
(after an historical study on Thessaly)

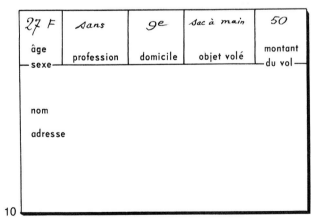

9

Descriptive library card

10

Research card
(After V.V. Stanciu, study of theft in department stores)

This type of card displays the data for numerous "contingency tables"
(see page 227)

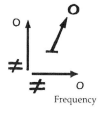
Three nonquantitative components

First example

Take the following information from the linguistics thesis of B. Quemada ("Le Commerce amoureux dans les romans mondains, 1640–1670" [Diss.: Université de Paris, 1949], quoted by G. Matoré in *La Méthode en lexicologie* [Paris: Didier, 1953], p. 85).

Let us look at the frequency of words selected from the "gallant" vocabulary of several seventeenth-century writers. Word frequency is determined approximately, not quantitatively; it simply involves three ordered categories:
frequent use (in black)
rare use (in gray)
exceptional or nonexistent use (in white).
This is an example of ordered evaluation. The problem can be considered in the following three ways:

≠ ≠ O (figure 1). Authors and words are merely considered as different and are thus reorderable according to their general frequency as determined by adding each column. To add nonquantitative evaluations, we give black a numerical value of 2, gray a value of 1. The total numbers merely establish an order.

The figure which is obtained shows the familiarity of certain words and suggests the richness of the different authors' vocabulary. Since the whole is ordered according to these two concepts, it is possible to group authors by type of vocabulary, and words by type of author.

≠ O O (figure 2). The words are still considered as merely different and reorderable according to their total frequency, but the authors are ordered over time (for example, by birthdates or by dates of works studied), with time running here from bottom to top. The interpretation of the image thus suggests literary "fashion" according to different periods.

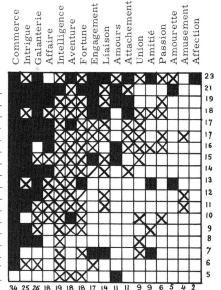

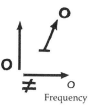
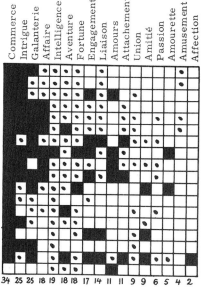

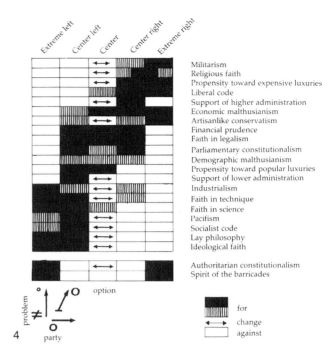

O O O (figure 3). Words and authors are considered as ordered, the authors over time, the words according to an a priori evaluation of their expressive power.

By using the temporal order we discover periods where certain words are most often used, and when their frequency diminishes. Thus the evolution in the meanings of the words and the stages of their disappearance emerge.

Obviously, the results would be even more interesting if words, and especially works and authors, were more numerous; this example is no more than a demonstration of methodology. Modern linguistics, incidentally, goes beyond the stage of graphic representation and establishes correlations by automated means, but the basic principles remain the same.

Second example

≠ O O (figure 4). Consider the dominant characteristics of five French political tendencies in 1956 (after C. Morazé, *Les Français et la République* [Paris: A. Colin, 1956]).

Three components:
≠ characteristics
O political tendencies
O evaluation on four levels: two "pro" (black and gray), a "changing" level (arrows), a "con" level (white).

Ordering permits us to class the characteristics, thereby defining the right and the left. As for the center, here is the commentary of C. Morazé: "By its very nature the Center is subject to changes of opinion on most essential points. These changes make for the possibility of governmental evolution and may occur either within one of the Center parties or by the replacement of one Center party by another."*

*Quotations and terms on figure 4 from translation of Morazé's book by Jean-Jacques Demorest, *The French and the Republic* (Ithaca, N.Y.: Cornell University Press, 1958), p.79 (translator's note).

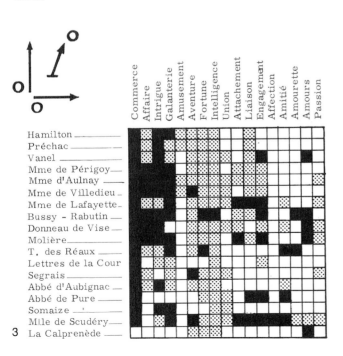

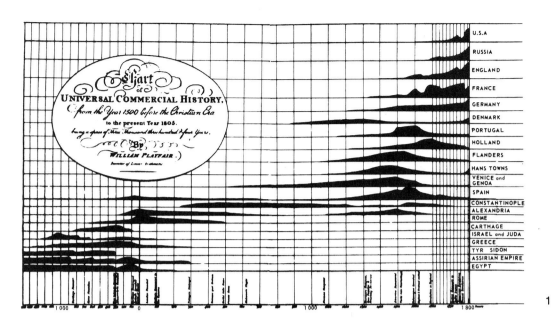

Third example

≠ O O (figure 1). The diagram representing "universal commercial history" (by William Playfair in *An Inquiry into the Permanent Causes of the Decline and the Fall of Powerful and Wealthy Nations* [London: Greenland and Norris, 1805]), heralds the era of historical "synoptics."

The three components are geographic regions ≠ considered as reorderable, the temporal order O, and a nonquantitative evaluation of relative importance O. The time scale varies according to historical knowledge; hence the divisions are larger for A.D. dates than for those B.C. Note the complete omission of information on the Asiatic powers, and the relative unimportance accorded to Muslim commerce in the later time periods.

Fourth example

≠ O O (figure 2). The same type of representation can be used in any historical problem when, in the absence of quantitative data, the author acquires a highly probable knowledge of the relative importance of the phenomena being studied. Figure 2 shows a nonquantitative evaluation of mining activity (after D. Kovacevic, "Les Mines d'or et d'argent en Serbie et en Bosnie," *Annales* 5, no. 2 [1960]: 248–258). It completes the diagram in figure 2 on page 198, in which the mines were ordered according to the date of their initial activity.

The question—should nonquantitative data be represented graphically?—can be answered affirmatively whenever the author judges that he or she is the foremost expert for the question being studied. In this case it is the author's duty to deploy all his or her knowledge, including an evaluation of the general tendency, which has emerged progressively through dint of work and constitutes new *information*. The reader must simply be warned of the approximate and thus modifiable nature of these evaluations if a detailed reading is to be undertaken.

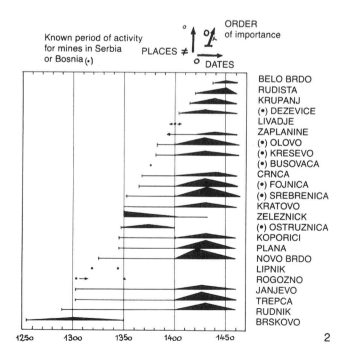

2. A quantitative component

We shall draw a distinction between:
contingency tables, which divide a set according to two components, $Q \neq \neq$;
comparisons of time series, based on a common order, $\neq Q\, O$ (figure 3);
And **comparisons of repartitions**, distributions, or concentrations, $\neq Q\, O\, (Q \neq)$.

$Q \neq \neq \quad Q \neq O \quad Q\, O\, O$
Contingency tables

In the analysis of a sociological survey a group of interviewed persons is divided according to two types of observation, which, along with the quantities, will produce a problem involving three components:

INVARIANT —*persons interviewed*
COMPONENTS —*Q of persons according to*
—*O four age-classes*
—≠ *five circumstances of visiting cafés (for business, with family, with friends . . .)*

Calculation permits us to reduce the information to a series of percentages by age-class. What information can be obtained from figure 4? We find that three circumstances are always at the top: for business, with friends, before meal. The totals convey this already. We also see that the item "after show" is last in three age-classes. This is anecdotal information. Questions of an intermediate or overall level are not answerable.

The standard construction (figure 5) permits a full exploitation of the information and indicates, in addition to what we just saw, that:
– the strongest "before meal" tendency corresponds to the 21–30 age-class and that this type of visit decreases regularly with age;
– the same is true for those who go to a café "with family," and the presence of relatives produces a variant;
– there is a slight tendency toward youth and toward the 31–40 age-class among those who go to a café "after show or game";
– business cafés have a majority clientele around 41–50 years, all other things being equal;
– the older one gets, the more one tends to go to a café "with friends";
– it is possible to class the visits as a function of age, which orders the figure, and thus justifies the categories defined at the outset.

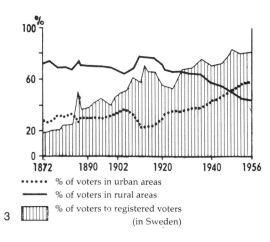

3 ······ % of voters in urban areas
— % of voters in rural areas
▒ % of voters to registered voters
(in Sweden)

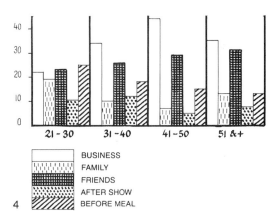

4 BUSINESS / FAMILY / FRIENDS / AFTER SHOW / BEFORE MEAL

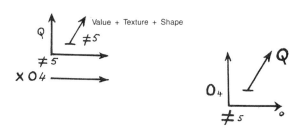

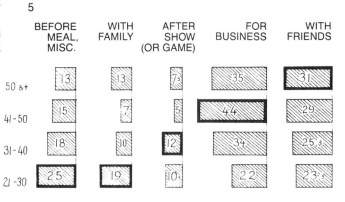

5

GRAPHIC PROCESSING OF CONTINGENCY TABLES

A defined set of objects or individuals, divided simultaneously by two intersecting components, produces a "crossed" data table or "contingency table" (figure 5). It enables us to perceive the information as a totality (rather than by successive divisions).

Two successive divisions

Imagine 100 persons, divided first by age-class: "30 years older" = 60 persons, "20–29" = 20 persons; "younger than 20 years" = 20 persons. From this we derive figure 1. The persons are then divided according to three favorite games: "tennis" = 20 persons; "cards" = 10 persons; "football" = 70 persons, from which we derive figure 2.

These divisions show that "on the whole" the population includes a majority of "older than 30" who prefer football, but such information does not tell us who plays cards: those older than 30, or those younger than 20? Who plays tennis? . . .

Proportional ("expected") distributions

We can only establish a distribution of the players (a possible but not probable construction) according to what we know about their distribution by age.

The twenty players are thus divided in the proportion of 60% to 20% to 20%, that is, "older than 30" = 12 persons; "20–29" = 4 persons; "younger than 20" = 4 persons. The same procedure is applied to the other games. We can plot this result (figure 3) and represent it as a table (figure 4). We see that this expected distribution is obtained for each box (cell) by multiplying the two corresponding totals and dividing by the general total (here 100), assuming independence of the two components. For example:

$$\frac{60 \times 20}{100} = 12$$

In fact, this table tells us nothing that we do not already know: distribution by age in the proportion 60 : 20 : 20, and by game in the proportion 70 : 20 : 10.

The intersection of the two components

Individual surveys should show to what degree the studied categories will cause a distortion of the expected distribution with respect to the observed distribution, assuming

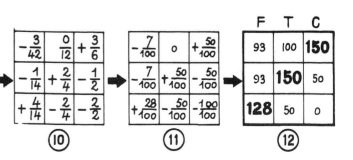

224

independence of the components. This distortion is seen as an interaction between the components. For example, in the illustration above, we suspect that age favors certain games, youth others. But which games and in what proportions?

Suppose that the individual questionnaires result in the table in figure 5. What we need to know is the difference between figure 4 and figure 5.

The tendency

In addition to the observed groups, provided by the totals (which can vary from one survey to another), this difference gives us the general tendency resulting from a comparison of the concept "game" with the concept "age." This tendency can be a constant.

In any case, the tendency is independent of the total groups observed. The differences, calculated in figure 6, are carried forward to figure 7. The table in figure 7 shows (according to the information in figure 5) that the tendency is to play football more when one is younger than 20 and cards more when one is older than 30.

Diagonalization favors the recording of this result (figure 8) by ordering the games ($\neq$) according to age (O). When one uses the age order to look at the games, the latter are ordered in the sequence football, tennis, cards. This is a general result, independent of the total quantities per class and of the contingencies of the survey.

Calculation of the tendency

This general tendency can vary from one game to another, from one age to another, from one survey to another. The absolute differences (+4, +2, +3) are not comparable given that they come from unequal groups of individuals. These differences are meaningful only if they express a certain percentage of distortion, in relation to the expected distribution (figures 4 and 9). Thus we are looking at a difference of four football players in relation to fourteen, of two tennis players in relation to four, etc.

It is sufficient to set up these relations as in figure 10 in order to calculate the percentage of distortion (figure 11). It can be interpreted, as shown in figure 12, in the following manner: We observe 128 football players younger than 20, whereas we would expect 100 football players younger than 20 from a calculation based on the expected distribution of the set being analyzed. This calculation makes all the surveys comparable.

Rapid determination of the general tendency

It is possible to discover the tendency rapidly, and thus to order a reorderable component, without undertaking a calculation of the expected distribution. In fact, a contingency table can be interpreted as a comparison of several sets: in figure 13, for example, as a comparison of the set "older than 30 years" to the set "20–29 years" and the set "younger than 20" on the subject of games (we can then speak of subsets).

For the three distributions to be comparable they must have equal or equalized totals. When each of the three sets is related to a total of 100, we obtain the table in figure 14. The numbers no longer represent real values or individuals, but frequencies, that is, the measurement of a characteristic belonging to each set of individuals defined by each cell. Thus, they provide an answer for the question: "Which of the three age-sets displays the highest frequency of tennis players, that is, the strongest tendency toward this game?" Answer: the highest frequency is 30, which is that of the "20–29 year olds."

When we calculate horizontal (row) percentages, thus forming sets which are comparable vertically, it is sufficient to circle the highest frequency in each vertical column. This frequency indicates the strongest tendency (figure 15). It also serves as a basis for the diagonalization (figure 16) and ordering of the games. The order discovered in figure 8 is obtained much more rapidly here.

Obviously the result is similar if we calculate vertical (column) percentages, thus forming sets which are comparable horizontally (figure 17). The question: "Which of the three sets of players displays the highest frequency of individuals of age "20–29," gives the answer 30, that is, the set of tennis players. Diagonalization achieves the same result (figure 18). See the applications of these procedures on pages 226 and 259.

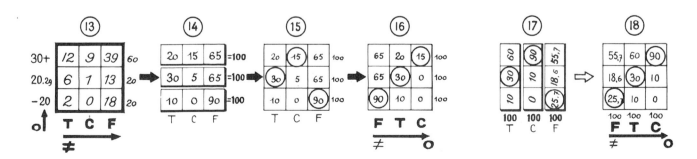

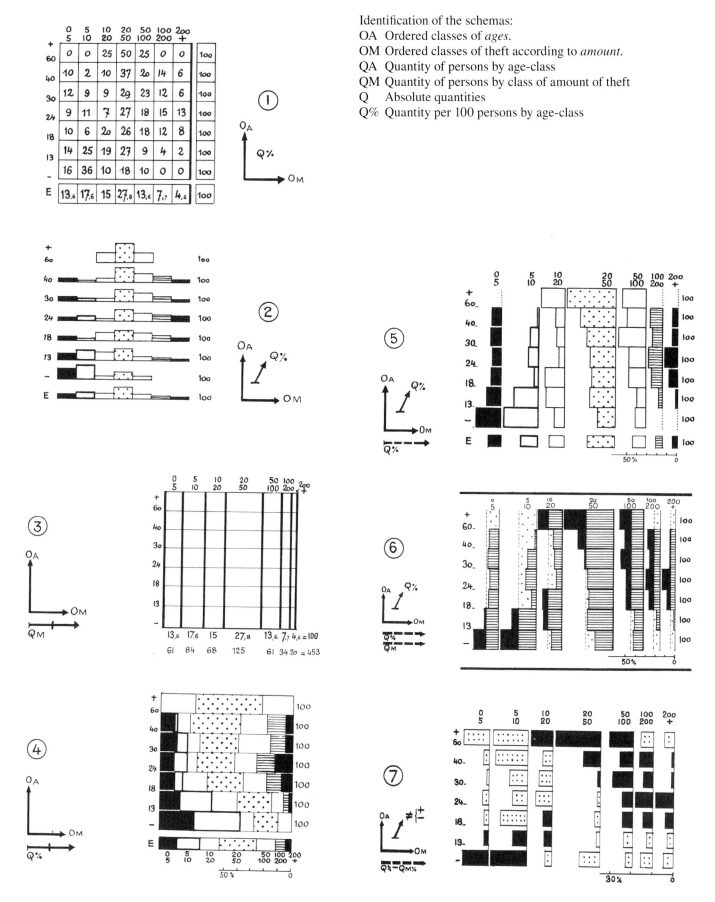

Identification of the schemas:
OA Ordered classes of *ages*.
OM Ordered classes of theft according to *amount*.
QA Quantity of persons by age-class
QM Quantity of persons by class of amount of theft
Q   Absolute quantities
Q% Quantity per 100 persons by age-class

THE GRAPHIC REPRESENTATION OF CONTINGENCY TABLES

Consider (after an unpublished study by V. V. Stanciu on theft in department stores) the distribution of thefts according to amount of theft and age of delinquent.

INVARIANT —*department store delinquency*
COMPONENTS —Q of delinquents according to
—O *age-classes*
—O *classes of amount of theft*

The numbers are the following:

AGE \ AMOUNT OF THEFT	0–5	5–10	10–20	20–50	50–100	100–200	200+	
60 +	0	0	1	2	1	0	0	4
40.59	4	1	4	17	9	7	3	45
30.39	4	3	3	9	7	4	2	32
24.29	4	5	3	13	8	7	6	47
18.23	7	4	13	19	12	8	5	67
13.17	29	53	39	56	19	8	4	208
12 –	13	18	5	9	5	0	0	50
	61	84	68	125	61	34	20	453

Transformed into percentages (figure 1), the numbers can lead to a rapid determination of the general tendency according to the method outlined on page 225. This produces table A and its graphic transcription B (see also page 37).

But can we represent the entire content of the information as a single image, which would permit:
– comparing various subsets with each other;
– perceiving the tendency, that is, the difference between the observed distribution and the expected distribution;
– perceiving the absolute quantities?

Comparison of subsets: percentages

The table in figure 1, given in percentages, establishes horizontal subsets of 100 persons per age-class, in which we observe the categories of theft.

It also permits us to construct a proportional (expected) distribution (figure 3), that is, the overall percentage E, all ages combined.

Finally, it allows us to construct figure 4, that is, the observed distribution. Each horizontal rectangle is equal to 100 persons per age-class, who are distributed horizontally according to the classes of theft. Since these are no longer aligned, a differential component (value, shape, texture) is necessary for their identification.

Although striking, the difference between figure 3 and figure 4 is difficult to interpret. To facilitate interpretation it would be necessary to superimpose the two images, rectangle by rectangle. This can be achieved by aligning the rectangles (figure 5). It is then possible to superimpose the already aligned rectangles from figure 3 onto figure 5, and we obtain figure 6, in which the positive differences (residuals) are shaded in black, the negative ones denoted by dotting.

Tendency (deviation from the mean)

The diagram in figure 6 represents, per 100 persons by age-class:
– the number of persons per class of theft;
– the distortion or exceeding of this number (in black) in relation to the number observed for 100 persons in all.

This excess is the *deviation* due to age, in relation to a distribution which ignores age: the *mean* distribution, all ages combined. We see this represented in figure B (with the extension of black to the entire rectangle).

The deviation, which expresses the general tendency, itself independent, as we have seen, of the absolute numbers and the contingencies of the survey, can be derived from figure 6, producing a representation such as figure 7, in which the common base (the mean distribution) has been eliminated. Positive and negative deviations can figure on the same side of a line, provided they are highly differentiated (black and white).

Such a representation is utilized on page 230 to determine a tendency and order for the eighty wards in Paris, according to the socioprofessional categories of the inhabitants.

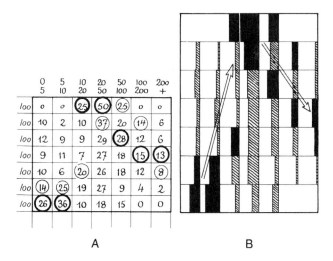

A   B

The representation of percentages and absolute quantities

Up to now, the absolute quantities have not been represented. However, they are useful in showing, for example, that those "older than 60" are not very numerous: four; whereas the "13–17 year olds" constitute the majority: 208. This is what appears on the table in figure 8; it can be represented graphically as in figure A or in the inefficient construction in figure 9.

The horizontal (row) totals can produce an expected distribution by age (all categories of theft combined), as shown by the horizontal lines in figure 10. Thus we transform the equal subsets of figure 4 from page 226 into subsets which are proportional to the total quantities.

Therefore, taking figures 4–7 from page 226, it is sufficient to give each horizontal subset a height proportional to the row totals in order that the areas represent the absolute quantities. Such is the case in figures 11–14.

The overall image

The *areas* of the rectangles in the diagram in figure 10 display the proportional distribution of the whole in absolute quantities. The *areas* in the diagram in figure 11 depict the observed distribution in absolute quantities.

The superimposition of these *areas* (figure 13) displays, rectangle by rectangle and in *absolute quantities*, the difference between the two: the *tendency*.

Studied *linearly*, on the horizontal scale alone, figure 13, like figure 6, page 226, shows the *percentages*, that is, per 100 persons by age-class:
– the number of persons per class of theft
– the deviation in black from the mean amount of theft (all ages combined) due to age.

Thus, the image in figure 13 resolves the problem posed on the preceding page.

Figure 14 is analogous to figure 7 on page 226, but it also depicts the deviations in absolute quantities.

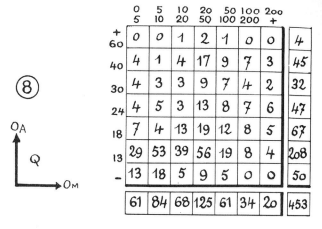

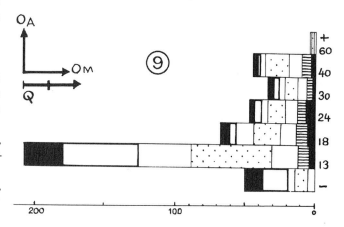

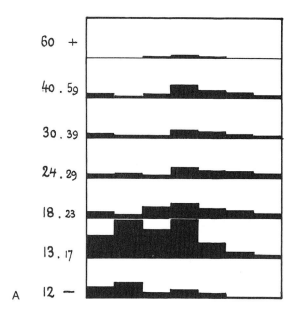

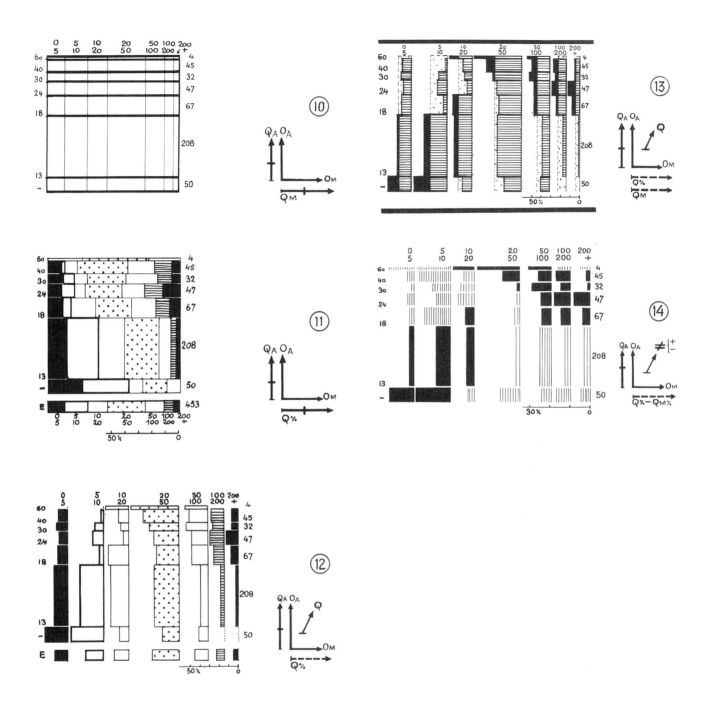

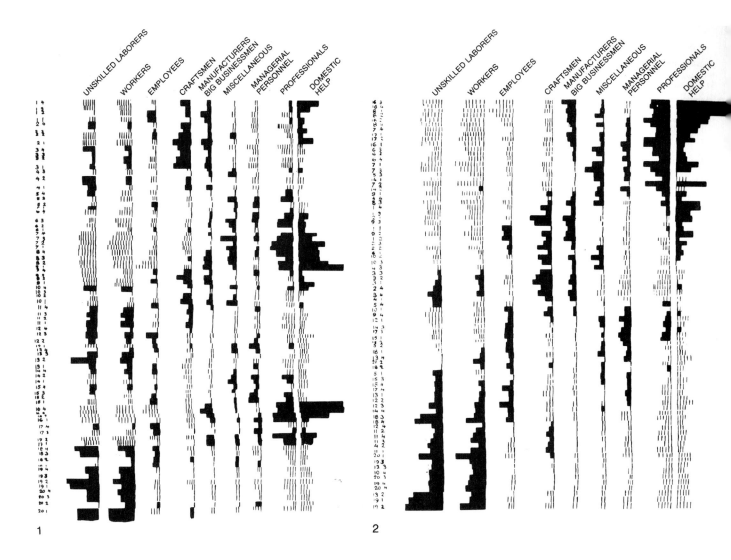

GRAPHIC PROCESSING BY MATRIX

Example: Socioprofessional tendency of Paris wards (1954 census) (information obtained from Germaine Belleville, *Morphologie de la population active à Paris* (Paris: A. Colin, 1962).

Components of the information:
Q of working population according to
≠ eighty different wards (geographic component)
≠ nine socioprofessional categories (one of which is "miscellaneous")

The total working population is divided by two intersecting components: wards and socioprofessional categories.

It forms a contingency table which can be interpreted as a comparison of eighty geographic subsets. However, a length of eighty prohibits a rapid determination of the tendency. The processing must be more subtle.

This information describes and characterizes each ward. It permits us to group them and reduce the geographic

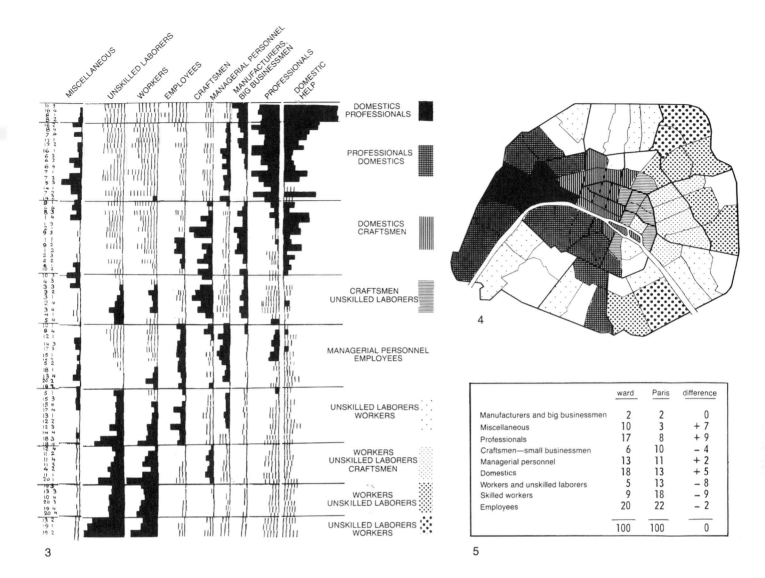

component from a length of eighty to eight or nine categories.

The character of each ward is evidenced by the differences which it displays in relation to the general mean of all the wards (percentage of Paris itself). It is these differences which are calculated. The table opposite (figure 5) gives an example of the calculations for the École Militaire ward.

These differences are then represented as in figure 7 on page 226, each ward on a separate card, so that we can group and class similar wards.

On the cards, the socioprofessional categories are ordered according to the principal groupings which evolve from a rapid preliminary observation. Thus, professionals and domestics, who seem to be closely linked, have been put together. The differences are represented proportionately by a black band (positive differences) or by dashes (negative differences).

The cards are filled out according to the administrative order of the wards.

Classed in this order (see figure 9, page 245), they produce the image in figure 1, whose oblique forms suggest a certain relation between the order of the categories and the administrative order. Since we know that the latter traces a geographic spiral, the oblique forms already bear evidence of radiating groups.

Classed by similarity, relying especially on the positive differences (in black), the cards produce figure 2, which can then be simplified by a reclassing of the socioprofessional categories. Practically speaking, one cuts up a photograph of figure 2 by columns, which are then permuted.

The "miscellaneous" professions, which do not seem to fit into the general picture, are excluded from the diagonalization.

Thus we obtain the image in figure 3, which has the simplest possible visual form. It indicates the most significant groupings resulting from the given information and constituting its originality. These groupings (whose number can be selected by the researcher) are defined by horizontal sections and can be mapped (figure 4).

231

SPECIAL CASE ≠ 3
TRIANGULAR CONSTRUCTION

A triangular construction (figure 1) is applicable whenever there is a component whose length is 3 and where the total of the Q is meaningful. Percentages must be calculated.

This type of graphic is used for internal processing and permits the length of the ≠ components to be "reduced." Consider the following example (after G. Th. Guilbaud, "Méthode d'analyse sommaire de la structure démographique," *Economie et humanisme*, [1946]: 515–525).

Q of population, in percent per age-group, in 1936, according to
≠ various countries
≠ three age-groups (J, young: younger than 20 years; A, adult; V, elderly: older than 60 years).

Problems of legibility

Note that the pertinent triangle (figure 2) can be cut out from the total triangle (figure 1), which is numbered from 0 to 100%, in order enlarge the scale.
- The most comprehensible identification consists in defining the corners of the total triangle (numbered 100) and the heights, rather than the sides.
- for reasons of legibility, we enhance *the visibility of the points*, which are the basis for potential groupings, rather than the names, which are arranged according to convenience (see figure 2).

Processing and discussion

Reduction of the component ≠ 3: G. Th. Guilbaud (in "Méthode d'analyse") comments on figure 2 in the following way: "The very elongated form of the cluster is determined by the proportion of young, which is the main characteristic in world demography. *When this proportion is known, the two others can be derived from it*, at least approximately."

Reduction of the extensive (long) component:
Again we turn to Guilbaud's commentary: "Within the area defined by the points, we can distinguish three groups: one corresponding to a high proportion of young (45% approximately) with Japan as an example; another to a small proportion (30% approximately) with France as an example; a third being intermediate."

Determination of groupings

Reduction of the extensive component raises the problem of different possible groupings. Three cases can discerned:
(1) Groups form themselves, with no ambiguity, as in figure 3. These are the groups that are retained.
(2) Groups are formed, but certain points imply a choice between two neighboring groups. In order to assign these points to a given group, we must rely on complementary information: for example, a department having an ambiguous position in the triangle will be assigned to the group which is closest geographically.
(3) The cluster displays no groupings. In this case we rely uniquely on the numbers and define groups geometrically, by drawing parallel lines from a central point to the sides of the triangle (figure 4). This center corresponds to the general mean in a departmental statistic, for example, to the percentages for the whole of France. Note that this amounts to symmetrically cutting out the pertinent triangle centered on the mean (figure 5). Several cuts are possible:
- from the center in six or nine categories (figure 6);
- by preserving a "mean" group, that is, by tracing a central triangle around the mean; this produces seven or ten categories (figure 7).

Tolerance zone

A triangular graphic permits us to take certain variations of percentage into account and define the limit of these variations. Consider, after Robert Satet and Charles Voraz, *Les Graphiques, moyen de direction des enterprises* (Tourcoing: G. Frère, 1944), the following examples:
- *an inventory*, according to percentage of raw materials, products being made, finished products;
- *a selling price*, based on percentage of raw materials, wages, general expenses;
- *a profit-bearing weight*, based on percentage of fuel, passengers, cargo.

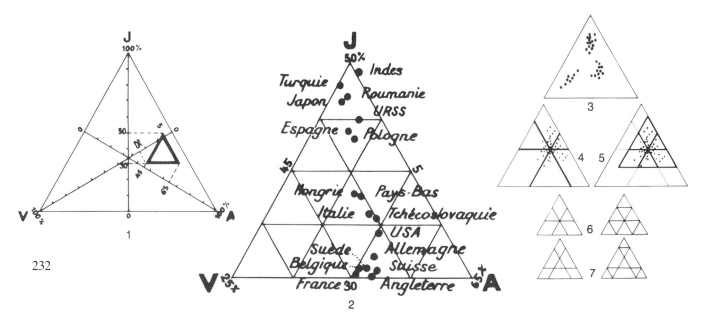

In each example, the percentage of the three terms cannot remain rigorously fixed. It can vary according to circumstance, period, place, type of aircraft, . . . But we can consider that there is an optimum position, around which the percentage can still vary, provided that these variations do not exceed certain limits defined by experience or by exterior constraints, and beyond which compensation impinges heavily on the other terms. The limits of each of the terms, or margins of tolerance, are placed respectively on each of the three heights of the triangle (figure 8). They allow us to see immediately:

(1) an increase in the range of one term, making it incompatible with the two others (figures 8 and 9);
(2) the maximum range of a term in relation to the two others (figure 10);
(3) the tolerance zones for the three terms (figures 10 and 11), and abnormal situations (figure 12).
(4) if one of the numbers is known (figure 13), it determines the range within which the two others can be inscribed.

Extensive ordered component

When an extensive component is ordered, as with a time component, for example, a triangular construction generally leads to a figure such as 14, which shows demographic evolution of a village. (After R. Baehrel, *La Basse Provence rurale* [Paris: S.E.V.P.E.N., 1960].)

Analysis of the extensive component permits us to define different demographic periods. They correspond to different directions of the drawing.

Introduction of the total or of a new component

A triangular construction (like the constructions ≠ 2, page 250) does not oblige us to utilize a third visual variable, which can then represent:
– either total quantities (figure 6, page 115);
– or a fourth component **O** or **Q**

(Indeed, nothing prevents us from superimposing onto figure 14 the population according to years or height of individuals or harvest value . . . provided that this new component relates to the total of the individuals and not to one of the categories identified by the triangle [mean height of *all* the individuals, harvest of the *entire* village, . . .]);
– or a fourth ≠ component, which permits us to superimpose several different series of points, representing, for example, the demographic evolution of several villages (see page 265).

Constraints added to a component ≠ 3

The burden which the elderly and the young represent for the work force obviously varies according to the proportion of young, elderly, and adult. But we can agree that the burden represented by an elderly person is greater than that represented by a young person.

The variation is therefore asymmetrical. A triangular graphic permits us to represent this asymmetry and establish a "chart" of the burdens. According to G. Th. Guilbaud ("Méthode d'analyse"), the graphic solution is as follows. Give, for example, the consumption of an adult a value of 1, of an elderly person a value of 0.8, of a young person a value of 0.6. This involves constructing a chart indicating the burden borne by an adult, as a function of this fixed weight and according to the different percentages of adults, elderly, and young.

Take, for example, a proportion of 55% adult, 15% elderly, 30% young.
The fifty-five adults support:
– their own consumption . . . 55 × 1 = 55
– plus the consumption of fifteen elderly persons . . . 15 × 0.8 = 12
– plus the consumption of thirty young . . . 30 × 0.6 = 18
Thus, the fifty-five adults support the burden of eighty-five consumptions, and an adult has a burden of 85/55 = 1.54 consumers.

We can say that the burden coefficient (C) = 154%.

On the graphic, all the points corresponding to the same value of C are on a straight line.

Simple calculations allow us to determine two points for each one of the characteristic values (150%, 200%, 300%, 400%) of C and draw figure 15. It enables us to compare several countries, several periods, from the point of view of the demographic burden incumbent on the work force.

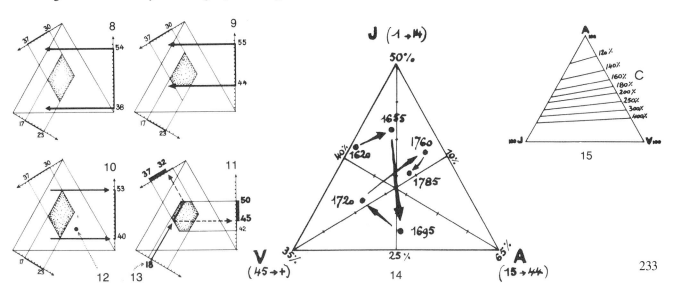

≠ Q O
Comparison of time series

Consider comparing different statistical evolutions, during a common time period, that is, different data sets **O Q** where the ordered time component is common to all the sets.

In addition to constituting an elementary inventory, the information has a general objective of reducing the length of the components, that is:
– ordering and categorizing the ≠ component;
– categorizing the **O** component according to the information being considered.

Consequently, one proceeds to a classing of the curves by diagonalization, which can determine groups of parallel curves and categories of the ≠ component, then to a vertical comparison of the elements of the curves, by addition, which can determine characteristic periods of the **O** component. How does one proceed graphically?

JUXTAPOSITION OF THE CURVES

The best means of treating the ≠ component is through a series of comparable curves A, B, C, . . . (figure 1). Here is a rapid way for ordering them, taking into account the overall correspondences expressed by each curve (and not only the two extreme dates).

One "blackens in" the highest peaks of the curves (columns are preferable) up to any horizontal level whatever, which is determined progressively and should reveal the general tendency of each curve (which amounts to creating a retinal visual variation and constructing the information according to the standard schema). It only remains to class the curves along a diagonal of "black." The ≠ component is classed: C, A, D, B, . . . (figure 2) and groupings are possible.

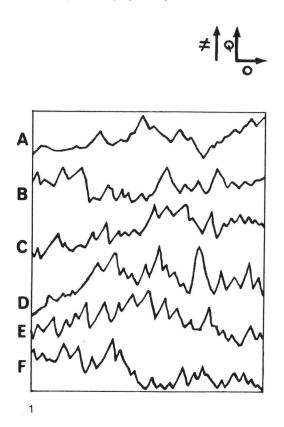

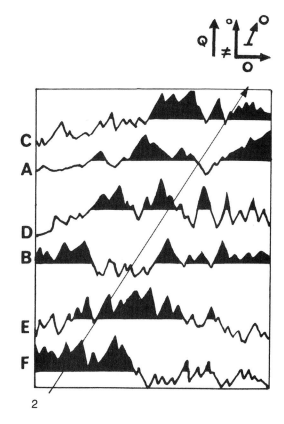

SUPERIMPOSITION OF THE CURVES

The categorization of the **O** component is based on an addition of elementary vertical coincidences (coincidence of peaks, of "lows") or averages (coincidence of increases, or decreases, . . .). The most natural construction is a superimposition of curves. But when the quantities are of a different nature, or simply of a different range, what scales should be chosen to represent them?

If the choice of scale is considered as "free," this can lead, for example, to the diagrams in figures 3 or 4, obtained from the same statistics. One can easily imagine the psychological impact of each! This example shows that **it is not the numbers but the slope of the curves which must be rigorously represented**. Any conscientious designer should realize that the choice is *not* free and that on the contrary such problems imply the strictest rules.

Take an example involving coal mines at Carmaux (department of Tarn) and compare:
≠ four different units:
    V value (sale price) of the coal
    S salary of the workers, in francs
    N number of workers
    P production in tons
Q quantities in each category, according to
O time.

In order to better appreciate the results of the different constructions, the same problem will be treated conjointly for the iron mines at Ariège (information taken from A. Armengaud, *Population de l'Est Aquitain* [The Hague: Mouton, 1961]).

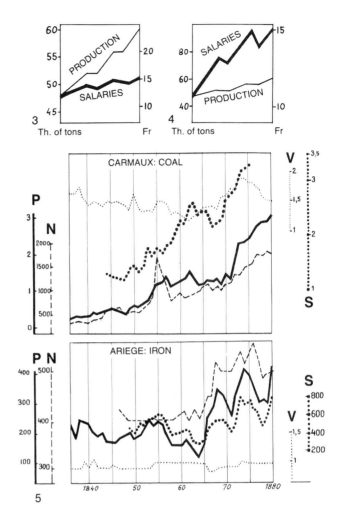

Scales constructed freely

One generally attempts to relate the curves, while avoiding a confusing superimposition (for reasons of legibility). The reader will deduce a "certain parallelism" of P, N, and S, in figure 5 as well as in figure 6. If we wish to compare figures 5 and 6 we note that the magnitude of each variation escapes us because we cannot see the base (the zero) of each curve. The information contained in these diagrams is reduced to the correspondence among the peaks of the curves. Any comparison on the intermediate level is ambiguous; it amounts to reading the numbers on the scales.

Visible zero

By constructing a zero point for each scale, we can perceive the magnitude of each variation. However, risk of confusion leads us to separate the curves, and the indication of zero does not resolve the problem of scale nor provide a basis for comparison. Why construct a given curve above or below another one? The reader will deduce from figure 1 that S increases more rapidly than P! And from figure 2 that P and S are parallel and different from N!

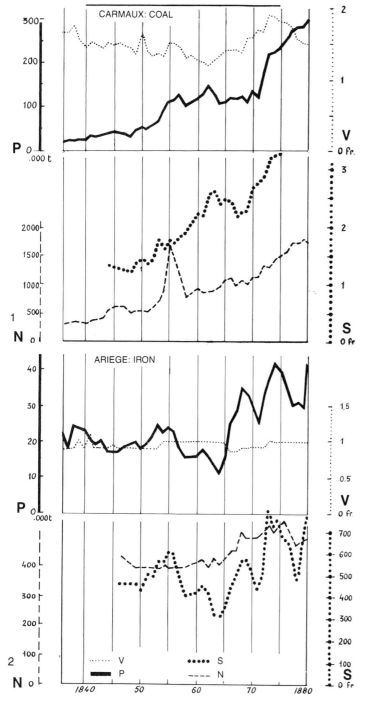

COMMON BASE: INDEXING

Through indexing, all the elements of a problem can be made comparable at a given moment, whatever their own numerical value. All the data points in each series can be indexed to a specified indexing point and scaled from a base of 100, according to the formula:

$$\text{Index at period } n = \frac{\text{data point at period } n}{\text{datum at period } i}$$

that is:

$$\text{Index } n = \frac{Q_n}{Q_i} \times 100$$

It is then possible to "describe" the different variations with numbers, since everyone knows the meaning of 100 and easily perceives what 80, 760, or any other index represents. All indices are comparable.

Graphic representation of indices

After the indices have been calculated, we only need to construct an index scale. This is common to all the curves, and its drawing is elementary.

When the indices are not calculated, *their calculation is not necessary*. In fact, what does indexing mean graphically? It means that all the statistical numbers at instant i are represented by the same magnitude on paper. In figure 4, for example, if instant i is 1836, the same base point can represent for Carmaux 20 000 t (P), 300 (N), 1.60 (V), etc.

In order to establish the scales, it is sufficient to make all the zeroes coincide on the same horizontal line A, and the numbers 20 000, 300, 1.60 . . . coincide on the same horizontal line B (figure 3).

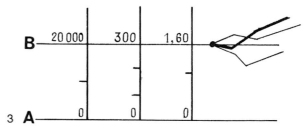

If one chooses AB = 2 cm, for example, the size of the unit (or of its multiple) for each scale will equal to:

$$\frac{AB}{20\,000} \qquad \frac{AB}{300} \qquad \frac{AB}{1.6}$$

The scales are inversely proportional to the magnitude of each variable at the indexing point. Once constructed, these scales allow us to plot the nonindexed numbers directly and obtain the indexed images, as in figures 4 and 5.

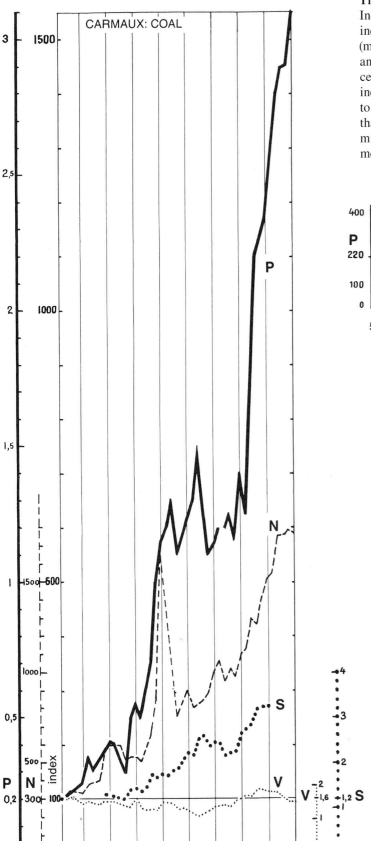

The indexing point

Indexing is still largely arbitrary, since one can choose the indexing point wherever one wishes. Chosen at the outset (most generally the case), it can lead to figures such as 4 and 5, that is, to amplifications which are erroneous in that certain factors have been indexed near the moment of their inception. The reader will note, for example, that from 1860 to 1880 the increase in coal production is much greater than that of iron! The indexing point can also be chosen at the middle or at the end of the period, or it can be based on the mean for several years.

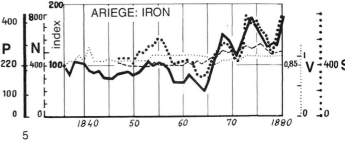

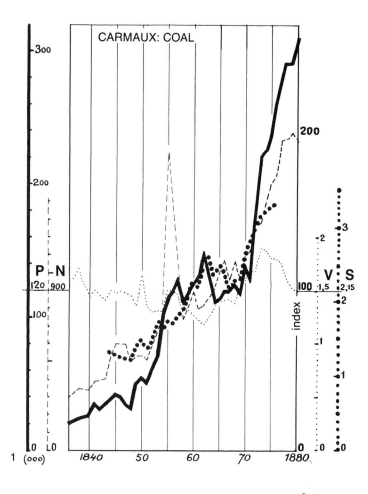

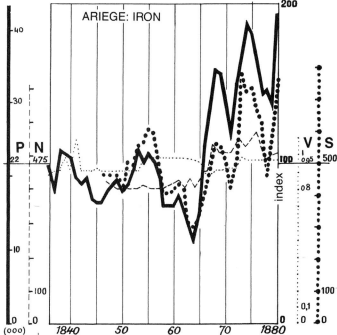

By extension, it can also be based on *the mean for all the years*, which corresponds to equalizing the totals and the flat areas subtended by each curve. The length of each unit on paper will be:

$$\frac{AB}{\text{mean P}} \qquad \frac{AB}{\text{mean S}} \qquad \frac{AB}{\text{mean V}}$$

(which resolves the problem of curves not having a known value for instant *i* chosen as index point.) This construction, figures 1 and 2, appears to be the best arithmetic solution for the superimposition of the curves. It avoids the exaggerations of the preceding solution (page 237).

Furthermore, the notion of a common base, easily explainable and conceivable for the reader, eliminates any arbitrariness in the choice of scales.

Its defects are that:
- the point common to all the curves disappears and along with it a visual classing of the different series;
- the construction leads to confusion among the curves towards the center of the graphic;
- as with any arithmetic solution, the slopes are not comparable; for example, coal production appears to increase more during the period from 1860 to 1880 than during the period from 1836 to 1860!

RELATIVE VARIATIONS

The only notion which is comparable among series of different units is the *slope of the curves*.

The most common solution (and the most satisfactory one) consists therefore in representing not each number but *each change*, that is, the ratio between one number and the preceding one. This amounts to constructing a permanent index, and translating into numbers not the quantities of units, but *differences in quantities*. This is what is achieved by a logarithmic scale.

The other elements of the comparison subsist in this solution—comparison of the peaks, comparison of the precise structure of the curves. A "rereading" of the data is possible, if one graduates the scale in sufficient detail, but it should be stated clearly that this is not the purpose of a graphic representation, which can never succeed in duplicating the precision of the data table. Only the "volume" represented by a series (whether it involves tens of tons or millions of tons) is not spatially visible. But this can be rectified by placing several reference numbers near the letters identifying each curve and repeating them on both sides of the diagram.

When constructed on a logarithmic scale (figure 3), the series related to coal and iron give us the most useful curves. All the slopes being comparable, one can easily assess the parallelism of the two curves, either in detail, or over a period, or on the whole. The "smoothing lines" (the straight lines drawn to emphasize tendencies) are comparable and even measurable, through the scale of the decennial variations.

No ambiguity persists. The relevance of the observations based on the preceding constructions can be rigorously commented upon here; we can speak with certainty of two very different evolutions and define all their aspects. Finally, the complementary information in figure 4 (whose purpose would have been difficult to understand in relation to the preceding constructions) now acquires its full explanatory value.

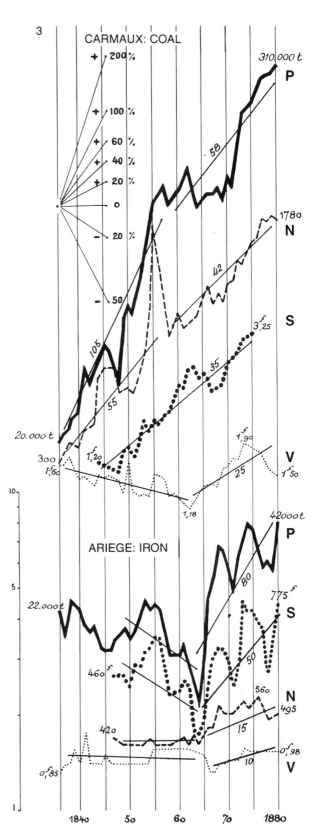

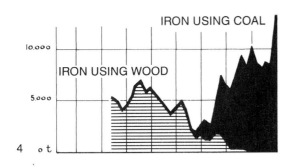

LOGARITHMIC CURVES

Terminology

Our analysis of the visual variables led us to consider the plane as having *two homogeneous and independent dimensions*. Consequently, we will speak of a *logarithmic diagram* when the logarithmic scale involves one planar dimension (the other being arithmetic), as in figure 1, and of *a double logarithmic diagram* when the logarithmic scale involves both planar dimensions, as in figure 2.

Construction of a logarithmic scale

The principle of this construction is summarized in figure 3, which displays the various properties of logarithms. For example, in order to multiply a number by 2, it is sufficient to add length AB to it. In order to multiply it by 3, we add length AC to it. In order to divide it by 1.5, we simply take away length AD from it.

Modification a logarithmic scale

One cannot move (translate) the numbers along a logarithmic scale, as in figure 5, without destroying its fundamental properties, but one can multiply them by a constant number (figure 4).

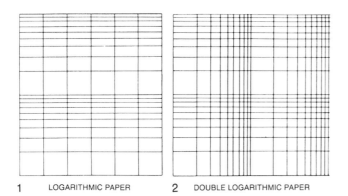

1 LOGARITHMIC PAPER   2 DOUBLE LOGARITHMIC PAPER

3

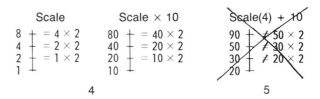

4

5

A logarithmic scale has no zero

In an arithmetic scale (figure 6) the curve passes through zero. In a logarithmic scale (figure 7) it tends toward zero, which is removed to infinity. When a statistical series involves null values, this will lead to a diagram such as figure 8. Since the zero is not representable, it is generally preferable to denote it by a convention (figure 9), which avoids useless lines and increases the legibility of the useful part of the curve.

A logarithmic curve visualizes the multiplier but not the product of a multiplication

Visually there is nothing constant in the arithmetic diagram (figure 10); the eye sees the absolute numbers (1, 2, 4, 8) and their differences (in black). However, one can derive from this figure that:

B = A $\boxed{\times 2}$ C = B $\boxed{\times 2}$ D = C $\boxed{\times 2}$

It is this constant $\boxed{\times 2}$ that the eye sees in the logarithmic diagram (figure 11), but without a visual notion of the absolute numbers nor of their difference. Conversely, when an arithmetic diagram displays a constant difference (figure 12), this corresponds to unequal variations:

B = A $\boxed{\times 2}$ C = B $\boxed{\times 1.5}$ D = C $\boxed{\times 1.33}$

These inequalities are what the eye sees in a logarithmic diagram (figure 13), and what the reader must understand and measure with the help of the legend. The legend for a logarithmic diagram cannot therefore be conceived according to the same principle as that used for an arithmetic diagram. The calibration of a logarithmic diagram is furnished by the base of the logarithm, that is, by the distance chosen to represent a tenfold increase.

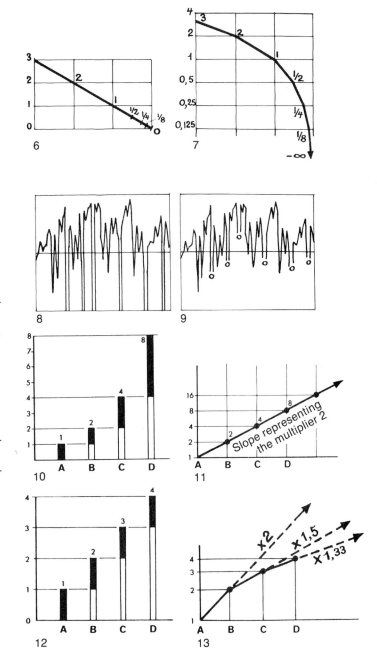

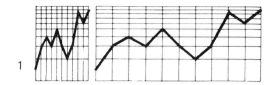

1

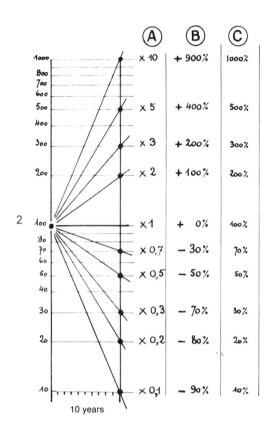

2

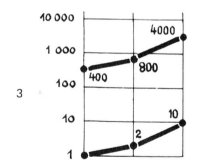

3

But the image also depends on the arithmetic scale, which modifies the slopes (figure 1). Accordingly:

The legend must visually define the slopes

We choose a base size (arithmetic)—for example, ten years—and we mark the slopes for ten years or decennial variation (figure 2), according to the multiplier A. We can also define the slopes by rates of increase B or by resulting numerical value C. Note, however, that an increase rate of 10% per year gives a decennial rate of 157% at the end of ten years. Conversely, a decennial increase rate of 100% gives a yearly increase which is less than 10%.

A logarithmic curve does not promote visualization of the absolute quantities

It expresses the multiplier, but not the multiplicand. The two series in figure 3, although very different in absolute value, are proportional. In a logarithmic diagram this means that they are parallel.

Consequently, except for the elementary level of reading (rereading of the numbers):
– only the slope is meaningful;
– the respective position of any two curves has no meaning in absolute quantities;
– the curves can be moved along the logarithmic axis, and the spacing between the curves can be chosen for better legibility;
– the curves can be superimposed in a meaningful order.

In other words:

The order of superimposition of the curves can express a useful classing

Take cultivated areas, in Sweden, related to date and different crops. Represented by an arithmetic scale (figure 4), they portray the numbers of hectares (area). Represented by a logarithmic scale, they show the trend of the crops, which can then be classed in different ways:
– according to the order of the quantities in 1937 (figure 5);
– according to their trend (figure 6);
– according to the order of dates of maximum area (figure 7) or any other useful order (alphabetic, geographic, dates of minimum area, . . .).

Finally, the logarithmic curves can be "indexed" at any date whatsoever (figures 8 and 9).

In order to benefit from this freedom, which transforms any homogeneous series of logarithmic curves into remarkable experimental material and remains the real objective of this construction, it is sufficient to trace each curve once on different pieces of transparent paper (along with two vertical reference lines), as in figure 10.

We can then proceed to all types of comparisons and useful experiments in classing.

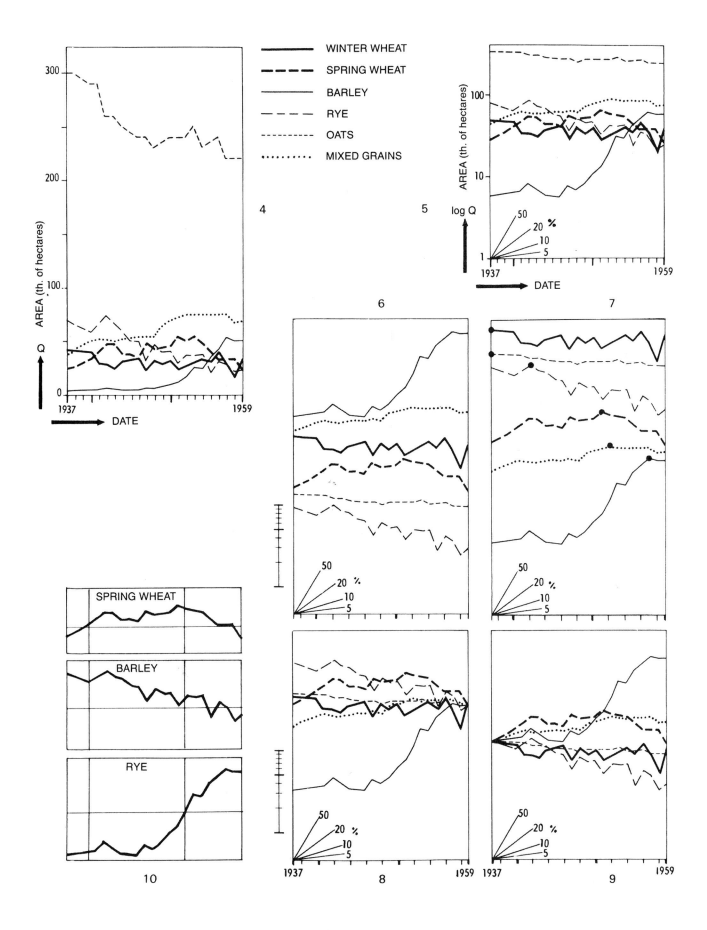

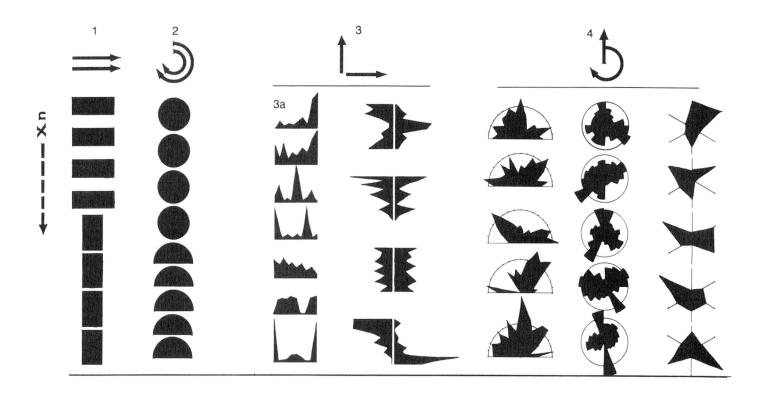

SPECIAL CASE ≠ "OPEN-ENDED" COLLECTIONS OF PROFILES

A problem involving three components can be represented by a series of images each accommodating two components. This construction is preferable in certain cases. Indeed:
(1) it is easier to draw than the standard construction;
(2) it is "open-ended"; new figures can always be added;
(3) we can rank the figures and order them in different ways, linearly or according to a double-entry table.

The collection of figures thus constitutes experimental material which leads to classifications.

Conditions to fulfill

For all the figures to be comparable, each one must form an image, that is, have a meaningful profile. This excludes figures 1 and 2, where the overall form is constant. To make them meaningful, it would be necessary to add a redundant visual variable, as on page 147. Only figures 3 and 4 create meaningful images and produce collections of profiles. Constructions such as those in column 3a are used in psychological tests, IQ tests, and professional aptitude tests.

A file card is constituted for each individual (figure 5). It includes in the first part (at the top) the elements of identification: name, age, family situation, profession.... In the second part, the notation corresponding to each of N questions is plotted on each line. We need only join these points in order to draw the characteristic profile of an individual. When numerous profiles are collected, their comparison makes "types" appear and enables us to place each individual into a given group, according to the sum of his or her characteristics. The profile is a summarized individual image. Its utilization (figure 6) involves:
(1) memorizing numerous images
(2) determining profile types
(3) classing the profiles according to these types.

The limits for utilization of collections of profiles

In order to fully utilize a collection, the reader must be able to easily memorize the meaning of terms A, B, C, D, ... of the analytic component, whatever they may be. In column 3a we need only juxtapose the images in order to find all the notations corresponding to a given term *on the same vertical straight line*. This is not possible with the other constructions, and for the polar constructions (figure 4) the reader would have to memorize the meaning of each orientation. Such a task becomes more difficult as the number of categories increases, and could only be accomplished by those few specialists familiar with a given classification. In fact, any collection of profiles has the disadvantage of including only half of the possible experiments, and consequently half of the information involved: the subsets corresponding to the intermediate level of reading are constructed on *only one* of the two identification components.

One can form a single image per tested individual, just as one forms a single image per branch of industry (page 144), but one cannot form a single image per test, any more than one can form a single image per size category of industry.

244

The collection of profiles is only utilizable if one is solely interested in the classification of the "open-ended" component (series of tested individuals, series of branches of industries, series of family budgets, series of housing options,...). It prohibits any discussion of the analytic component (elements of the test, size categories of industry, elements of the budget, elements of housing evaluation,...).

Returning to the standard construction: the image-file

Like any series of images involving two components, a collection of profiles can be constructed as a single image involving three components, which completes, or even replaces the collection.

The three components of this image are, in the example shown in figure 5:
O (or Q) of notations according to
≠ a series of test questions
≠ a series of individuals.

In order to apply the standard schema (figure 7) we need only: (a) transcribe the notations graphically on a single line (figure 5), in a way which is visually ordered, that is, by an amount of black corresponding to the notation recorded for each question on the test (A, B, C, D, . . .); and (b) transcribe certain ordered elements of identification, such as number of children (e), ages (a). . . .

Arranged in an appropriate file and separated from each other by the thickness of a piece of corrugated cardboard (figure 8), the cards construct an image such as that in figure 9, or any other image, according to the basis which we adopt for ordering them.

In this completely hypothetical example (figure 9), the cards have been ordered according to the age of the individuals (a). An image of this nature would provide the opportunity for numerous commentaries, for example:
- the test questions are not independent of the age of the individuals;
- their recorded order A, B, C, D. . ., is related to age;
- the test defines three main age-groups;
- the third group is characterized by the best but the most irregular answers. . . .

Note that the individuals who represent characteristics in relation to the whole (black spots in the white areas or conversely) are easily detected.

The "image-file" enables us to class the profiles without resorting to a preestablished typology, or by using different typologies, each adapted to a particular problem. The standard construction augments the collection of individual images with the possibility of creating an overall image, which is simplifiable in several ways. The file thus becomes complete information, capable of contributing answers to questions of an overall level (comparison of files). These answers are applicable to the very elements of the test, to their pertinency, to their mutual relationships, and to the pertinency of the order in which they were inscribed on the cards, etc. . . . The image-file introduces possibilities for factor analysis.

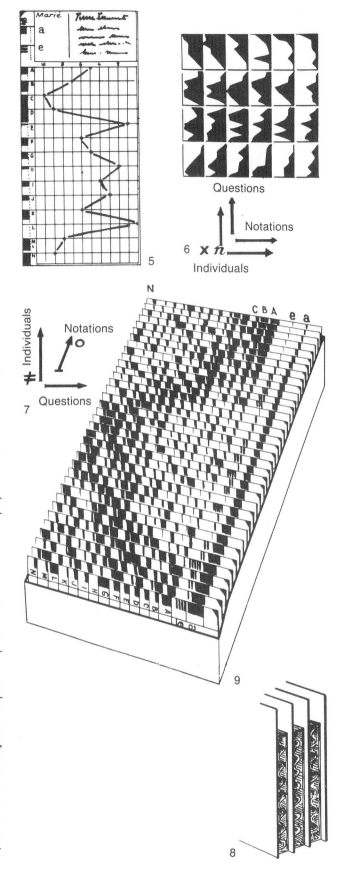

≠ Q O (Q ≠)
Comparison of repartitions, distributions, concentrations

Take two statistical series (tables A and B) by classes of quantities:

Ⓐ Classes	Pupils f	Mean number per class f		Numbers of pupils (cumulative) f	
0.10	41	× 5 =	205	205	41
10.20	12	15	180	385	53
20.30	4	25	100	485	57
30.40	2	35	70	555	59
40.50	1	45	45	600	60
	60		600		

Mean $(M_A) = 600/60 = 10\ f$

Ⓑ		m	m	m	
0.2 m	10	× 1 =	10	10	10
2.4	4	3	12	22	14
4.6	3	5	15	37	17
6.8	5	7	35	72	22
8.10	15	9	135	207	37
10.12	3	11	33	240	40
	40		240		

Mean $(M_B) = 240/40 = 6\ m$

(A) sums (in francs) held by a group of sixty pupils
(B) length of twine (in meters) held by a group of forty pupils.

Comparing the two series poses a problem involving three components:
– number of pupils
– classes of numbers of units (francs, meters)
– different groups.

This means comparing two repartitions, two distributions, or two concentrations (see pages 205–208).

The differences between the groups to be compared are not great enough to be evaluated in separate images. The curves must be superimposed, and in order to be comparable, they must be related to comparable sizes, that is, equally scaled by their totals. In **repartitions** and **distributions**, these totals are expressed by the *area defined by the curve*. How can these areas be equalized?

SUPERIMPOSITION OF REPARTITIONS

When they are constructed separately on millimetered paper, the two series produce the diagrams in figures 1 and 2. To agree that the two groups are comparable, *it means agreeing that "sixty pupils" is comparable to "forty pupils."*

This is accomplished by relating the real value of the two groups to a frequency per 100, and accomplished graphically by using the same size to represent 60 and 40 (figures 3 and 4).

It also means agreeing that "600 francs" is comparable to "240 meters."

This is accomplished by representing the two means by the same size (figures 5 and 6). The two areas are equal and represent an equalization of the total of the statistical elements (francs and meters).

Calculation of the scales
Let 1 cm of length on the paper represent each of the means.
The size O–50 (or C_A) representing the classes will be equal to:

$$1\ \text{cm} \times \frac{50}{10} = 5\ \text{cm}$$

The size O–12 (or C_B) will be equal to:

$$1\ \text{cm} \times \frac{12}{6} = 2\ \text{cm}$$

The lengths representing the numbers of elements are inversely proportional to the mean.

$$C_A = \frac{K_A}{M_A} \left(\frac{\text{upper limit of class A}}{\text{mean A}} \right)$$

$$C_B = \frac{K_B}{M_B} \left(\frac{\text{upper limit of class B}}{\text{mean B}} \right)$$

The two repartitions are comparable (figure 7).

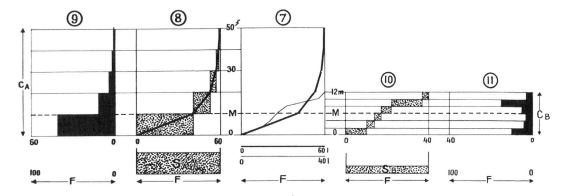

SUPERIMPOSITION OF DISTRIBUTIONS

Take figures 9 and 11, which are the distributions obtained from figure 7.

To be comparable, the numbers of pupils must be equalized, which is accomplished by equalizing the areas S_A and S_B represented by the distributions (figures 8 and 10).

Calculation of the scale

The class scales C_A and C_B cannot be modified. They are linked to the common mean, previously positioned.

Therefore we will modify the scale of the frequencies F, after having verified that the areas S_A and S_B are:
– proportional to C (size of the classes) (figure 12)
– inversely proportional to N (number of classes) (figure 13).

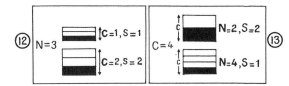

Consequently, for S_B to be equal to S_A the length F′ must be modified in relation to F according to the formula:

$$F' = F \times \frac{C_A}{C_B} \times \frac{N_B}{N_A}$$

(F being taken as base, it is generally replaced by 100). And if, in the example opposite, a frequency F (= 50%) = 2 cm (figure 14), then

$$F' \, (= 50\%) = 2 \text{ cm} \times \frac{5}{2} \times \frac{6}{5} = 6 \text{ cm (figures 15 and 16)}$$

The areas S_A and S_B are equalized and the two distributions are superimposable (figure 17). The scales are marked in real values.

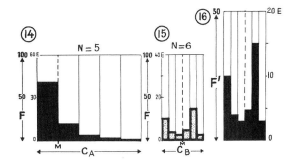

Superimposition of a third distribution

We have seen that $C_A = K_A/M_A$ likewise $C_B = K_B/M_B$
From which we obtain:
$F' = F \times K_A/K_B \times M_B/M_A \times N_B/N_A$
And for a third distribution D:
$F' = F \times K_A/K_D \times M_D/M_A \times N_D/N_A$
That is:

$$F'' = F \times \frac{\text{(upper limit of class A)} \times \text{Mean D} \times \text{N of classes D}}{\text{(upper limit of class D)} \times \text{Mean A} \times \text{N of classes A}}$$

The comparison of age pyramids

Here we have a superimposition of distributions, in which C is constant (0 to 100 years). It is therefore a visual error: (a) to juxtapose two separate images, men and women (they must be superimposed); (b) to construct absolute quantities whenever the total populations are dissimilar. The frequencies must always be compared and the totals provided in another way.

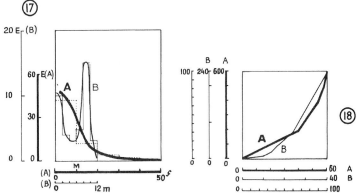

SUPERIMPOSITION OF CONCENTRATIONS

The totals will be represented by the sides of a square and, graphically, it is sufficient to divide:
– *the base of the square* according to the total number of pupils in each group (total of the observations) (figure 18);
– *the height of the square* according to the total number of units counted (francs or meters).

In order to make the points correspond, we should calculate the cumulative numbers of the observations, then the units counted (last two columns of tables A and B).

When the graphics are finished, a common scale, in percentages, can replace the proper (individual) scales.

SPECIAL CASE ≠ 2
COMPARISON OF ORDERS

Information ≠ 2 ≠ n O is a comparison of orders. It can be analyzed as O O ≠ n and constructed according to the schema in figure 10.

If *n* does not exceed approximately thirty, the information can be constructed according to the schema in figure 1.

Consider the information:
comparison of the industrial population (II) and the tertiary population (III) according to the order in which they class the ninety departments of France.

INVARIANT   –*a French department*
COMPONENTS –≠ *two sectors of the work force (II) (III)*
            –≠ *ninety French departments*
            –O *numbers of persons in each sector*

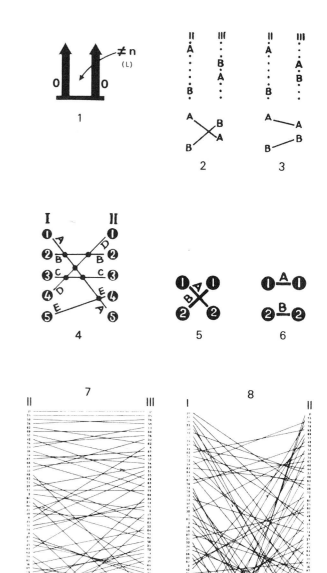

COMPARISON OF ORDERS

Our information consists of ordered lists of departments (ordered according to population II, then III, but these quantities are not given to us).

Question—Are the two orders different, or are they related? What is their "distance"? In fact, what meaning can we give to the term "distance" between two rankings, given that only the orders matter here, not the quantities? A simple way of answering these questions is to count the number of the "inversions" between the two classings. "Inversion" occurs whenever a pair of departments (A and B) is classed in a certain order in relation to II and in inverse order in relation to III (figure 2). In the opposite case (figure 3) we can say that there is "agreement" between the two rankings for this pair. Two rankings are said to be more "distant" from each other when the number of inversions is greater. The number of inversions between two rankings will be called their distance. To count these inversions and represent them graphically involves two possible solutions.

GRAPHIC CONSTRUCTIONS

Parallel alignments
Each ranking is translated by an alignment of points. The two alignments are parallel. We use a line to join the two points representing department A, then another line for those representing B, etc. If there is an inversion, and in that case only, the two lines (A and B) will intersect each other. There are, for example, six intersections and thus six inversions in figure 4.

It is therefore sufficient to draw the ninety lines corresponding to the ninety departments and count (two by two) the number of intersecting points between scales II and III (figure 7). If there are many inversions and many classed objects, counting by this graphic method will very quickly become laborious. At least the muddled impression in figure 8 permits visualizing a substantial distance between I and II and comparing it to the proximity of II and III revealed by figure 7. In general this construction should be limited to a component which does not exceed approximately thirty categories.

Rectangular alignments
In the preceding construction (figure 4) the ranks are represented by points on a scale, the departments by lines.

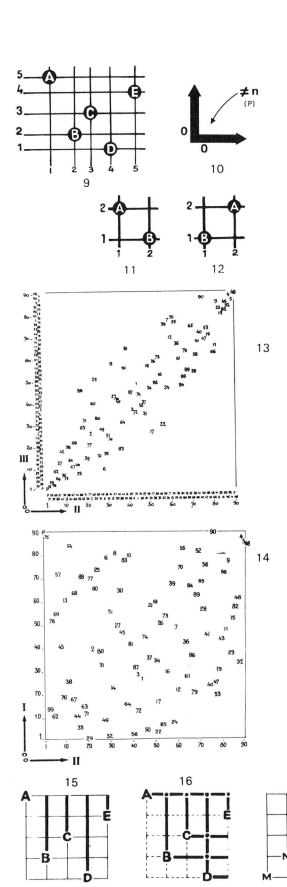

In the rectangular construction (figure 9) the ranks are represented by lines, the departments by points. The two constructions are "duals" of one another. Of course, since order alone matters and not quantity, the lines do not have to be equidistant, and we can use whatever is the most convenient grid—equidistant, logarithmic, or other—provided that it preserves the order of the ranks.

Inversions between rankings will be represented as in figures 4 and 11. Agreements by figures 6 and 12. If there are few inversions, that is, if the "distance" is small, the points will appear as a cluster elongated in a "Northeast" direction (figure 13). If there are many inversions, the points will appear as a "Northwest-Southeast" axis. Incidentally, we have perfect agreement, null distance in figure 17 and maximum inversion, ranking in inverse order (ten inversions) in figure 18. Between the two extremes we can have an absence of relationship, that is, no relationship between I and II, as shown in figure 14.

MEASUREMENT OF INVERSION

The distance N between two rankings is the number of inversions. In figure 16, it is the number of "summits" (nodes) situated to the right of each point, each node being the intersection of the vertical lines erected above each point (figure 15) with the horizontal lines drawn to the right of each point (figure 16). There are six inversions in both figure 16 and figure 9.

The maximum distance D between two rankings (figure 18) corresponds to maximum inversion, that is, two opposite rankings. In a network, it is the number of intersections situated to the right of the diagonal. For n objects, it is the total number of intersections $n \times n$, minus the points of the diagonal, that is, n. We retain only half of the result (half of the network), that is: $\frac{n^2 - n}{2}$. This can be expressed as

$$D = \frac{n(n-1)}{2}.$$

The relative distance N/D. To measure the distance between two rankings and compare rankings not involving the same number of objects, we can relate N to a standard scale (0–10, 0–100). It is sufficient to divide the number of inversions N by the maximum distance D. In the case of five objects one could have ten variations at most:

$$\frac{5(5-1)}{2} = 10.$$

In this way, a distance 6, as in figure 16, represents 60% of the maximum.

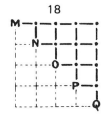

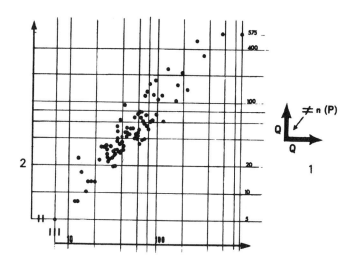

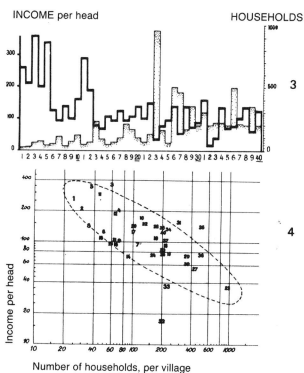

COMPARISON OF QUANTITATIVE ORDERS

Comparisons of order require that the elements be numbered or ranked prior to the graphic operation: In figure 13 on page 249, the departments are considered "already ranked." This prior ranking is not necessary when quantitative series are involved.

$$Q \neq \neq 2 = Q\,Q\,\neq$$

First example
Q of population according to
≠ ninety departments
≠ two sectors (II and III).

As previously, **information $Q \neq 2 \neq n$ can be analyzed as $Q\,Q \neq n$ and constructed according to the schema in figure 1.**

It is sufficient to construct two rectangular scales of quantities and plot each element at the intersection of the two quantities which define it (figure 2).

Since the graphic retains only the order of the points, the scales of the quantities can be arithmetic or logarithmic.

The latter stretches the scale and thus separates points which would otherwise be squeezed together when the distribution involves groups very distant from one another (figure 2).

Second example
For forty villages in the vicinity of Madrid during the sixteenth century we are given the number of households and the average income per head, that is:
Q (of households and income) according to
≠ two types of quantities
≠ forty villages.
Analyzed thus, the information leads to a fastidious and inefficient superimposition of curves (figure 3). Analyzed as:
Q of population
Q of income
≠ forty villages
it leads to the construction in figure 4, which is rapid and efficient; it immediately translates the only internal relationship contained in the information: the larger the village, the more the average income tends to diminish (figure 4), which distinguishes the Madrid area from other areas of Spain.

Retinal variables
In these constructions, retinal variables are not employed. All the points are considered as similar (≡) in figure 2, and their identification is not necessary for treating the problem. They are identified by numbers (shape variation) in figures 13 and 14 on page 249 and in figure 4 here (for purposes of external comparison).

The points are thus free to convey a new variation, which can be, as in figure 5, the total population.

The correlation among the populations of II, III, and the total appears clearly.

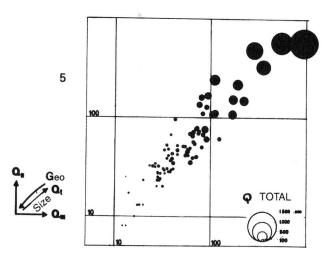

3. Several quantitative components
Q Q ≠ Q Q O

Each point on the plane has a different meaning from all the neighboring points. However, within a component **O**, all the points of the same category have the same meaning; that is, they share a common characteristic. Thus when occupied by nonreorderable components, the plane forms a constant reference system; consequently, diagrams such as these are sometimes called "maps" (color, sound, or vocalic maps, page 266).

Q Q ≠. "Ombrothermal" (rainfall/temperature) distribution of different categories of vegetation (figure 6) (taken from P. Rey, *Phytocinétique biogéographique* [Paris: C.N.R.S., 1960]).

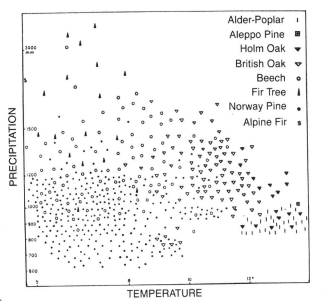

A sampling procedure defines a "population" of sites, in a given region (here the region covered by the map of le Puy at 1/200 000). For each site we know: **Q**, annual temperature; **Q**, total annual precipitation; ≠, the characteristic vegetation. When the plane shares the **Q** components, each category of vegetation forms a reduced, characteristic cluster. The diagram thus defines the optimum conditions (center of the cluster) and the extreme conditions for a given category of vegetation, as concerns precipitation and temperature.

Since each grouping of points is relatively homogeneous, small visual differentiations among the types of signs are sufficient to circumscribe each zone.

Q Q ≠. Map of audible sounds (figure 7) (taken from A. Moles, *Théorie de l'information et esthétique* [Paris: Flammarion, 1958]).

This problem involves three components: ≠ different perceptible domains according to:
Q frequency of sound (height) in periods per second
Q level of sound (amplitude) in decibels.

The characteristics of frequency and amplitude enable us to situate all the sounds and divide them into categories, into "perceptible domains," which form different zones on the diagram. The zones are differentiated visually by applying value to both lines and zones, with a redundant addition of orientation to the zones.

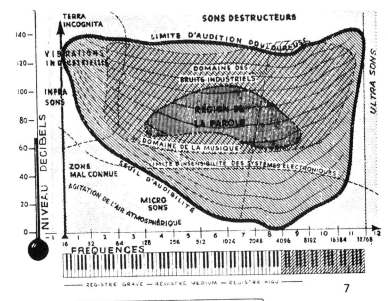

Q Q O. Evolution of "ombrothermal" conditions (figure 8), based on P. Rey, *Phytocinétique*.
O series of zones (climates) over time, according to:
Q of precipitation
Q of temperature.

The study of pollen gathered in a bog lets us reconstruct the appearance of different vegetation categories over time. Knowing that each category implies a precise and practically constant zone in the ombrothermal diagram, we can construct a sequence of zones (climates) based on that of the vegetation categories. It reveals variations in temperature and precipitation during the last millennia. Rey defines a standard evolution characterized by four climatic stages:

P Preboreal (−8500), followed by increasing temperature (2–3°) and decreasing precipitation (200–300 mm);

X Xerothermal (−5000), followed by a temperature decrease (1°) and great increase in precipitation;

M Mesohygrothermal (−1500), followed by a temperature decrease (1°) and precipitation decrease (200 mm);

A At present.

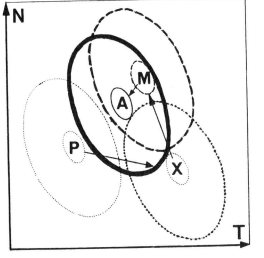

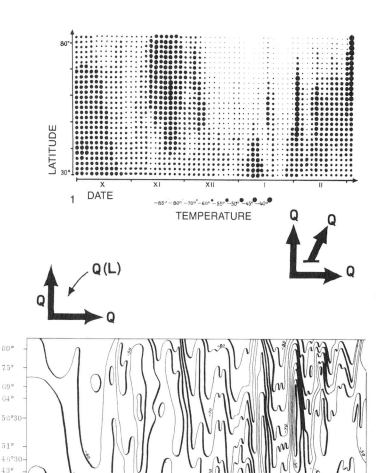

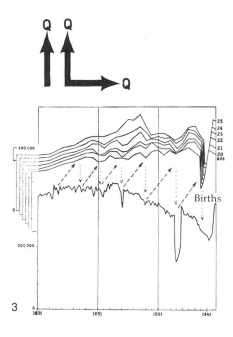

QQQ

THE STANDARD CONSTRUCTION

(Figure 1)
Example: Temperature variation in the stratosphere (25 km altitude) (after F. Kenneth Hare, "The Stratosphere," *The Geographical Review* 52, no. 4 [October 1962]: 525–574). Three components:
Q of degrees centigrade according to:
Q latitude (latitude in degrees north of observatories situated along the meridian 80 W)
Q date.

The standard construction produces a good image for expressing the temperatures. However, the graphic should emphasize the difference between the placid regime of summer, still perceptible in October, and the brutal temperature variations of winter.

SPECIAL CONSTRUCTIONS

Isarithms (figure 2) (see also page 385)
Obtained by joining all points of the same temperature, the number of isarithms increases with temperature variation, and this construction leads to the following visual result: the more black, the greater *the slope*. This construction is therefore preferable here.

Furthermore, isarithms suggest the perceptible continuum of the phenomenon, since this procedure allows each point of the paper space to be defined numerically in terms of the three components. Note, however, that the *sense* (up or down?) of the variation only appears through a complementary procedure, here through shading the curves.

Juxtaposition of curves (figure 3)
Take the information:
Q total population in France according to:
Q age (annual age-classes)
Q time (1800–1947)
(after C. Morazé, *Les Français et la République*).

The age pyramid is a familiar construction. Compared over time, several pyramids cause the well-known phenomenon of population aging to appear; here, it is translated by a "bulge" which rises progressively.

The pyramids also show another phenomenon, just as important to the general economy of a nation: *the "wave movement" of the "low" spots*. But in order to see it, the time continuum must be restored visually. With a juxtaposition of curves (figure 3)—births, individuals aged 20, 21, 22 years, . . . —the low spots appear, along with their successive repercussions. But more numerous curves would overlap, and the image would become confusing.

Another construction must be used.

The standard construction, which would lead to an image of the type in figure 1, would not be legible, because we are especially concerned with variations of relatively small quantities.

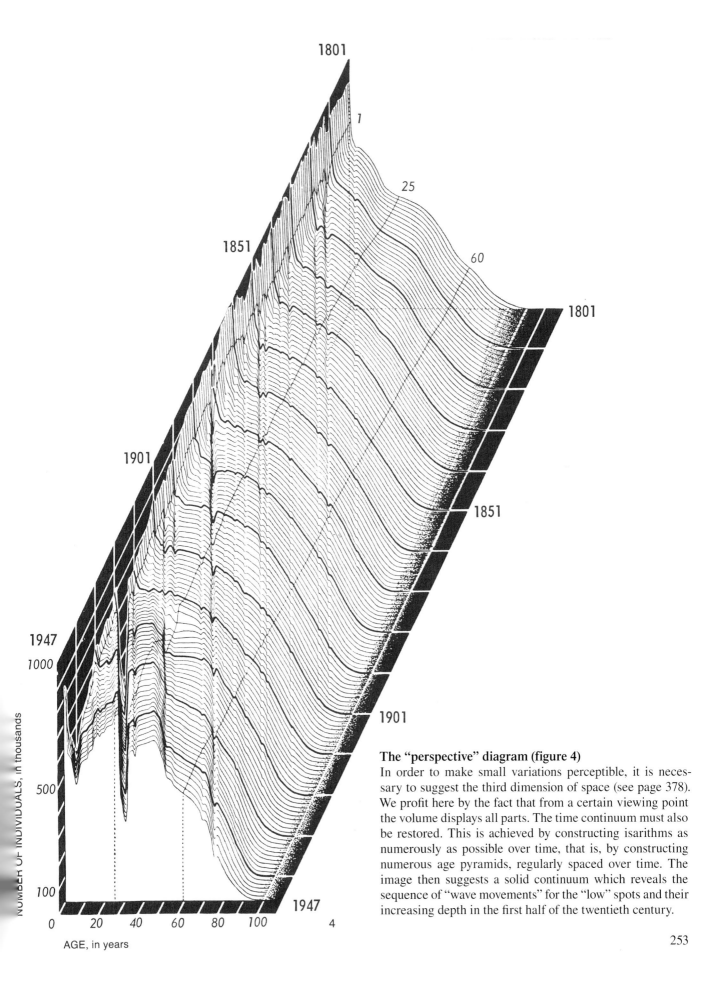

The "perspective" diagram (figure 4)

In order to make small variations perceptible, it is necessary to suggest the third dimension of space (see page 378). We profit here by the fact that from a certain viewing point the volume displays all parts. The time continuum must also be restored. This is achieved by constructing isarithms as numerously as possible over time, that is, by constructing numerous age pyramids, regularly spaced over time. The image then suggests a solid continuum which reveals the sequence of "wave movements" for the "low" spots and their increasing depth in the first half of the twentieth century.

C. Problems involving more than three components

1. Graphic information-processing

Modern research generally involves numerous components. It will constitute a unit, however, only if the information on which it is based can be thought of in the form of *one* double-entry table—albeit of very large dimensions. In fact, any homogeneous set of information can be conceived as a series of *objects* AB . . . (or of individuals, places, moments, concepts, . . .) which is observed across a series of 1, 2, 3, . . . n *indicators* (or characteristics, or variables, . . .).

If we enter the objects along *x* (the abscissa) and the indicators along *y* (the ordinate), the response *z* (either numbered or yes–no), marking the intersection of each object with each indicator, is inscribed in the corresponding cell.

This double-entry table must result from the totality of the information involved in the study. It constitutes the point of departure for information-processing, since it reduces any problem involving n components to three components.

The number of rows in this table—1, 2, 3, or n rows (see opposite page)—determines the basic graphic constructions and permutation procedures.

They are a function of the *length* (number of objects and number of indicators) and *level* (reorderable [≠] or ordered [O]) of the components. Very limited lengths (1, 2, 3) permit direct constructions, without requiring permutation. Extensive lengths will entail the graphic operations of classing and permutation.

Problems without permutation

(1) *The table involves only a single row (see entry 1 opposite)*. This is the simplest kind of information. If AB . . . is reorderable (≠), the standard construction is a repartition diagram (see page 203). When AB . . . is very limited we can use a circle; when AB . . . is very extensive it leads to a distribution diagram (see page 205). If AB . . . is ordered (O), the standard construction is a time series (histogram, polygon, curve, or cluster; see page 212).

(2) *The table involves only two rows (see entry 2 opposite)*. The standard construction is the rectangular scatter plot (see pages 248–249). The objects AB . . . become points which are classed automatically in relation to the two indicators entered along x and y. If AB . . . is ordered (O), a row can manifest this order and be meaningful.

(3) *The table involves only three rows (see entry 3 opposite)*. (a) Addition by column is meaningful. Percentages must then be calculated. By analogy with the preceding case the standard construction is the *triangular scatter plot* (see page 232). (b) Addition by column is not meaningful (for example: price + tonnage + inventory). Such information enters into the following general case.

Graphic classing and permutation
(see entry n opposite)

Whenever the table involves more than three rows, the standard construction consists of entering the series AB . . . of objects along *x* and the series n of indicators along *y*; this means graphically representing the quantities by an ordered visual variable in *z*, and simplifying the image by permutation.

Let us recall the principle of permutation (figures 4, 5, and 6). Take seven objects A, B, C, . . . across the presence or absence of ten indicators 1, 2, 3, 4. . . . This information produces the table in figure 4. If one can permute first the rows (figure 5), and then the columns (figure 6), the image is simplified and its comprehension, which at the outset necessitated assimilating $7 \times 10 = 70$ elements, only requires retaining three groups, whose originality is revealed at the same time.

This type of permutation is possible only if objects and indicators are reorderable (≠). The problems of graphic permutation can therefore be reduced to three basic situations, according to whether the imposition involves two reorderable dimensions of the plane (≠ ≠), or a single one (≠ O), or two ordered dimensions (O O).

Basic constructions

(≠ ≠) *Objects and indicators are reorderable.*

The standard construction is the **reorderable matrix** (figure 4 opposite, and see also page 256). Permutation requires technical means: permutator, domino apparatus (page 169), or computer terminal. These means permit processing information up to approximately 150×400 elements. Beyond that, it is necessary to proceed to a prior reduction of the information, either by random sorting or by procedures of automated classification.

In certain conditions of unequal lengths (30×1000, for example), a *matrix-file* (page 262) permits compensating for the absence of specialized equipment.

(≠ O) *One of the two planar dimensions is ordered.*

Two constructions are possible: An **image-file** (page 258) represents the quantities by a variation in amount of black; an **array of curves** (page 263) portrays the quantities in the form of curves.

(O O) *The two planar dimensions are ordered.*

It can be useful to construct an ordered table. This is the case, for example, with the "vocalic map" (page 266) which transforms a song into an image. It is also the case with thematic maps utilizing one image per indicator (pages 398, 402). By definition, these tables are not permutable, either in *x* or in *y*.

Freedom of permutation can be regained by considering a **collection of tables** (page 266) or a **collection of maps**

INFORMATION IMPOSITION

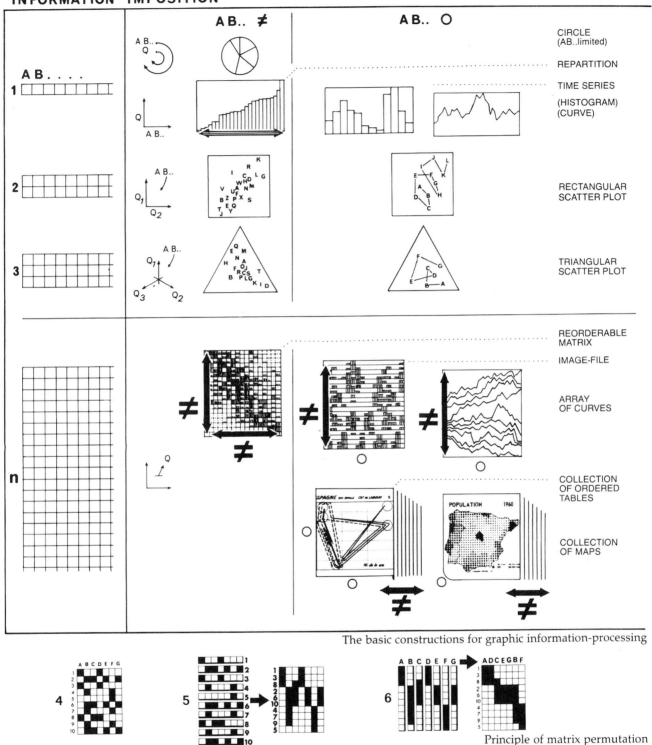

The basic constructions for graphic information-processing

Principle of matrix permutation

(pages 268, 398, 402). It is the series of images formed by the collection which constitutes the reorderable component (≠) and which allows groupings, classings, and trichromatic superimposition.

A collection of tables or maps affords the means of processing extremely broad information. In fact, we know that the eye perceives the resemblance between two shapes no matter what the number of elementary constituents of these shapes. To class 200 communal maps of France (38 000 communes) is relatively easy, and yet this represents the processing of 7 600 000 pieces of information!

2. The reorderable matrix

Figure 4 on page 255 is a reorderable matrix whose capacity is 7 × 10. The example below involves a matrix of 82 × 80 and also contains quantitative indicators.

Typology of ionic capitals
(After D. Theodorescu, "Le Capital ionique" [PhD Diss: University of Paris I, 1973].)

The problem. The eighty-two specimens being studied represent the essential aspects of known "capitals" (tops of columns; see figure 1). They differ notably, and various specialists have developed different typologies, based subjectively on one or another criterion. An objective approach is to inventory all the possible variables and to consider them initially as equivalent indicators. Next, the similarities which emerge from the whole are studied, and finally the most representative indicators of this stylistic typology appear.

The information and its graphic representation. The capitals are entered along x, the indicators along y. Each capital is the object of twenty-eight measurements involving ratios or dimensions. These are defined in figures 1 and 2A and are represented on the matrix (figure 3) in nine steps running from white to black. The qualitative model (figure 1B) suggests fifty yes–no indicators, which are recorded in figures 2B and 3B. Missing data are represented by the sign (=); doubtful data by (◁). The graphic procedure utilized the "domino" apparatus (figure 6), which accomplished all the necessary permutations.

Processing and interpretation. By successive permutations, similarities emerge and lead to figure 4, which reveals three systems of relationships between indicators in y and three major types of capitals in x.

The system in figure 4A displays a series of indicators which evolve chronologically (chr.). A second system (figure 4 B) reveals, on the contrary, indicators which are only present or important (k, l, m) at the middle of the chronological order, or conversely (o, p) at the extremities. Finally, the indicators C and D in figure 4 seem unrelated to chronology. These observations are schematized in an "interpretation matrix" (figure 5). System A from figure 4 can be characterized by the design of the capital (b), which becomes less rectangular over time, and the shaft of the column (c), which becomes less narrow.

System B from figure 4 can be characterized by the size of the volute (m), small at the outset, then large, then becoming small again at the end of the period, whereas the overall size of the capital (p) follows an inverse evolution. These two systems account for 85% of the indicators and thus correctly represent the whole of the stylistic characteristics of the ionic capital.

Typology. These results enable us to divide the capitals into three main types, each containing three subtypes. Their essential characteristics appear in each column of figure 5. Note in figure 4 that the chronological order (chr.) is not strict; certain capitals are out of place. The explanation is simple. It is obvious that at any given time, a sculptor can be inspired by former models, or, on the contrary, be innovative. As a result, the order of x in the matrix (figure 4) is that of the style and not of the construction dates.

This typology can be verified. One need only rearrange the matrix according to the typology proposed by any given author. In no case, does the number of criteria corresponding to these subjective typologies exceed 5%!

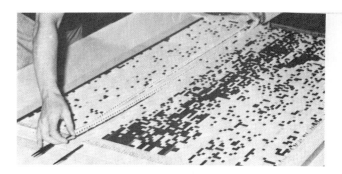

6 Above, permutation of column using manual "domino" apparatus. Below, keyboard permutation on a cathode-ray screen.

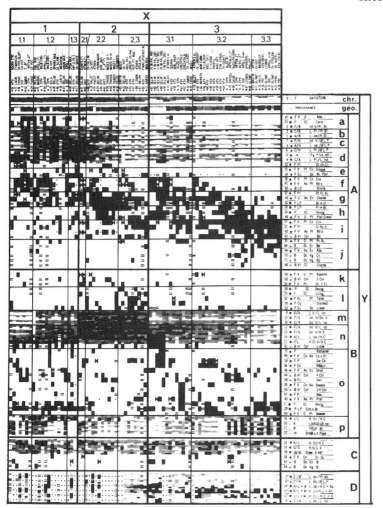

4

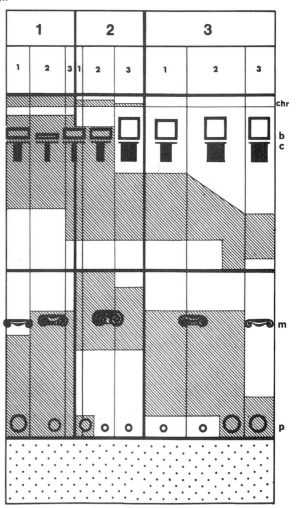

5

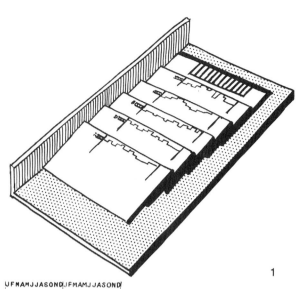

3. The image-file

When the component entered along *x* is ordered, it is generally not useful to permute the columns; simple paper cards suffice for the graphic representation. They are aligned along a sheet folded on one edge (figure 1) and are permutable in *y*.

A file 20 × 12

A promotion campaign for a large hotel* requires studying the monthly trend of several indicators (figure 2). Each of these is recorded on a separate card. The numbers are transcribed in the form of a histogram with a maximum column height of 1 cm. For all the numbers above the annual mean, the column is shaded in black. The twelve months are recorded twice, along *x*, in order to reveal potential cycles. No more is required later than to put similar cards together (figure 3). This causes four types of indicators to emerge—two in a semiannual pattern (at the bottom of figure 3) and two in a quarterly pattern (the first five at the top and their opposites, numbers 12 and 5).

Studying the incidence of each indicator in the "(s)low" periods (numbers 14 and 20) defines the nature of the efforts necessary to increase activity: advertising orientation, types of services to develop during these periods, organization of supplies, phasing of conventions. . . . This image-file led to business increasing by 10%.

A file 1000 × 9

Within the framework of the *Atlas de Provence-Côte d'Azur*, the Laboratoire de Géographie d'Aix-en-Provence charged G. Peugniez with designing a map involving one of the main variables of rural economy: the size of farms. An image-file is utilized. Along *x*, it represents seven categories of size, ordered from the smallest (0.5 to 1 hectare) to the largest (more than 100 hectares). A commune (line) is characterized by the percentage of "useful agricultural area" (S.A.U.) in each category, that is, by seven indicators, to which are added the total S.A.U. and the number of farms. The series of communes is recorded along *y*.

The objective is to discover the types of communes resulting from this analysis.

The atlas covers 1000 communes. Whenever the number of cards exceeds several dozen, it becomes necessary to use cards of about 1 mm in thickness, and to draw on the top edge (figure 4). The latter is divided: 3 mm for the yes–no marks; 15 mm for the quantities. These dimensions are constant. The purpose is, in fact, to display the profile of each indicator equally, whatever the numbers that characterize it. The cards are all supported in the same way along a guide strip and are permutable.

Figure 5 shows the file at the outset. In figure 6 it is classed by departments and, within these, by categories. The departments themselves are ordered according to their tendency toward small farms (Alpes-Maritimes) or large

---

*For a detailed study of this example, see *G.I.P.*, pp. 1–11, 242–245 (translator's note).

ones (Basses-Alpes). The classing in figure 7 defines types of communes, independent of their departments. The main types A, B, C, . . . are divided into subtypes, whose characteristics appear despite photographic reduction. These types will be recorded on the map, and the file will serve as the legend.

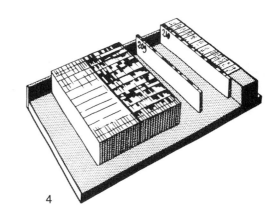

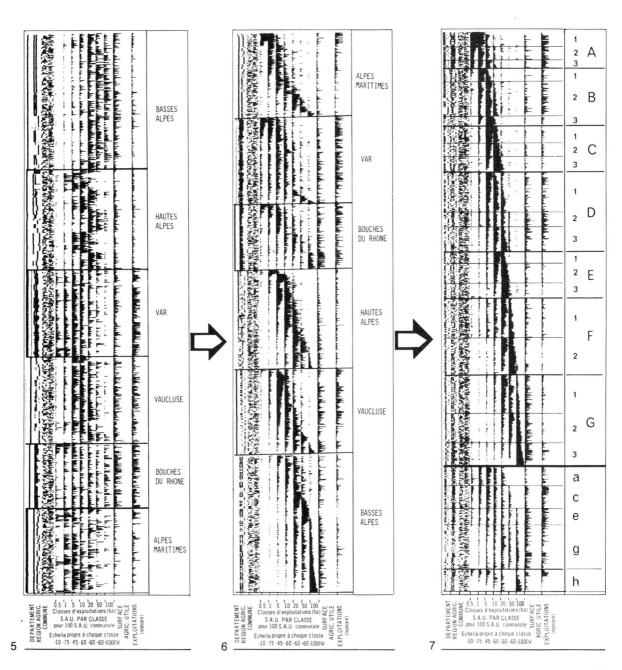

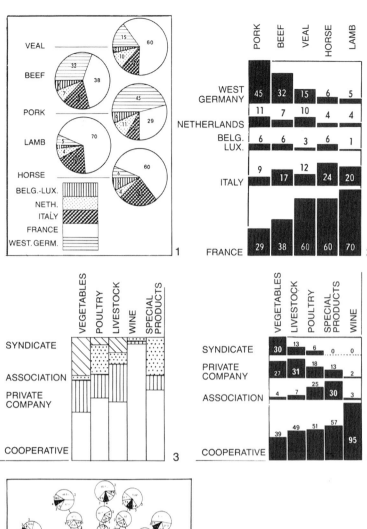

EFFICIENCY AND UNIVERSALITY OF THE MATRIX CONSTRUCTION

Showy but useless graphics are still encountered everywhere, even in scholarly journals and works devoted to graphic representation. Their main result is to hide the content of the information from the user, thus doing a real disservice to graphics. Experiments have shown that readers do not even look at them. Since a reader seeks to understand and retain information, he or she instantly evaluates the time which will be "lost" in attempting to discover the relationships which are not readily displayed by the graphic. Consequently, the reader turns the page and seeks the anticipated "reduction" (any message involves a reduction) in the written text. And what is reduced? Quite simply the lengths of x and y, and this cannot be achieved when the two components are separated by the representation. To test this statement, the reader is invited to cover the matrices opposite with a sheet of paper, to look only at the odd-numbered figures, and to seek the answers that a graphic construction should provide: what are the objects and indicators dealt with? How are they related in this information?

Figure 1: The meat problem in the common market (1966)

Are Common Market ministers seeking to evaluate the meat problem greatly assisted by figure 1? Only a few numbers are obtainable from it, and a simple table would have been more efficient. How are the different nations grouped or opposed?

The same information, represented correctly (figure 2), reveals a problem of a political nature. The nations form two groups—Germany, the Netherlands, and Belgium–Luxembourg, on the one hand, France and Italy on the other—whose policies would appear to be dictated by their group affinity.

Figure 3: The legal form of production groups

This form varies with the products. Are certain forms concurrent or particular? which ones? in which branches? In figure 3, only the importance of cooperatives, especially for wine, is immediately shown. Although this information is still visible in figure 4, the distribution of the other forms, the concurrence of the first two, and the resemblance of various products suggest directions for further research.

Figure 5: Employment trends in physics in the U.S.

At least that is the so-called subject of figure 5, which appeared in a professional review, as a full-page spread and without commentary! Can we even determine what is involved?

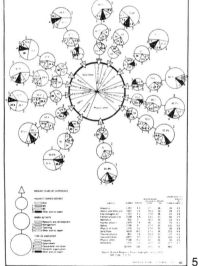
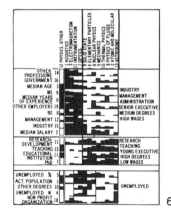

The matrix construction (figure 6) shows that the information deals with the distribution across different branches of physics, of types of professions (P), types of employers (E), university degrees, age, average salary, and finally unemployment. This information divides the indicators into two systems. The most important further divides the branches into two categories: acoustics, electromagnetics, instrumentation, and optics are the province of industry, management, high salaries, and older men. The other branches are in the areas of research, teaching, advanced degrees, . . . and low salaries! But another system of indicators, at the bottom, disturbs this order and shows that unemployment follows quite another pattern.

Figure 7: Property and farming

Fifteen types of person are supposed to be defined and represented by the sectors of figure 7. This is a complex problem which only a double-entry matrix (figure 8) can elucidate. We have to define types of owners, and types of farmers, as well as the number of individuals and amount of area for each type. We also know that an individual can be an owner or not, a farmer or not, that there are different types of farmer and that all combinations can occur. For a given geographic canton, the information has four components. (1) Different types of person. Two main categories, owners (P) and farmers (E), produce three groups: the owners-non-farmers (P nonE), the owners-farmers (P E), the farmers-nonowners (E nonP). Rentals are of three kinds: tenant farming (f), sharecropping (m), or both. This makes fifteen types. (2) The numbers of persons per type (in percentages). (3) The mean areas per type. (4) The total areas per type.

If we represent along x the different types, in proportion to the numbers of persons per type, and along y the mean areas, the area ($x \times y$) represents the total area, and all the components are represented. Note that the content of rectangle (P) represents the total area of the canton. What is outside (third row) represents the redistribution, across the farmers, of lands not farmed by their owner (first row).

Figure 9: Foreign labor in France

Belgians (B), Polish (PL), North Africans (NAF), Portuguese (P), Spanish (E), and Italians (I) have furnished foreign labor to France in chronological sequence. Can we determine in figure 9 the main periods of this sequence, the nations which characterize it, the uniqueness of the modern situation and the probability of the next situation? In figure 10, black depicts the quantities which are greater than the mean from 1921 to 1968.

Figure 11: Crop cycles

This is a standard example of the image-file. In this African region, the agricultural year is divided among several crops: squash (C), beans (H), manioc (M), different types of corn (m), tobacco (T), and peanuts (A). The precipitation curve (P in figure 12) is of fundamental importance here. Various types of work are noted: (1) groundbreaking; (2) sowing; (3) hoeing; (4) harvesting. Need we stress the difference between the two figures? We see in figure 12 that:

– the various phases of the crop are grouped;
– the crop is conceived as a whole which has a beginning, the groundbreaking, and an end, the harvest. This unity is not visible in figure 11.
– in order to discover the cycles, it is necessary to represent at least eighteen months in x and to repeat the crops at the beginning (H) and at the end (M_3) of the cycle in y.
– in figure 11 attention is monopolized by deciphering the graphic, whereas figure 12 makes one regret the lack of complementary information: cultivated areas, amount of work, feast days, market days, demography, which could easily be represented, thus furnishing a true reconstruction of communal life.

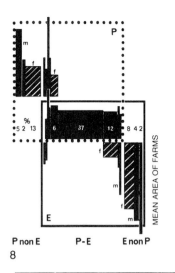
8

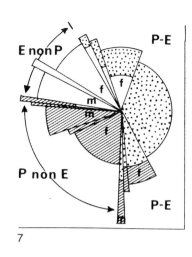
7

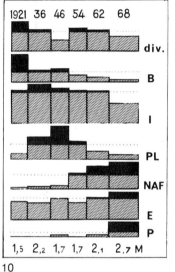
10

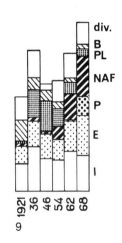
9

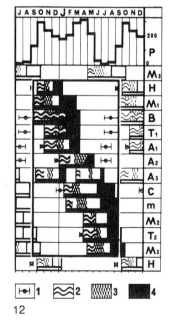
12

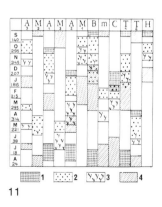
11

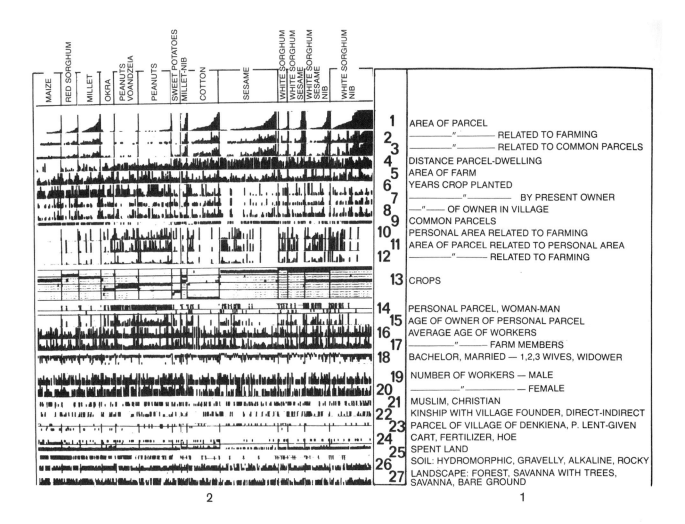

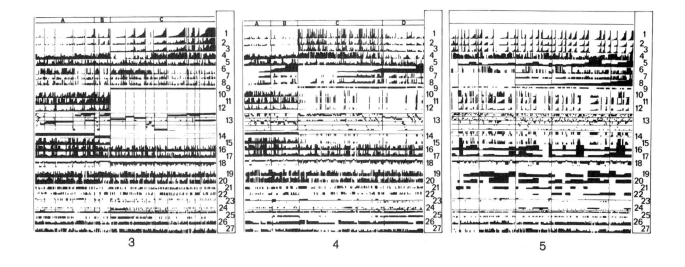

4. The matrix-file

The technique of the image-file (page 258) enables us to compensate for the absence of permutation equipment and to process information ≠ ≠, which is normally submitted to a reorderable matrix. The construction in figures 1–5 here is called the "matrix-file." It has the simplicity and capacity of the image-file, but the impossibility of permuting in x causes a substantial increase in the number of classings. As a result, some thirty indicators is a maximum.

Mossi colonization in the Bwa country (Upper Volta)

(After Michel Benoit, "La Genèse d'un espace agraire mossi en pays bwa [Haute-Volta]," *L'espace géographique* no. 4 (1972), pp. 239–250.) In figure 1 the image-file is applied to the study of agrarian land. Along x are 350 parcels of land; along y twenty-seven indicators. Conceived on the very site of the study, this file permitted numerous classings, of which four examples are given here.

Figure 2: The first-order classing is by crop (row 13), the second-order by area (row 1), then the crops are classed by mean distance from dwelling (row 4). This classing permits characterization of the crops.

Figure 3: The first-order classing separates common parcels from personal parcels (row 9), then the personal parcels by sex (row 14). The second-order classing is by crop (row 13), and the third-order by area (row 1). Here the personal parcels are characterized, which refines the characteristics of the preceding classing.

Figure 4: In first-order classing are the women's personal parcels (row 14); in second-order, the number of years the crop has been planted (row 6); in third-order, the type of soil (row 26). Here a first typology of the parcels, taking account of numerous indicators, can be defined.

Figure 5: The first-order groups are parcels from the same farm (rows 16, 17, 19, 20); the second-order, years spent by the owner in the village (row 8); third-order, area of farm (row 5); fourth-order, area of the parcel (row 1). Here we reconfirm the great homogeneity of farms which characterizes the colonial system.

5. An array of curves

This construction (see also pages 239–243) accommodates extensive lengths in x and y (several hundreds) and is particularly useful in historical studies. The example in figure 6 treats twenty "rates of exchange," across 200 dates. The classing in y is of a geographic nature and displays a financial regionalization and its periodic characteristics.

For certain information, the array of curves can be permuted, not only in y but also in x. When the curves represent the stages in the life of an individual or an object, for example, they can be aligned on various bases, independent of the first reference system.

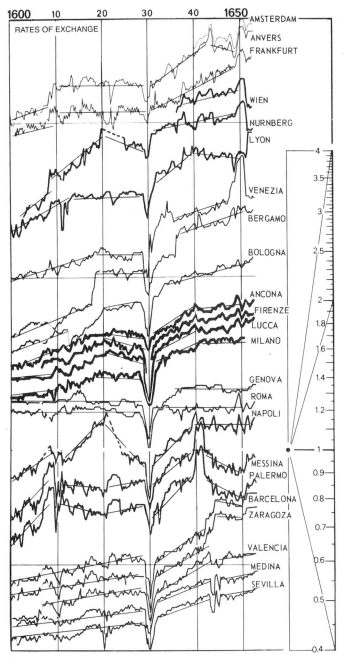

6

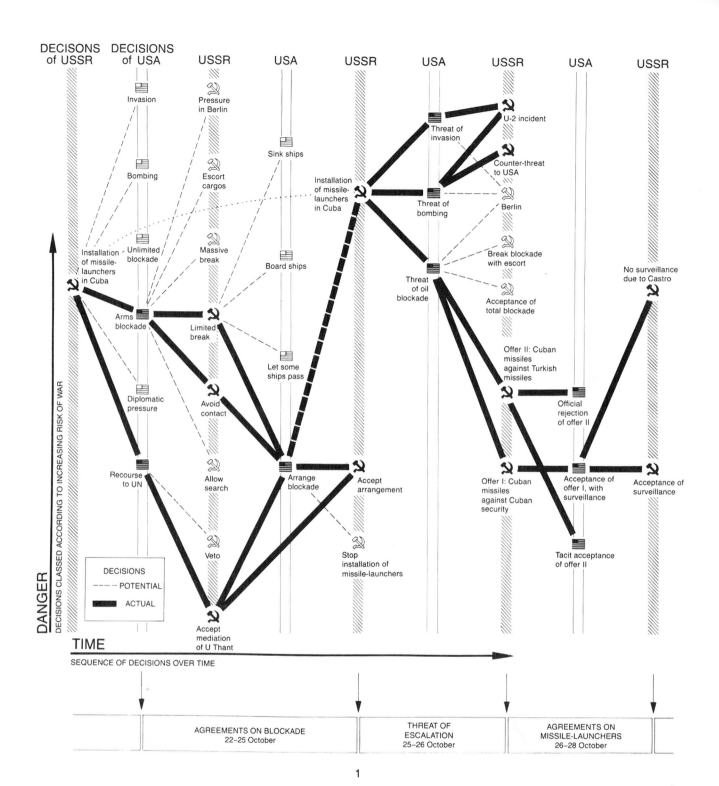

6. An ordered table

An ordered table shows how various elements are arranged in relation to two orders which the reader can easily grasp.

A table can have only two or three components (see rectangular or triangular scatter plots, page 254, and also pages 251–252). Here is an example involving five components.

The Cuban crisis
(After A. Joxe, "La Crise cubaine de 1962, Eléments principaux de décision, au cours de la crise 'chaude,' " *E.P.H.E.: Groupe d'études mathématiques des problèmes politiques et stratégiques* [February, 1963], Document 3E.)

The information can be analyzed as a series of decisions which are: ≠ potential or actual; ≠ American or Russian; ≠ of different kinds; O as to time; O as to risk of war. The two ordered components are universal, easy to grasp, and will thus form an efficient table (figure 1). It only remains to diversify the decisions according to three components:

- ≠ 2, potential or actual decisions, are represented by value: light-dark.
- ≠ 2, American or Russian decisions, alternate along *x*. To differentiate them more clearly, we add a value difference and two shape differences—flags and titles—to the vertical bands.
- ≠, kinds of decisions, are denoted in writing and are readable only on the elementary level.

The elements of this crisis are obviously more numerous and complex (one need only imagine the strategic and logistic problems posed by each eventuality), but it is the aim of "communication" and languages to consider this complexity as "reducible," and to attempt to display a choice of elements such that one can become "informed" of the principal characteristics of the crisis.

Put into image form, this choice has the advantage over a written report of opening the discussion at any point, and any objection, any different choice, immediately implies a visible chain of consequences.

7. A collection of ordered tables

A collection of comparable tables (that is, where the reference plane is similar) permits us to reduce the component formed by the series of tables through classings and groupings. When this component is ordered, as in figure 2, we can define "periods"; when the component is reorderable, we can define "types" (see page 266).

Demographic periods
(After R. Baehrel, *La Basse Provence rurale* [Paris: S.E.V.P.E.N.,1960].)

This information has four components: Q of population according to ≠ three age groups (young, adult, elderly, ≠ villages, O of dates.

≠ 3 authorizes a triangular construction, since the total of the three categories is meaningful. Two solutions are possible:

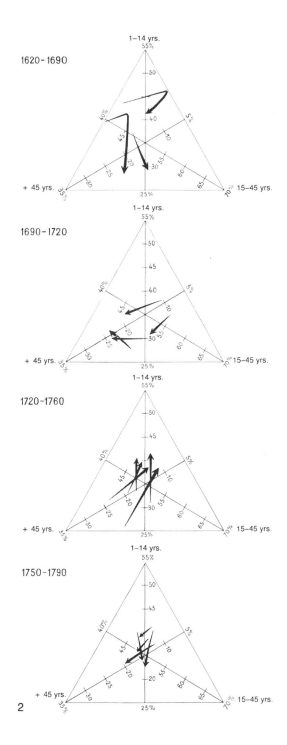

(1) Construct one image per village (see figure 14, page 233), thus forming a collection of villages. The time order is represented by a line joining the dates. In this case, we perceive that all the villages tend to resemble each other in their evolution, whereas graphic processing is more efficient when the type of construction produces different images.

(2) Construct one image per period, as in figure 2, by superimposing all the villages. The differences among the images characterize the periods.

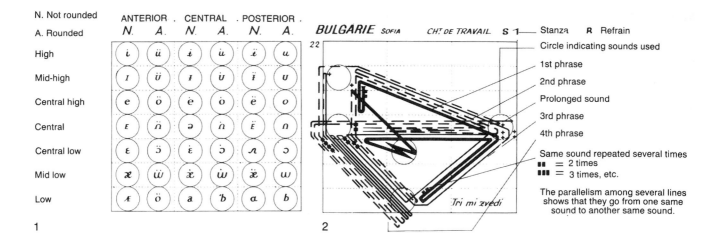

Comparison of languages

(After Alan Lomax and Edith Trager, "Phonotactique du chant populaire," *L'Homme* 4, no. 1 [January–April, 1964]: 5–55.)

Tape recordings of innumerable types of "folk song" form an inoperable inventory if not transformed into a sign-system which will facilitate comparisons. The graphic image is an instrument which enables us to exploit these riches.

It should be noted from the outset that the collection assembled here consists of folk songs recorded uniquely by local singers, that the analysis does not concern the music but only the words, and at that only the vowels. To make use of this collection is to pose a problem involving seven components.

This is a "population" of sounds: **O** ordered in time and ≠ different according to: ≠ phonetic differences, ≠ language differences, ≠ regional differences, ≠ differences in kinds of song, ≠ differences of stanza or refrain, ≠ different titles of songs.

Construction of an ordered table

First the sounds must be represented. Numerous graphic solutions are imaginable, the best in this case being the *"vocalic map"* (figure 1), which situates all the vowels in relation to three variables: the openness (or in English, height of the tongue), the anterior-posterior variation (sounds from the throat or the mouth), the form of the lips (rounded or not).

This solution adds two additional variables to the problem, since it transforms ≠ phonetic differences into **O** difference of height of tongue, **O** anterior-posterior difference, ≠ form of the lips. This enables us to construct an ordered reference table which reserves a particular place for each possible sound.

The time order is represented in the following way (figure 2).

Each sound is linked to the following one by a thick black line tracing the first phrase, then by black dashes tracing the second, then by a red line for the third, and red dashes for the fourth (all these are reproduced in black and white here, which makes the sound order less visible).

Certain very long stanzas have been divided into two images (see example 18 in figure 3). In spite of this initial distinction, the number of lines still produces confusing images.

However, in general the number of paths followed (number of linkings between two sounds) is relatively small. This constitutes a second element of distinction, provided each path is perceptible.

This task has been entrusted to the visual notion of straight parallel lines, and the impact point of each sound can be enlarged (see examples 16 and 36).

A collection of tables

The images thus obtained allow us to compare all the recordings. All the other variables of the problem are simply recorded on the image, always in the same place, and the collection of tables can be classed in different ways.

The presentation in figure 3, for example, tends to class them from top to bottom by country and complexity (except for the Bushman, which should be at the top), and from right to left by kinds of song: (1) refrains; (2) lullabies, blues, melodies; (3) the rest.

This classing shows that the Romance languages do not contrast with the others, but that the Mediterranean languages (Spanish, Italian, Greek, Serbo-Croatian) form a much simpler group of images, as opposed to the complex images of English, French, Irish, ...

Note the simplicity of all the refrains (R) and the tendency of lullabies toward a horizontal line, which is also found, incidentally, in the melody, the religious chant, and the blues!

Many other observations are possible, and we come to regret a lack of more examples for each kind and each language (do all Spanish songs end as here in a triangle?) and that other characteristic languages (German, Russian, Arabic, Asiatic languages) have not yet been "mapped."

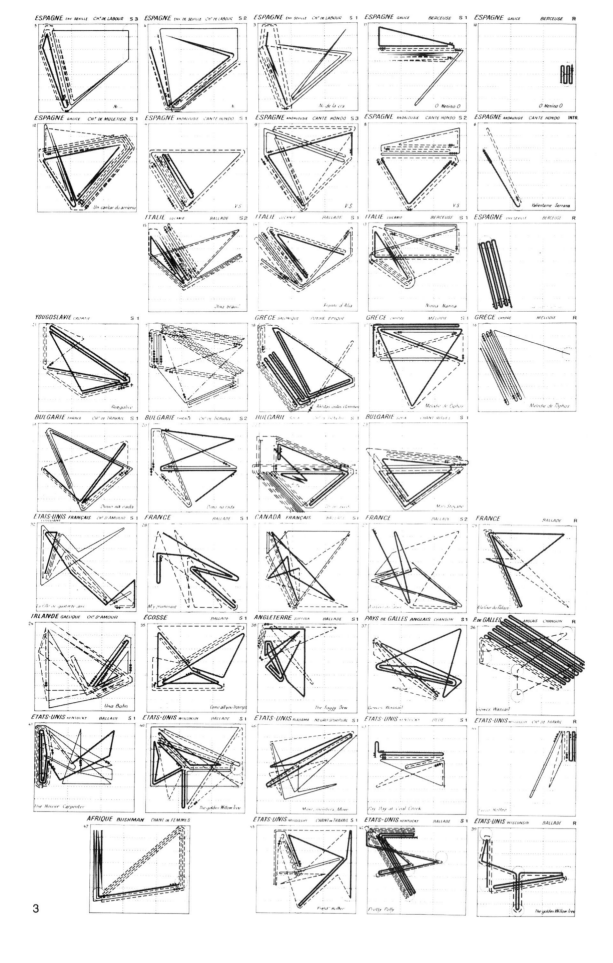

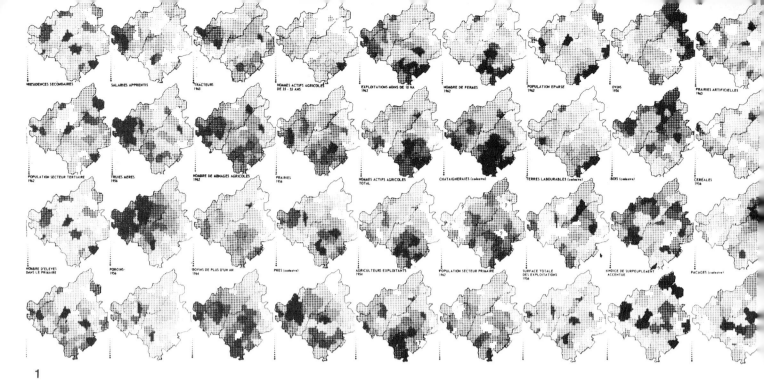

1

8. A collection of maps or a matrix permutation?

The properties of a collection of maps are dealt with beginning on page 397. We should note here that within certain limits of information length, the collection of maps and the matrix permutation are both possible. These limits are the following:

GEO 400 × 150 indicators: matrix
GEO 150 × 400 indicators: matrix
GEO 500 × 30 indicators: matrix-file
GEO 1000 × 10 indicators: image-file

The image-file on page 259 treats 1000 communes across nine indicators. It permits us to discover a typology that is more precise and more certain than we would find from the comparison of nine maps.

Figure 1 above, the study of rural planning for the Tulle region in France, deals with sixty-two communes across 108 indicators. These lengths authorize both the collection of maps (automated) and the matrix permutation.

This extract shows how the maps are grouped in order to characterize certain regions, and how the matrix (figure 2) can lead to a geographic synthesis (figure 3) and a map (figure 4) based on all the indicators.

But when the geographic component exceeds the matrix lengths (a study involving the 38 000 French communes, for example), the collection of maps becomes the only possible form of graphic processing. However, we can utilize matrix processes by calling upon mathematic "routines" (factor analysis, clustering typologies) by which extensive lengths can be reduced to dimensions which are approachable by graphics.

All modern problems lead, in the final analysis, to conception of a "processing chain," in which the computer affords the means of utilizing both automated algorithms and visual manipulations, whether with matrix or map.

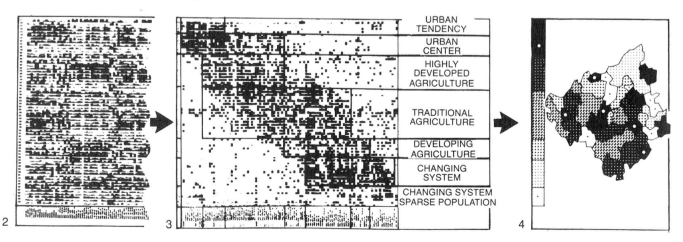

II. Networks
(flow charts, trees, inclusive relationships)

Definition
When the correspondences on the plane can be established among all the elements of the same component, as in figure 1, the graphic is a network.

Process of construction:
The transformation of a network
The new, pertinent information stems from the *observed correspondences* (figure 2), which must trace the *simplest and most efficient image possible.* Accordingly, each set of information poses a particular problem and entails a process of construction which distinguishes a network from a diagram.

In a diagram, one begins by attributing a meaning to the planar dimensions, then one plots the correspondences.

In a network, one can plot the figures on a plane which has no meaning, and then look for the arrangement which produces the minimum number of intersections, or the simplest figure. After this *transformation*, the graphic will yield maximum efficiency, based on the discovery of a meaningful order expressed by the plane.

Meaning of elementary figures
In a network, the size of the points, the length and width of the lines, the size and shape of the areas, theoretically have no meaning on the plane. *Their presence merely signifies presence* of an element or of a correspondence between two elements. On the other hand, the correspondences may be oriented in one direction or another, which can be expressed either by an "arrow" or by a meaning attributed to certain parts of the plane, or by both.

Unity of the image
Since the network occupies the two planar dimensions, any other components must be represented by retinal variables. As a result, *a network can only be perceived as a single image when two components are involved:* one forming the network, and a second represented by an ordered retinal variable.

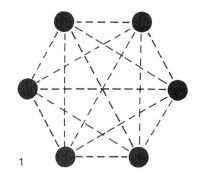

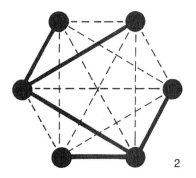

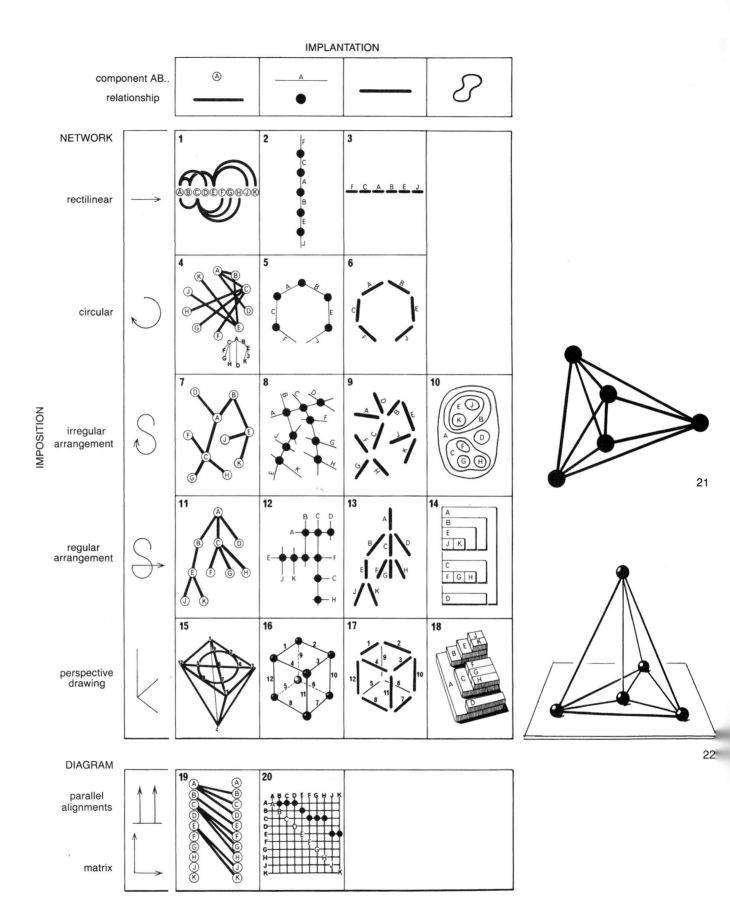

CONSTRUCTION AND TRANSFORMATION OF A NETWORK

Consider the data: A is father of B, C, D; C is father of F, G, H; B is father of J and K. A genealogical tree depicts the set of correspondences (kinship relations) linking the members of a family, that is, the elements AB . . . of a group of individuals. A flow chart represents the set of relationships linking a series AB . . . of preestablished functions. These datasets are constituted by the relationships among the elements AB . . . of a single component. When such information is transcribed onto the plane, it produces a NETWORK. For the same information, various constructions are possible.

Available graphic means

We have seen that graphic representation *(implantation)* involves three elementary figures: the point, the line, the area. The elements AB . . . of a component can be represented by points and the relationships by lines, or conversely. In certain cases, the lines alone can represent both elements and relationships. The same is true for areas, when the relationships are inclusive. Furthermore, the utilization of the planar dimensions *(imposition)* enables us to organize these figures in a rectilinear or circular manner, or order them along one of the two planar dimensions. A perspective drawing can suggest depth and situate the network in a "three-dimensional" space. Finally, any network can be constructed in the form of a diagram, provided the component AB . . . is represented twice. In the set of figures opposite, implantations and impositions are combined to illustrate the various possible constructions of a network.

Types of network construction

A rectilinear construction (figure 1). This type of construction orders the elements. The relationships are curves and can be distributed from one part to another on the line. This construction is useful when AB . . . has an ordered characteristic (page 273) or when the nature of the relationships justifies a distribution in two groups.

The constructions in figures 2 and 3 are possible only in a series without ramifications.

A circular construction (figure 4). By arranging the elements AB . . . on a circle, any relationship can be transcribed by a straight line. This is the construction which produces the least confusing image, whatever the number of intersections stemming from the raw data. Consequently, it is useful for a first graphic transcription, enabling us to pose visually the problem of simplification. The constructions in figures 5 and 6 are governed by the same principles as those in figures 2 and 3.

Irregular arrangements. One can forsake rectilinear or circular alignment and use the entire space to arrange the elements. In figure 7, the relationships are represented by lines, the component AB . . . by points. In figure 8, the opposite is true. In figure 9, lines alone represent both. In figure 10, the example chosen utilizes the properties of area representation. By expressing the notion of inclusion, areas enable us to transcribe all the relations in the information being considered. They can either, as here, express both the element and all the successive groups which it engenders, or group elements among each other (see page 282).

Regular arrangements. In the preceding examples, neither of the two planar dimensions was meaningful. If one considers the vertical direction to represent the order of generations, one arrives at the classic form of the genealogical tree (figure 11). The ordered meaning of the plane facilitates comprehension of the image in contrast with figure 7. A line-point inversion leads to figure 12, in which the series of generations is represented successively on one or the other of the two planar dimensions. Lines alone are utilized in figure 13, which, in this case, appears to be the simplest solution (see also page 276). Areas can be constructed in an ordered manner and can trace out images which are easily accessible, as in figure 14.

Perspective drawings. Whatever the arrangement of five points on the plane, their correspondences will produce at least one meaningless intersection (figure 21). However, if we suggest three-dimensional space, it is possible to avoid any intersection (figure 22). If the drawing creates a sense of volume (figures 15–18), it will also suggest that the lines do not cut across each other. The impression of depth is obtained by utilizing various perceptual properties (see page 378). In figure 15, the elements 1, 2, 3, . . . of the component are represented by points. The set of relationships is simplified considerably when these same elements 1, 2, 3, . . . are represented by lines (figures 16 and 17). Areas can also be situated in three-dimensional space (figure 18), illustrating the stratification of generations, already suggested in figure 14.

Diagrams. Any network can also be constructed in the form of a diagram. One merely represents the component twice and considers the elements AB . . . as starting points for relations leading to arrival points AB . . . . Two constructions are possible. Parallel alignments (figure 19) are useful for comparing orders (see pages 248 and 260). A matrix such as that in figure 20 allows for permutations in rows and columns and can thus lead to the simplification of complex information by diagonalization.

The transformations of a network

The simplest, most efficient construction is one which presents the fewest meaningless intersections, while preserving the groupings, oppositions, or potential orders contained in the component AB.... In the absence of a simple and general calculating procedure, which would permit us to define the optimal construction and arrangement of the elements for given information, it is necessary to pose and resolve most problems graphically. When the information is not too complex, experience shows that it is the circular construction which affords the best visual point of departure. For example, it permits us to discover that the order ABEJKDHGFC eliminates meaningless intersections (figure 4) or to see that an arrangement (page 278) produces a greater simplification. It also enables us to reconsider the conceptual relationships contained in the component AB... (see figures 5 and 6, page 274).

When the information is highly complex, the permutable matrix (figure 20) affords the means of proceeding to an initial simplification prior to construction of the network.

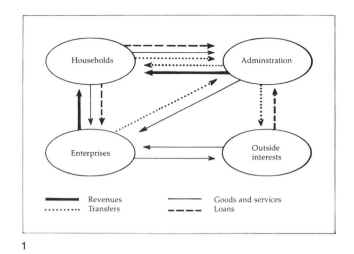
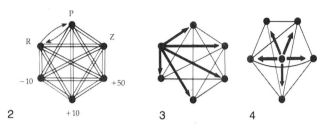

EXAMPLE OF STANDARD CONSTRUCTION

Figure 1 shows the schema of an economic circuit (after Edmond Malinvaud, *Initiation à la comptabilité nationale* [Paris: Imprimerie nationale, 1957]).

Analysis of the information:
≠ four groups of economic entities (households, enterprises…) and
≠ four types of vectors (directed relationships).
Each group (element) is placed in a circle. The correspondences are represented by straight lines, which are differentiated by a redundant combination of value, shape, and texture.

FIRST EXAMPLE OF TRANSFORMATION

Elector movement in 1953 (after P. Vieille and P. Clément, "L'Exode rural," in *Etudes de comptabilité nationale*, no. 1 [Paris: Ministère des Finances, 1960]).

Information:
O six categories of communes (Paris [P], suburbs [Z], rural communes [R], communes with more than 50 000 inhabitants [+50], more than 10 000 [+10], less than 10 000 [–10]), joined by vectors, according to
Q of moved electors (net movement), which weights the relationships;
Q of total population, which weights the categories of communes;
≠ two age classes (21–29 and 45–59).

If the objective is to compare the two age classes, it is advisable to construct one image for each.

The standard construction (figure 2)
This construction gives us an initial look at the structure of the weighted relationships (figure 3) and leads us to place the rural population in the center (figure 4). We then utilize a regular arrangement, producing two images (figures 5 and 6), which are in stark contrast.

But, the attentive reader is not fully satisfied by these constructions, since the central position of the group of rural communes makes it a privileged group, overwhelming all the others. The concept "categories of communes" is, after all, a concept which is ordered from the largest to the smallest commune, and the reader unconsciously feels the need to compare this order to the quantitative order of elector movements and their directions.

A linear construction (figure 7)
Here one can order the categories from top to bottom by commune size (not by the total population of each group), thus giving a meaning to the plane. It is then possible to divide the relationships into "ascending" (toward the top) and "descending" (toward the bottom) ones. The comparison of the two images is striking, and each group of communes receives a visual characterization which is no longer determined by a slighted or privileged position on the plane, but by the total content of the information. Thus we can see from figure 7 that the direction of emigration for young persons (21–29 years old) from all groups of communes (except Paris), and not only from rural communes (as it appeared in figure 5), is directed towards the larger communes.

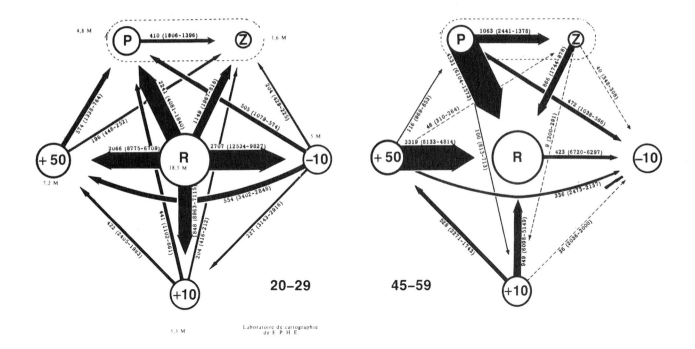

5   6

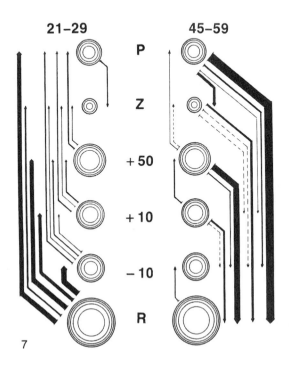

7

273

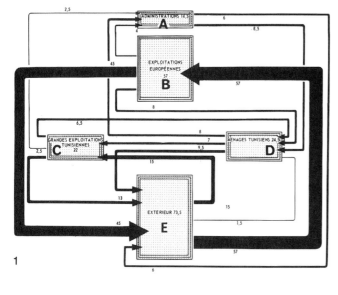

SECOND EXAMPLE OF TRANSFORMATION

Financial exchange among five main groups of economic entities in a market economy (after J. Cuisenier [Laboratoire de Cartographie, E.P.H.E.]).

Analysis of the information:
≠ five groups of economic entities (administration [A], European farming [B], Tunisian farming [C], Tunisian households [D], and outside interests [E]), linked by vectors (directed relationships)
Q of francs (millions) weighting entities and relationships.

In figure 1, the groups are arranged in such a way that the inputs are situated on the same side for each group. This construction produces no visual simplification. In order to reduce the number of intersections and discover the simplest construction, it is first necessary to study the network of relationships. The *circular construction* (figure 2) is generally the one which permits the best visual posing of the problem. It leads to a first simplification (figure 3), from which we discover that *the arrangement* in figure 4 eliminates all intersections. It is then necessary to consider the elements being represented and determine whether their meanings create distinct groups which planar position could highlight. This is the case here. The five elements are distributed in two groups: farming operations and general economic entities.

Consequently, in spite of its two intersections, figure 5 is superior to figure 4. The arrangement in figure 5 reveals the two groups and allows us to compare the two types of farming operation (figure 6). Figure 4, which combines the elements of the two groups and does not allow a clear comparison of the two types of farming operation, is not so accessible, even though it involves no meaningless intersections.

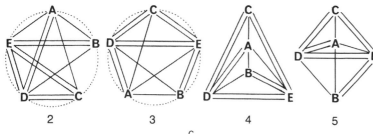

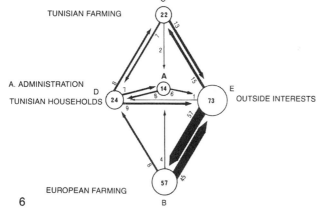

EXAMPLE OF AN IRREGULAR ARRANGEMENT

Figure 7 depicts a squadron with good morale (A) and a squadron with poor morale (B), after a study on leadership and isolation by John G. Jenkins (1945).

Analysis:
≠ personnel hierarchy (commander, second in command, men, individuals outside the squadron)
≠ two types of relationship (friendship: black line; enmity: broken line)
≠ two squadrons.
The objective is to compare the two networks of relationships.

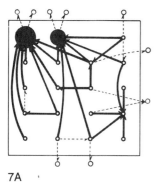
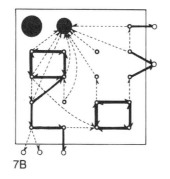

The arrangement of the personnel is similar in each squadron and occupies the entire plane. The contrast between the two squadrons is perceptible only if we can *distinguish the networks* of friendship from the networks of enmity.

This visual differentiation is obtained by *retinal difference of value.*

The contrast between the networks of friendship is striking and underlines the broken nature of squadron B. One also notices an equally perceptible contrast between the networks of enmity.

APPLICATION OF NETWORKS TO CLASSIFICATIONS

As we saw on page 248, the distance between two rankings can be defined by the number of their "inversions." We can visually portray the distances among *n* different rankings by a network in which each ranking is represented by a point, and the distances by lines, one line representing a distance of 1, two lines (end to end) representing a distance of 2 . . . . Figure 8, for example, shows a network of the six possible rankings of three objects. The distances are added: from ABC to CAB, the distance is 2. Note, however, that the distance is represented by the number of links. The length and shape of the fines are not meaningful, so figure 9 has the same meaning as figure 8.

Reading a network

The above example forms a simple image, but whenever the number of objects to be ranked exceeds three, the figure can no longer be interpreted on the overall level. It becomes necessary to read it on the elementary level, to follow the shortest path point by point in order to pass from one ranking to another. The construction of such networks requires several precautions, since it is easy to arrive at erroneous figures. For example, four objects A, B, C, D, furnish twenty-four possible rankings. Represented as in figure 10, certain lines join rankings which are distant by five inversions!

In order to construct the correct representation (figure 11), we adopt the following method: Take a ranking. Place its three immediate neighbors at a distance of 1, then the immediate neighbors of these, etc. The distances are to be added on the shortest course, provided we consider the distance between two rankings to be represented by the number of links on the shortest chain (based on the number of links) between two rankings.

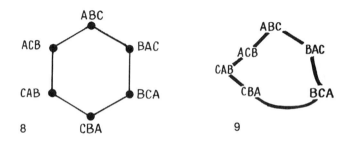
8 9

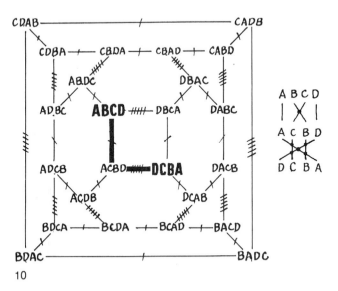
10

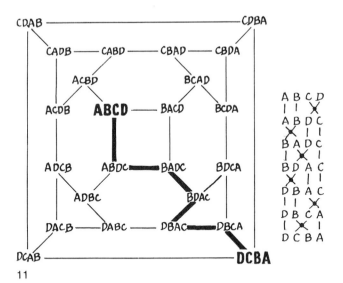
11

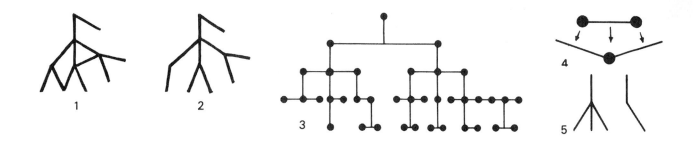

TREES

A network in which there is only one possible path to go from one point (node) to another is a tree. Figures 1 and 2 are both networks, but figure 2 is also a tree. The genealogical tree is the most common example of this geometric figure, but we also find it in classifications, structural analyses of property and language, etc.

Points and lines

We have examined (page 270) the main possible constructions of a tree. The classic construction (figure 3) has the advantage of simplicity in drawing. However, it is not the simplest to evaluate visually, and any increase in the number of relations limits its possibilities.

Replacing points by lines, as in figure 4, simplifies the figure and permits us to eliminate the points, provided the lines are delimited by changes in direction (figure 5).

This principle is often utilized in kinship networks. The conventional form (figure 6) can be simplified, as in figure 7, where a man is represented by a dark line, a woman by a broken line (or a thin line). Complex forms, such as that in figure 8, finally become readable (figure 9). (After J. Cuisenier, "Pour l'utilisation des calculatrices électroniques dans l'étude des systèmes de parenté," in *Calcul et formalisation dans les sciences de l'homme* [Paris: C.N.R.S., 1968], 31–46.)

Circular trees

A substantial increase in the number of relations leads to a tree spread out over all directions of the plane (figure 10). However, it has the disadvantage of not making the various stages of the tree perceptible. A first ordering is provided by the construction in figure 11, which uses an ordered circular web. This same principle is applied to the genealogical tree of Genghis Khan and his descendents (after M. Toptchibachy, "Rachid-ud-Din, la réunion des chroniques," unfinished work). Figure 12, which uses the form depicted in figure 3, produces a good deal of visual confusion; whereas figure 13, based on the form shown in figure 11, displays the genealogical sequence more clearly. Figures 12 and 13 are reductions of the original drawings, 90 cm in diameter, which include the identification of 1230 persons. Genghis Kahn is singled out by a circled point.

However, the order of generations is only detectable on the elementary level of reading (it is difficult, for example, to locate the father or the brothers of Genghis Khan).

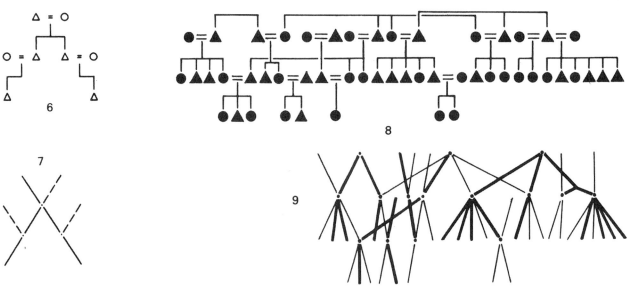

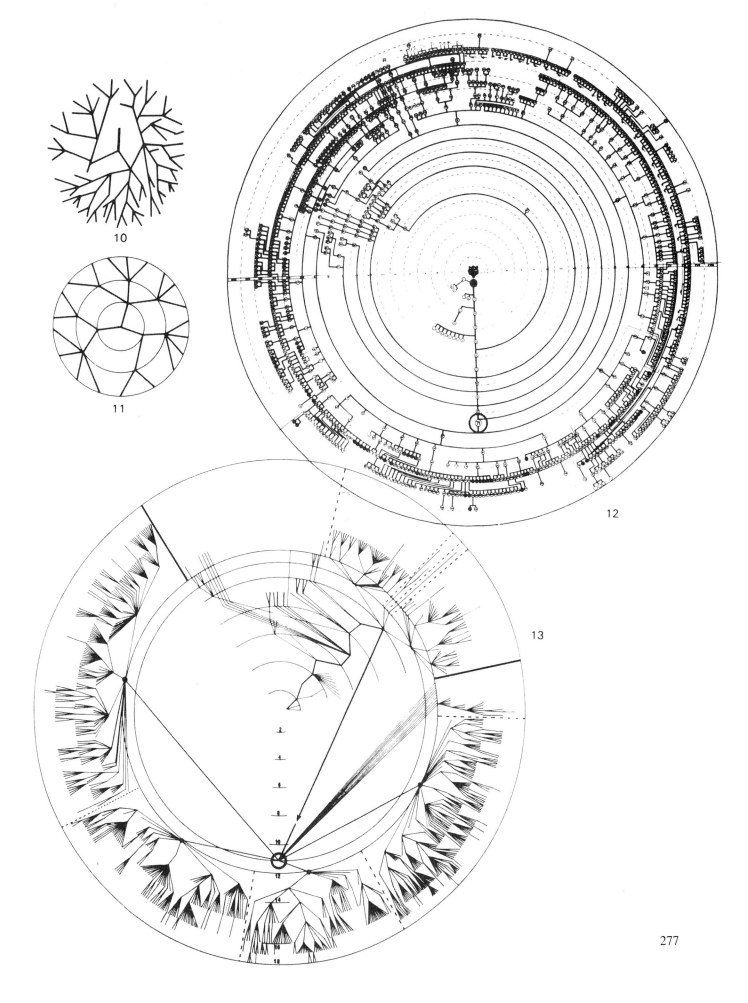

Seeking a perceptible order

Figure 1, which represents only one sector of the circle, renders the succession of generations more perceptible. It approximates a plane ordered from top to bottom, but A appears to be more recent than B!

In the final analysis, it would seem that figure 2, in which the plane is ordered and the lines denote individuals, represents the most efficient construction for an ordered genealogical tree (A is visibly anterior to B). This form can depict the length of each individual's life and can be extended to numerous generations.

Very numerous populations can lead to a square (figure 3), which preserves all the properties represented in figure 2.

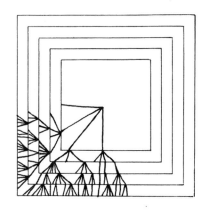

3

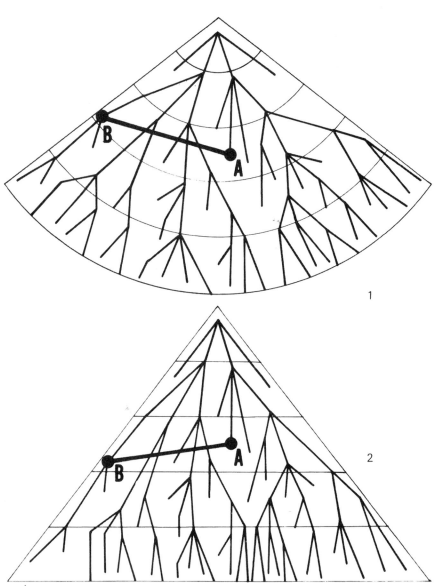

1

2

Application of trees to linguistic analysis

A visibly ordered tree has also been applied in recent methods of linguistic analysis. One such example is the "stemma" of L. Tesnière (*Eléments de syntaxe structurale* [Paris: Klincksieck, 1958]).

All thought is multidimensional and results from the convergence of several concepts. The nature of verbal language resides in the linearization of this convergence and its transcription in a temporal sequence.

L. Tesnière attempts to break down the linearity of the sentence and rediscover the basic thought, whatever the language. He proposes constructing a sentence in the form of a "stemma," that is, according to a tree which portrays, not the sentence extended over time (where the grammatical sequence of elements changes from one language to another), but the concepts expressed, in their universal attributional relations. He takes a ≠ component, which categorizes the words in four species—substantives, adjectives, verbs, adverbs—and represents them by signs—O, A, I, E—that is, by a variation in shape.

He uses the two planar dimensions to construct the relations constituting the thought to be expressed. The stemma will thus be practically constant for a given thought, no matter what the language (figure 4). The stemma is a means of analyzing a thought and comparing it to another thought, but it does not permit comparing different languages or grammars in terms of the same thought.

In order to compare languages, it would be necessary to visually restore the linear regime characteristic of each grammar. This is the tendency of studies like those of Y. Lecerf and P. Ihm (*Eléments pour une grammaire générale des langues projectives* [Brussels: Euratom]). These authors give an ordered meaning to the horizontal dimension of the plane—the order of the words in the sentence—and Tesnière's stemmas become trees which are ordered along one dimension.

However, the properties of the image would no doubt allow even more. All the images of the same thought can be compared, provided they are constructed on a plane which has constant meaning. Consequently:
– if one gives to the vertical dimension of the plane the meaning ≠ (verb, adverb, substantive, adjective);
– and since the order adopted vertically is preserved across all the languages, the signs become unnecessary;
– and if one also adopts a constant spacing between the words on the ordered horizontal dimension of time;
then all the images become comparable, since the two planar dimensions have a homogeneous and constant meaning (see figure 5).

The transformation of the images and their particular shapes represent the differences in grammatical structures for the expression of the same thought. The images, and therefore the grammars, can be grouped by types of resemblance and thus foster numerous experimental classings.

Incidentally, one can compare these methods with those outlined on page 265 and observe that, in fact, we have a network constructed on an ordered plane, which amounts to having a "map." The drawings in figures 4 and 5 here are "grammatical maps."

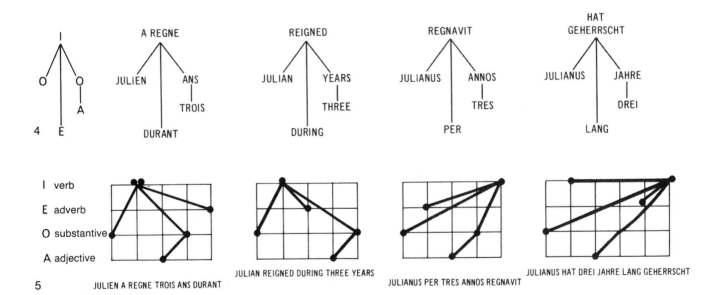

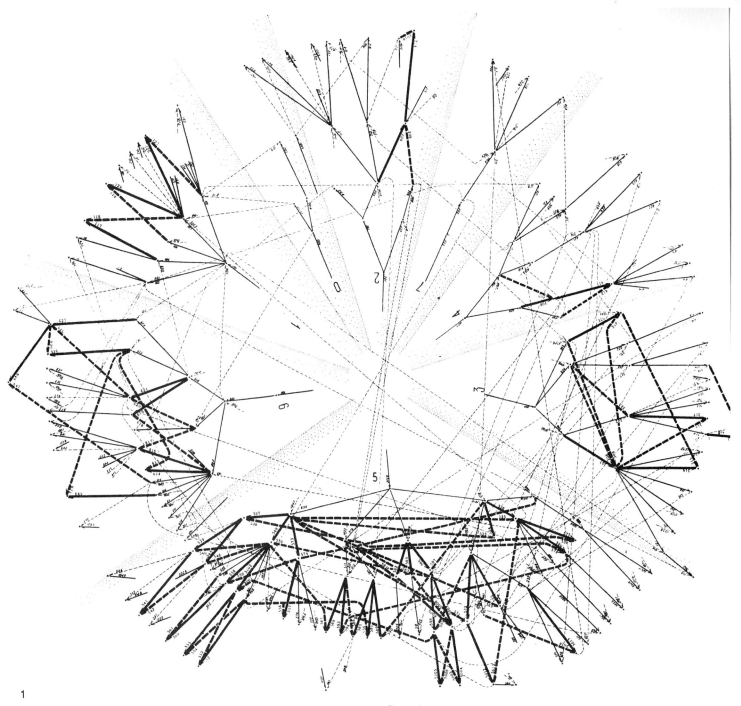

1

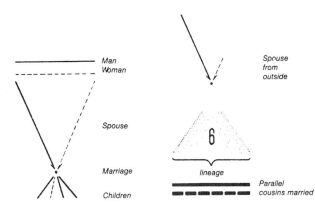

Superimposition of two trees

Consider the genealogy of the populations of Djebel Ansarine (after J. Cuisenier, "Endogamie et exogamie dans le mariage arabe," *L'Homme* [May–Aug. 1962], pp. 80–105). The representation of both sexes in a genealogy involves two trees. In order to superimpose them, the designer must construct one and depict the other by a "trellis."

In figure 1, the male tree constitutes the base of the representation. This is consistent with the tribal conception of the society being studied and highlights the different lineages (note the visual differentiation of the lineages obtained by shading, which creates an area rather than an additional line). The women form a complex trellis which, to avoid confusion, must be differentiated visually, by being

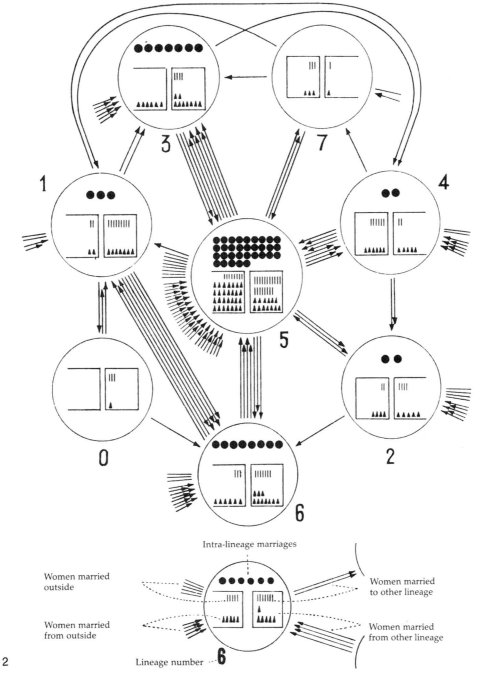

transcribed in a light value. In spite of its complexity, this tree highlights the importance of alliances between parallel cousins (underscored by lines of darker value). This trait characterizes the matrimonial regime of the population and distinguishes it, for example, from the Eskimo regime, as described by J. Malaurie in collaboration with Léon Tabah and Jean Sutter ("L'Isolat eskimo de Thulé," *Population 7*, no. 4 [Oct.–Dec. 1952]: 675–692). Thus, the objective of genealogical trees can and must go beyond the elementary level of reading where it constitutes an inventory of all the kinship relations. As an overall image, the tree can be used either to compare family structures as they differ in space and time, or to compare family structures with socioeconomic structures that are based on the possession of wealth.

Network and exterior relationships: Supplementary components

The preceding information, no longer considered over time, but for a given period, permits us to identify the nature and extent of relationships among given lineages, as well as between these lineages and the outside world. The information becomes:

≠ eight lineages plus the outside world and relationships, according to

Q of directed relationships (women whom marriage causes to leave or enter the lineage).

This is a network (figure 2). Each group (lineage) contains various notations, translated by conventional signs, which recall and summarize different aspects of the relationships for each group.

281

AREAS, INCLUSIVE RELATIONSHIPS

Consider the constitutions of 1795 and 1875 (after C. Morazé, P. Wolff, and J. Bertin, *Nouveau cours d'histoire* [Paris: A. Colin, 1948]).

The constitutional organization of a country, with its different seats of power and their relationships, forms a complex network of relations of incidence or inclusion which are more or less hierarchical. Graphic representation allows us to emphasize their characteristic traits, by combining inclusive relationships (elements in areas) with oriented relationships (vectors).

The constitution of 1875 (figure 1) appears as homogeneous and hierarchical. Without our needing to define it, the plane suggests a meaning which is ordered from top to bottom and at the same time forms categories on equal levels horizontally.

At a single glance, the constitution of 1795 (figure 2) illustrates the politics of laisser faire characterizing the *Directoire*. This example shows that the plane can visibly suggest a meaning of "nonorganization" or disorder.

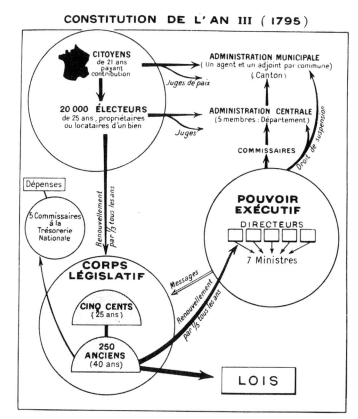

PERSPECTIVE DRAWINGS

The lattice of tetrahedrons (figure 3), developed by Geneviève Guitel ("Compte rendu," *Académie des sciences* 5, no. 235 [November 1952]: 1274), is a spatial figure which allows us to pass from a tetrahedron of type O (six acute elements) to any other type of tetrahedron by the successive addition of a single obtuse element. This is a "three-dimensional" network.

The "totemic operator" of Claude Lévi-Strauss *(La Pensée sauvage* [Paris: Plon, 1962]) presents a microcosm of the problem of categorization. It involves two contrasting trees, whose relationships intersect two by two (or three by three, four by four, according to the number of paths at each node).

Inscribed on a two-dimensional plane (figure 4), it produces numerous meaningless intersections.

Imagined in "perspective" (figure 5), it avoids intersections. In both figures 3 and 5, the suggestion of volume results from a variation in the width of the lines, which creates the illusion of different planes in depth (see also page 378).

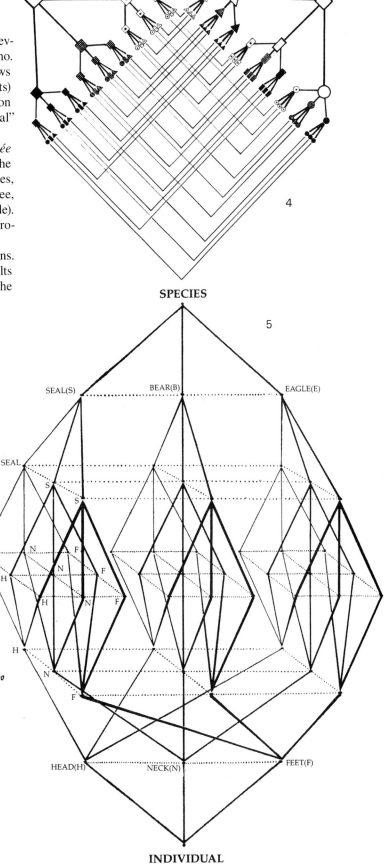

3   Lattice of tetrahedrons

III. Maps

Definition
A graphic is a geographic "map" when the elements of a geographic component are arranged on a plane in the manner of their observed geographic order on the surface of the earth. Let us recall that a geographic component can always be constructed in a linear manner, thus taking the form of a diagram (see page 51).

Since geographic space is continuous, one can always imagine a more precise survey, a more refined enumeration, which means that geographic information always results from human choice. It cannot, therefore, be a question of "accurate maps," but of a degree of cartographic accuracy, involving two levels: the degree of precision of the information, provided by the topographer or researcher; the degree of precision of the representation, produced by the designer.

The construction and meaning of elementary figures
A map is an ordered network. One can either represent this order (geodesy, topography) or reproduce the survey's data. The geographic elements can take the form of POINTS, without theoretical area (geodetic points, junctions, crossroads), LINES, also without area (coast lines, river axes, natural or human boundaries), or AREAS, which *do* have perceptible dimensions (cities, lakes). Since any representation requires a minimum amount of planar area in order to be visible, it is obvious that:
- only areas are theoretically representable;
- lines and points are always conventions. However, since they signify a position area, they can be made to vary in planar area when representing additional qualitative or statistical components. Of course:
- positional meaning naturally relates to the center of the point or the axis of the line;
- areas cannot vary in size without varying in position. Thus in order to express a "thematic" component, their space must be filled in with points or lines which will portray the variations within that component.

Unity of the image
The geographic network occupies the two dimensions of the plane. Consequently, the other components of the information must rely on "retinal" variables for their graphic transcription. Thus a map can only be perceived as a single image when two components are involved: the geographic component and a second one, represented by an ordered retinal variable.

In any problem involving more than two components, a choice must be made between the construction of several maps, each one forming an image, and the superimposition of several components on the same map.

A series of diagrams (graphics involving two components or more) superimposed on a geographic base will be called "CHARTMAPS." They are formed of as many images as there are diagrams (see page 119).

Graphics which deform the geographic base in order to display a nongeographic component will be called "CARTOGRAMS." They can form an image, but at expense of deforming the geographic shape with which the reader is familiar (see page 120).

CHARACTERISTICS OF THE GEOGRAPHIC COMPONENT

The reading of a map
The irregular shape displayed on the graphic immediately indicates the presence of a geographic representation. Without reading the title, we know that a map is involved. But in order to assimilate its content, we must be able to recognize the components of the information and IDENTIFY:
- the space represented: What region is involved?
- the invariant: What does each line, each point depict?
- the "thematic" components: What is the difference between the red lines and the black lines?

In a diagram, the WORDS of the title and the terms written along the planar dimensions provide these identifications.

In a map, it is the title and legend which enable us to identify the invariant and the thematic components; these have a standard arrangement and formulation, as we have seen (page 19).

But we must also be able to identify the *geographic order* represented by the plane.

Geographic order
Represented by a straight line (diagrams, tables) the geographic elements have no particular properties. Like other components, the geographic component is characterized by its level of organization and its length (number of categories). It lends itself to classing, permutation, and simplification.

Transcribed according to the geographic order as observed on the surface the earth, the geographic elements trace out a map, that is, an ordered network.

Cartography, the planar translation of geographic order, is the sole means of simplifying the geographic component as a function of spatial relations. There is no other system which can accomplish this "regionalization" of space.

Properties of geographic order

Among the ordered concepts on which knowledge is based, geographic order has unique properties:

It is visible, which means that the same sense, sight, governs its perception as well as its graphic translation. This fact, the transcription of a space (geographic) onto a space (sheet of paper), has made the map the simplest, the most comprehensible (and thus the most ancient) of graphic representations.

It has two dimensions, and for this reason it provides the most practically divisible component, that in which we can identify the greatest number of categories without having to use complex apparatus and definitions. For example, on the most elementary level, time affords three unambiguous categories: before, during, after. On this same level planar space offers five: in front, in the middle, at the back, on the left, on the right.

It is constant on the scale of human time. This property, linked to the preceding one, makes geographic order the *component which offers the greatest number of constant practical reference points*.

It is universal; that is, any identification based on its reference points, whoever the author, gives nearly the same visual shape.

Whatever the number of diagrams previously observed, a complete identification must be undertaken for each new diagram.

On the other hand, cartographic observations, by repeating themselves, by fitting into a universal structure, progressively facilitate the effort of identification, and construct a unique mental totality, which is homogeneous and ordered: the geographic reference system.

Thus, whenever problems of memorization are posed, geographic order remains the most privileged domain, even though spatial relations might appear, particularly in human contexts, to be less significant than age relations, or social and professional relations.

These properties of identification also make geographic order the most privileged domain for problems of "visual communication."

Recent experiments seem to indicate that, due to automated means of drafting and especially of graphic documentation, graphic processing is at least as efficient as purely mathematical processing, whenever a large number of concepts is involved and whenever all known relevant information can potentially be introduced into the problem.

Cartography is fast becoming the most practical comparative basis for integrating and reducing the vast reaches of modern information. If the *geographer* seeks to define a domain in terms of space in order to discover "regions," and if the *historian* seeks to define a domain in terms of space and time in order to discover "civilizations," the *cartographer* can be said to use space in order to "inform" us about all conceivable domains. The cartographer is the "mathematician" of a nonmathematical system of signs.

A. External geographic identification

External identification is the first stage in the reading process; a map is only useful when the reader can situate the space which it represents within the field of prior geographic knowledge.

What space is involved? This question entails two answers: a situation and a dimension.

Situational identification: projections

A situational identification enables us to relate the space represented by the map to a known geographic shape.

This can be achieved through several means.

Words

A noun figuring in the title can characterize a city, a river, a region, or a sea. For the reader, words vary greatly in familiarity: France; America; Dakar; Szechwan; Sidobre. . . . Geographic indexes allow an unfamiliar word to be related to a shape (region) which is already familiar.

The system of meridians and parallels

This is a universal convention, well known to the average reader. It can be used to relate a point on the map to a larger region that is already within the reader's field of knowledge.

Identification by hundredths

This system is based on the geographic coordinates, but, within the broadening framework of modern research, it enables the reader to situate a given map in a more simple and rapid manner. It is analogous in principle to the rectangular grids used by the military on topographic maps.

The terrestrial sphere is cut up into 432 areas of 10° latitude and 15° longitude. These areas are numbered in latitude from 1 to 9 in the northern hemisphere and from 01 to 09 in the southern hemisphere. In longitude they are identified from A (30° west of Greenwich) to X. Each one of the 432 areas is thus identified by two signs (for example, 7B) or by three signs in the southern hemisphere (for example, 07E). Two additional numbers can designate a rectangle corresponding to a hundredth of one of these areas (for example, 7B–28).

Another two numbers can specify a hundredth of the preceding rectangle (for example, 7B–28–44). Two more numbers would permit situating a point down to the kilometer!

In practice, the focal point of a given map (one involving the northern hemisphere at 1 : 5 million, for example) can be situated in a rectangle whose average size will be about:
22 cm for an identification having two signs
2 cm for an identification having four signs
2 mm for an identification having six signs.
This is the most convenient system for rapidly and precisely situating the great diversity of data stemming from modern research.

A table of correspondences (or a simple map) would allow the reader to pass from the classic system of geographic coordinates to an identification by hundredths.

The situational map

This is a map which is on a smaller scale than the one in question and situates it in relation to an easily recognizable space.

Shape

In the final analysis, all identification of a strictly "cartographic" nature amounts to the recognition of a shape. Certain characteristic shapes are sufficiently familiar to furnish in themselves all the elements of identification. Well-known images, such as the map of the British Isles, Japan, or France, have a degree of universality which elevates them to a symbolic level.

However, the shapes to the average reader are not all that numerous, and the smaller and more distant the regions, the more precise the external identification will have to be. In fact, we still encounter sketches and even publications in which a situational identification is impossible. Such a map is devoid of "information."

On the other hand, the larger the space the more the exterior shape of the map suffices for its identification. The largest space, the ultimate reference shape, is, of course, the terrestrial map or "planisphere." Unfortunately the transposition of the globe onto the plane has no single solution; it involves a convention which is always debatable. This means that the planisphere is the only map which cannot produce an image of constant and universal shape. A succinct study of the **planisphere and its projections** is therefore indispensable to any further discussion.

Peirce's periodic projection

PROJECTIONS

The transposition of a sphere onto a plane is impossible without bending or tearing; the planar representation of a spherical surface will always lead to a *deformation* of the relative arrangement of the elements of this surface (figure 1). The study of the nature of this deformation and the means of reducing its impact fall within the domain of mathematical cartography.

However, since this deformation can be a source of error in the interpretation of geographic images, we will discuss it here. It is well to be forewarned of its potential consequences and to know the principles for making an eventual choice among several projections (among several planispheres, for example) in relation to future use.

Deformation

This notion can be summarized by three propositions. To arrive at these we take a plane that is tangent to a sphere at point P (figure 2). Then, aside from point P, any circle of very small spherical area:

(1) can be represented on the plane by a circle, but of larger area,

(2) can be represented by an equivalent area, but of an elliptical shape,

(3) and any large-sized circle (except for those centered on P) will be deformed on the plane.

A system corresponding to the first proposition is called "conformal"; at any point on the map, the angles are similar to their corresponding ones on the sphere, but the area deformations are considerable. In these maps, the meridians and parallels cut each other at right angles (this can also occur in nonconformal systems: for example, figure 10, page 294).

A system corresponding to the second proposition is called

"equivalent"; any area on the plane is equivalent to the corresponding spherical area, but the angular deformations are considerable.

A system which establishes a balance between areal and angular deformation is called a "compromise" (or aphylactic) projection.

But in all cases where a considerable portion of the sphere is considered, as in proposition 3, there is always deformation of *directions* and *distances* (except for those originating from point P).

Consequences for small spherical areas
For "plans," such as regional maps, these deformations are perceptible only in a precise survey of positions (in geodesy, for example). They are not important in the comparison of images, when these comparisons involve the same region. It is sufficient that the projections be of the same nature and on the same scale. For medium-sized regions, the use of two different systems can render comparisons difficult, as with the Mercator and conic projections in figure 3. The difficulty of comparing different regions on the same map increases in proportion to the deformations inherent in certain systems; for example, the Cameroons, West Germany, and Finland on a Mercator map (figure 4) are in fact of similar size and should be arranged in a different order (figure 5).

To control the equality of scales between two regions, it is sufficient to measure the distance between two parallels. A difference of 4° in latitude, for example, must have the same length in each region (figure 6). In effect, at any point on the globe:
1° in latitude = 111.11 km (40 000 km/360); 1′ in latitude = 1852 m (111.1/60). This is the nautical mile or marine mile (not to be confused with the statute mile or terrestrial mile, which is equivalent to 1609 m).

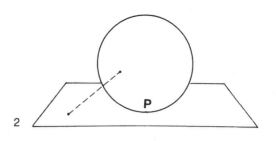

2

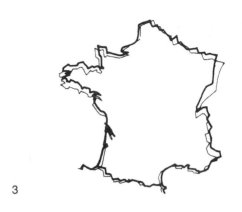

3

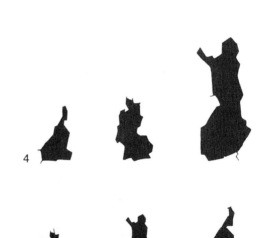

4

5

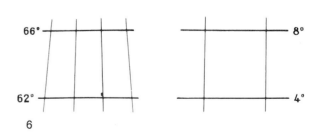

6

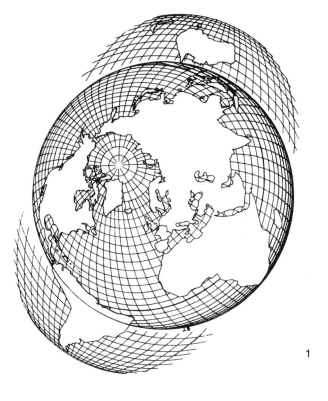

Consequences for large regions

For a continent, an ocean, or a hemisphere, the deformations become appreciable and will have a direct effect on any problem involving a precise geographic enumeration. Depending on the purpose of the map, one can choose to retain either the angles (as with airway or maritime navigation systems based on radio wave fixes, such as the "Loran" and "Decca" systems), or the areas (as with point or area enumerations of various "populations"). Linear problems, such as those involving distance or flow, will have no satisfactory solution and will require the adoption either of compromise systems, or of perspective projections. Indeed we know that the eye restores the normality of shapes seen in perspective, provided the perspective is natural and its laws obvious.

A correct perspective can thus restore a large spherical body and give an acceptable, though not measurable, perception of it for shapes and distances. However, the technical conditions required—construction, shading, reproduction—are difficult to fulfill (figure 1).

Moving the point of view can create striking positions and define interesting systems, but it will not fulfill the conditions of natural perspective perception.
- Placed at the center of the globe (figure 2), the point of view defines a *gnomonic* projection, in which any great circle arc is represented by a straight line.
- Placed on the globe (figure 3), the point of view defines a *stereographic* projection, which is conformal.

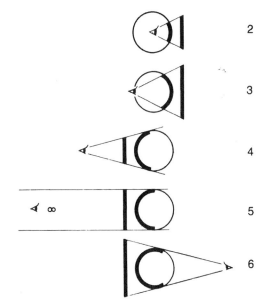

When the point of view is exterior to the sphere, the perspectives become natural, "photographic," as in figure 4. The perspective becomes orthogonal when the point of view is removed to infinity, or "seen from the moon" (see figure 5).

To go beyond the hemisphere, one can use a transparent globe and proceed as in figure 6. If the point of view is placed on the side opposite to the observed side, then the geographic shapes need only be inverted on the globe to be seen in normal fashion. The transparent globe permits us very rapidly to trace an outline for any of these perspectives, whatever the geographic center (figure 1).

The shape of the regions

If the region to be represented is of an elongated shape, as is the American continent or an airway route, one can profit from the fact that the plane can be transformed into a cone or a cylinder without tearing. Point P (figure 7) becomes a line, along which there is no deformation (figures 8 and 9). Systems based on the principle of a point are called "azimuthal" or "zenithal" (figure 7). All perspectives are azimuthal. When systems are based on the principle of a cone, they are called "conic" (figure 9). Those systems based on the principle of a cylinder are called "cylindrical" (figure 8). The tangent can be replaced by the secant (figure 13), providing two contact lines for conic and cylindrical systems. In this case these systems are also qualified as "secant" (figures 14 and 15).

Of course, these lines are independent of the position of the poles and the equator, and all the above constructions can be either conformal, equivalent, or compromise.

Thus, a different system can be chosen according to the shape of the region being considered.

A *region inscribed in a circle or a square leads to an azimuthal system*. This is the case with maps representing large bodies, in atlases, for example (figure 10).

When the line involved is a great circle arc, cylindrical systems are the most adaptable. For example, the Mercator projection, which is cylindrical and conformal, suits the air route in figure 16 (map by Kahn), the American continent (figure 17), the map of the Pacific (figure 18) (after L. Strohl), or the circular zone of classical civilization (figure 19). But the Mercator system, which removes the poles to infinity, cannot represent the entire sphere, and by any logic the Mercator map cannot be termed a "planisphere."

A cylindrical system is utilized for the map of the ancient world (figure 11), which also incorporates a "secant" cylinder and several compromises in the marginal regions.

When the line involved approximates a small circle arc, conic systems are generally used. For example, they are the bases for numerous topographic maps of the temperate zone countries and for the atlas maps derived from them (figure 12).

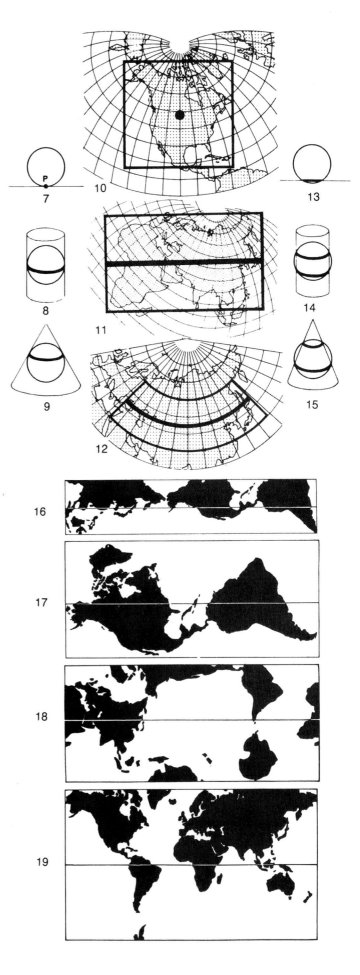

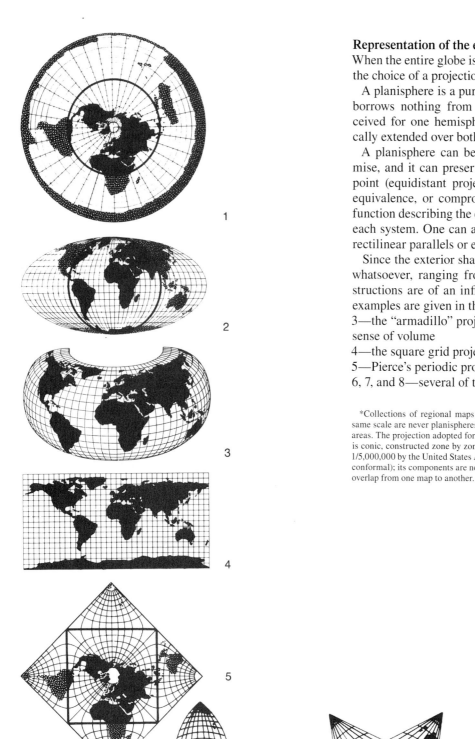

Representation of the entire globe

When the entire globe is to be represented in a single image,* the choice of a projection is of fundamental importance.

A planisphere is a pure mathematical construction, which borrows nothing from normal perception. A system conceived for one hemisphere can, for example, be geometrically extended over both hemispheres (figures 1, 2, and 5).

A planisphere can be conformal, equivalent, or compromise, and it can preserve directions or distances around a point (equidistant projections). The laws of conformality, equivalence, or compromise along with the mathematical function describing the exterior shape, are the parameters of each system. One can also add further constraints, such as rectilinear parallels or equidistance from a point.

Since the exterior shape of a planisphere can be anything whatsoever, ranging from a circle to a star, possible constructions are of an infinite number. Several characteristic examples are given in the following figures:

3—the "armadillo" projection E. Raisz, which introduces a sense of volume
4—the square grid projection (or *plate carré* projection)
5—Pierce's periodic projection, which is given on page 288
6, 7, and 8—several of the many star projections.

---

*Collections of regional maps that are meant to cover the entire world on the same scale are never planispheres, and the maps can only be connected for limited areas. The projection adopted for the international map of the world at 1/1,000,000 is conic, constructed zone by zone (polyconic). The projection of the world map at 1/5,000,000 by the United States Air Force is a regional conic projection (Lambert's conformal); its components are not meant to be connected, but have large areas that overlap from one map to another.

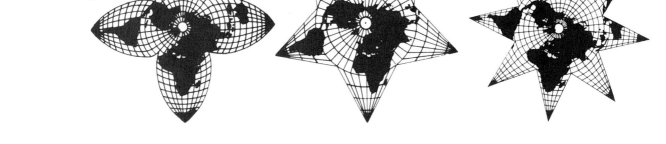

The favored region

The various deformations become more accentuated from the central spot toward the periphery, reaching a maximum at the boundary of the image. It is easy to calculate these deformations for each system and to represent their progressive increase.

Figure 9 shows the area deformations found in Gougenheim's conformal system, and figure 10 gives the angular deformations of Mollweide's equivalent system (after A. Robinson).

One normally tries to make the inevitable point (or points) of maximum deformation coincide with the least pertinent geographic area(s) or, stated another way, to reserve the zone of least deformation for the geographic area which is particularly germane to the purpose of the representation.

It is most often the populated regions which provide the focal point of maps, so we usually place the areas of greatest deformation in the immense void of the Pacific (figure 12).

As a result, we can detect two "generations" of planispheres:
– classic planispheres, in which the mathematical axes of the projection coincide with the geographic system of coordinates—the terrestrial equator (or the pole) being considered as a privileged area. An example is Mollweide's projection (figure 11). According to the particular case, these projections are called "polar" or "equatorial."
– a second "generation" of planispheres, in which the zone of least deformation coincides with the continents, or with any region of the globe relevant to the purpose of the map. An example is Brisemeister's projection (figure 12). These projections are called "oblique."

Obviously the obliquity of a projection is independent of its mathematical formula, that is, of its system, and any system can produce a classic projection (equatorial or polar) or an oblique projection.

To summarize:

Conditions of preservation	Shape of the region to be represented		Exterior shape of the planisphere
conformal equivalent compromise	azimuthal conic cylindrical	tangency or secancy	

determine the mathematical SYSTEM
↓

the tangent point of the system on the earth
polar
equatorial
oblique
determines the PROJECTION (drawing).

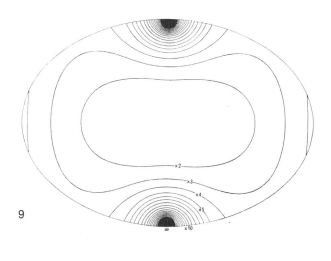

9

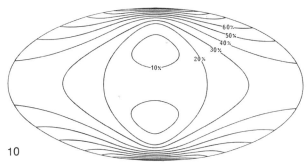

10

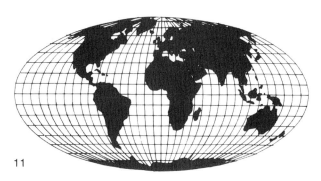

11

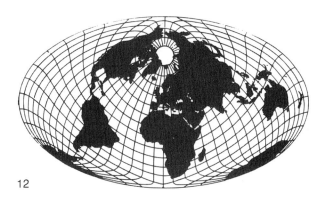

12

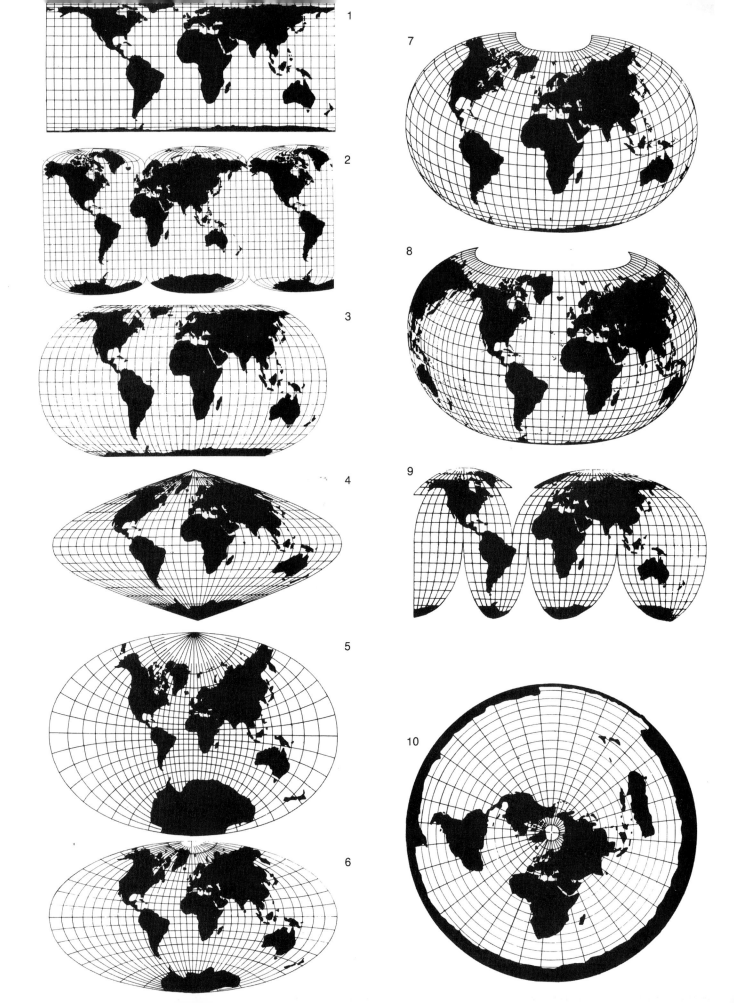

Examples of several "classic" planispheres

In addition to Mollweide's equivalent planisphere (figure 11, page 293) several typical projections are given in the following figures:
1—equivalent cylinder;
2—Gall's cylinder (modified for the poles by J. Bertin, 1950);
3—Eckert's IV projection;
4—Sanson-Flamsteed's equivalent projection;
5—Gougenheim's congruent projection;
6—Guillaume Postel's compromise projection
7—orange rind projection (J. Bertin, 1950);
8—orange rind projection extended, giving a closed Pacific Ocean (J. Bertin, 1951);
9—Goode's equivalent projection, by juxtaposition of sectors from the Mollweide projection.
All of the above projections are "equatorial."
10—"polar" azimuthal equidistant projection. The whole of the outer circle corresponds to the South Pole.

Examples of several oblique planispheres

Along with Brisemeister's equivalent projection (figure 12, page 293), several oblique projections are given in the following figures:
11—Guillaume Postel's compromise projection. This is similar to the projection in figure 6, with the axis through the Pacific Ocean (J. Strohl);
12—"Atlantis" projection. Mollweide's equivalent projection, where the main axis corresponds to the meridian 30° west;
13—projection with regional compromise (J. Bertin, 1953), in which the compromise is no longer homogeneous, but is modified for a larger deformation of the oceans, to give a lesser deformation of the continents;
14—azimuthal equidistant projection, centered on Khartoum; this is like the projection in figure 10, centered to group the set of the continents, including the South Pole, in the zone of least deformation;
15—split projection with regional compromise (J. Bertin, 1952);
16—projection with regional compromise achieved by juxtaposition of azimuthals (J. Bertin, 1954).

The principle of compromise justifies this latter planisphere, in which nearly all of the deformations are placed in the oceans. This is a juxtaposition of compromise azimuthal projections, one per large continental mass (with the exclusion of the South Pole), and each continent can be considered, with few exceptions, as equivalent and conformal. The vast majority of readers relate geographic identification to the shape of the continents. Thus we contend that figure 12 on page 293, and figures 7, 13, 15, and 16, here, provide the most useful images for general geographic knowledge.

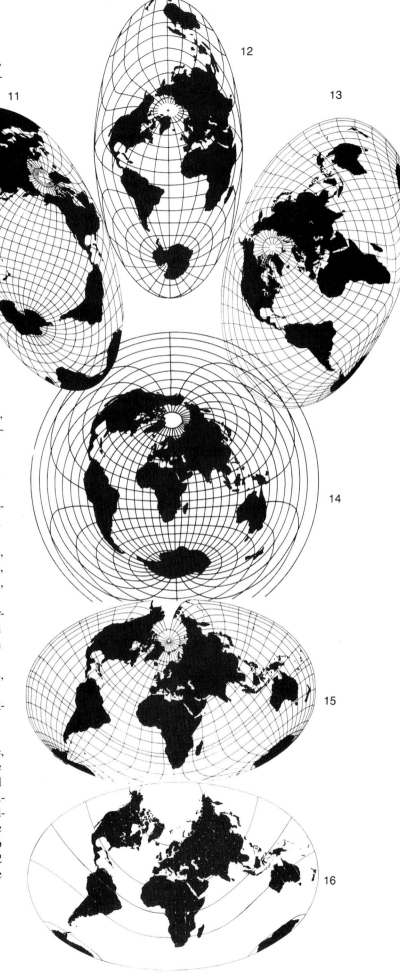

Dimensional identification: The scale

A dimensional identification enables the reader to evaluate the extent of the space represented by the map.

A situational identification, by relating the map to a region whose size is familiar, can lead to an approximate evaluation of the dimensions of the space being represented. But the most precise and rapid evaluation is provided by the scale.

Note that the image is capable of representing any space whatsoever, from the infinitely small (atomic and crystalline structures) to the infinitely large (charts of the heavens and the solar system), including the "life size" image (industrial and architectural drawings). What is properly called "cartography" covers the range from the village map to the celestial chart, and involves having to reduce the size of natural elements. In contrast with this infinite field of possible representations, the sheet of paper is nearly always the same relative size, and:

THE SCALE expresses the relationship between the linear dimensions of the sheet of paper and the dimensions of the space being represented.

However, this relationship can vary up to infinity, and there is a point beyond which human imagination can no longer follow, where the scale becomes without perceptual meaning. In extreme relationships, it is often omitted. This "disjuncture" point varies according to familiarity with maps, and it must be recognized that for the average reader fractionary scales, for example, have practically no meaning. The dimensions of a space must be identified by all possible means:

A graphic scale

This utilizes an image of known length, the meter, the km, the mile.... It should be drawn in the simplest possible way (figure 1). With worksheets, where space is lacking, it can be placed on the neat line (figure 2).

A fractionary scale

For example, 1:5000 ("large" scale) or 1:1 000 000 ("small" scale) are called "maps at five thousand" or "maps at one million"; the terms signify that any length on the map represents a length 5000 times or one million times larger in space. In addition to the correspondence given in figure 4, the following formulae should be retained:

at 1/10 000   1 mm on the map = 10 meters;
at 1/80 000   1 mm on the map = 80 meters;
at 1/126 720   1 mm on the map = 126.72 meters;
at 1/10 000 000, 1 mm on the map = 10 000 m = 10 km.
(1/126 720 is 1/2 inch to the mile; 1/253 440 is 1/4 inch to the mile.)

The fractionary scale can be written in three ways:

$$\frac{1}{80\ 000} \qquad 1/80\ 000 \qquad 1:80\ 000$$

One should not put commas between the thousands (80,000), but leave a visible space (80 000).

Rounded scales greater than 1:1 000 000 can be written 1:1 M, 1:3 M, 1:20 M.

Remember that when a map is to be reduced by the printer, the scale will change proportionately; for example, a map at 1/100 000 becomes a map at 1/200 000 if the drawing is reduced 50%. The fractionary scale must be calculated for the dimensions of the printed map.

A grid

The notion of spatial dimension can be furnished by a grid of squares having sides of a known length, such as a km or a mile, extended over the entire plane. In series of maps relating to the same area, but involving studies of specific phenomena, this system permits an immediate identification of the dimensional relationships among the maps. The grid must be extremely discrete; it will be readily visible due to its regularity.

Known shapes

On a given map, involving an unfamiliar area, we can also include a more familiar area "at the same scale." We can, for example, represent "Paris at the same scale" on a map of Athens (figure 3) or choose any other area, provided that its shape and dimensions are familiar.

Distance numbers

These are commonly used on highway maps to inform us of real distances ("curves included") which must be travelled for a given itinerary. Although they provide very useful elementary information, they do not easily produce a notion of the overall scale of the map.

Spacing of parallels

This is a geographic constant which can always be translated into km or miles (see page 289). In latitude:
10° = 1111.111 km = 600 nautical miles
 1° = 111.11 km = 60 nautical miles
 1′ = 1852 m = 1 nautical mile

SCALE UNITY IN A SERIES OF MAPS

The notion of scale acquires particular importance when a study involves several maps.

If each a map represents a different region, an overall map on which each of the individual maps in the set can be located facilitates their identification. The unity of the scale makes these regional comparisons possible. If one excludes details of villages, a set of regional maps can always be simplified to, at most, three scales, and usually one or two. Different scales must be immediately perceptible as such, and the reader must also be informed of similar ones. The overall map or the index must visibly show this.

Scale unity is even more important when the reference space is constant, and the maps differ only in theme or time. It is then necessary to eliminate any variation in scale. Any exception works against geographic comparison and points to a methodological error coupled with an incomplete knowledge of cartographic capability. It would be better to have maps which are different in style and precision, but comparable, than maps of the same visual "density," but different scales.

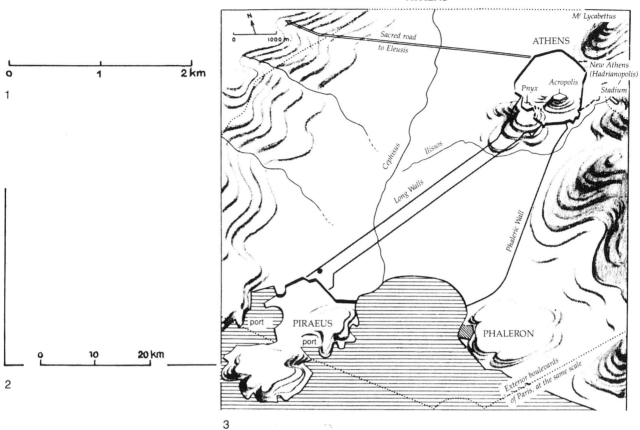

B. Internal geographic identification

Reading on the intermediate and elementary levels involves internal identification. The notions of ACCURACY and GENERALIZATION must be defined in order to construct an appropriate BASE MAP for such problems.

Cartographic accuracy

We should specify once more that we are dealing here with degree of accuracy in the graphic representation, not in the information furnished to the designer.

It is duty of the cartographer to use the most precise and reliable information available (and consequently to take all appropriate steps to ensure that these conditions are met). However, this duty does not involve doing field work or doubling as a geometer, topographer, or any other researcher in order to obtain better information. Our problem concerns the "container" rather than the "content," which allows us to assume that the information to be transcribed is accurate.

GRAPHIC ERROR

The material concerns of graphic representation and the limits of visual perception contaminate any length measurement with a certain amount of unavoidable error. This is graphic error.

With two points 555.55 mm apart, a correct drawing should enable any reader to measure this distance and report it as 555.55 mm. This is rarely possible.

Graphic error is the difference between the length which is read and the length which should be read. This difference is itself a length, which can be measured on the sheet of paper, but it is independent of the *meaning* of this length, which depends entirely on the scale.

For a reader to read 555.55 mm it is necessary that: (1) the drawer plot the distance with that degree of precision; (2) the signs utilized be as precise as possible, yet remain visible; (3) the reader be capable of a precision comparable to that of the drawer, which poses the question of human precision, its limits, and its average. This is *human graphic error*.

But it is also necessary that: (4) the material not have varied between the two measurements; (5) the rule used be similar in the two readings, which poses the question of the dimensional stability of the material throughout the various graphic operations of drafting, reproduction, storage, and reading. This is *technical graphic error*.

A precision draftsman can plot, with a needle, on stable material, a series of given distances: 55.55 mm; 18.20 mm; 210.35 mm; 85.25 mm; etc.

Another draftsman, just as precise and utilizing the same rule, can measure and report these lengths practically without error. This obviously involves an exceptional situation, which could only be improved by mechanical means. It enables us to define the human limits of dimensional evaluation as 1/20 of a millimeter.

It is normal to require of a draftsman a precision of 1/10 of a millimeter, but the reader's precision cannot be counted on. It varies greatly with each individual, and 1/2 of a millimeter is the best we can expect of the average reader. But this still implies that 1/2 of a millimeter is useful, that technical error is not greater than human error. However, it is quite common for a sheet of paper of 0.80 m to vary by more than 5 mm in length!

Reduction of technical error

This error results from a variational difference between the measuring object and the object being measured. It is therefore proportional to the length of the measurement. It can be remedied in several ways:

– by reducing the length measured, that is, by avoiding documents which are too large, measurements which are too long;
– by reducing the variations, that is, by using the best technological means, such as stable material and calibrated rules;
– by making these variations equal, that is, by ensuring that any variation in the measuring object matches that of the object being measured. This means, of course, placing the measure on the map itself, that is, utilizing a precise graphic scale. But the length and width sheet of paper often have different degrees of variation. Thus, when linear precision is imperative, the graphic scale should be placed on all four sides or, better, over the entire surface of the map (i.e., grids).

Reduction of human error

This error results from visual faculties, and it can only be reduced by appropriate education. It is a personal constant, which can be measured and defined as an absolute quantity: 1/20 mm; 1/10 mm; 1/2 mm; etc. It applies to the extremities of the measurement and is independent of the length being measured. Since approximately the same amount of individual error occurs with each measurement, the total error is proportional to the number of measurements. It is impossible to remedy human error completely, however:

– The useless addition of several errors can he avoided, and one should never proceed by addition of partial measurements when the total can be measured.
– Increasing the error can be avoided by utilizing the most precise possible marks (points, dashes). However, this precision works against the visibility of the marks. The minimum acceptable mark width is 1/10 mm, which corresponds to minimum human error.

– It can be reduced at the moment of drafting, by using a larger drawing, followed by a photographic reduction. But this supplementary technical operation introduces a risk of deformation and has no influence on the human error committed at the reading stage.

Reduction of the total error: Graphic precision

For any question involving a precision measurement, it is standard procedure to take all the above technical precautions, in which case technical error is less than the average human error.

In this case, graphic precision is a constant which characterizes a given map. It is irreducible and can be defined as an absolute quantity: 1/10 mm; 1/4 mm; 1 mm; etc. However, when it is not possible to take all the appropriate technical precautions, graphic error will vary with the length of the measurement, the draftsman and the reader.

We will now look at the meaning and effect of this error.

DIMENSIONAL ACCURACY

What is the distance between two given steeples? Any cartographic question leading to a linear metric evaluation (in meters, kilometers, miles, etc.) involves the representative fraction, the ratio of reduction, that is, the scale.

At 1/2000, 1/4 mm represents 0.5 m. At 1/200 000, it represents 50 m.

Dimensional accuracy is the metric meaning of graphic precision.

It is inversely proportional to the scale and can be expressed as an absolute number; one can say that a map is accurate down to 10 m, down to 2 km, etc. Consequently:

(1) Linked to graphic error, accuracy has an absolute limit for a given scale.
(2) The dimensional accuracy of a representation can be determined when one knows the scale and the degree of precision (figure 1).

	Scale	Graphic precision		
		2mm	1mm	1/10mm
		DIMENSIONAL ACCURACY		
	1/ 1 000	2 m	1 m	0,1 m
	1/ 10 000	20 m	10 m	1 m
	1/ 50 000	100 m	50 m	5 m
	1/ 100 000	200 m	100 m	10 m
	1/ 200 000	400 m	200 m	20 m
	1/ 500 000	1 km	500 m	50 m
	1/ 1 M	2 km	1 km	100 m
1	1/ 5 M	10 km	5 km	500 m

(3) When one knows the accuracy required by the information and the graphic precision afforded by the drafting conditions, the scale of the map can be based on these factors. If one knows, for example, the respective position of villages down to 1 km, and if the probable precision (determined by the researcher's experience) is not high, the scale must be drawn at 1/500 000 or 1/250 000. It is always preferable to plot on a larger scale; in fact, any plotting done on a scale smaller than that indicated in the table in figure 1 will lead to a loss of information.

(4) It not possible to represent "at scale" a ground distance which is smaller than maximum dimensional accuracy.

A maximum graphic precision of 1/10 mm represents a distance of 20 m on standard highway maps at 1/200 000; as a result, any meaningful distance smaller than 20 m cannot be represented in true size on this scale. For example, a road 10 m in width will have to be represented by a ground distance of at least 20 m, and, in fact, often 200 m. This is a convention. In the above case, any quantitative observation leading to the evaluation of a ground distance will involve a certain amount of error. One can do no better than to reduce it to an admissible magnitude, by reducing graphic error and increasing the scale.

RELATIONAL ACCURACY

Fortunately, distance measurements amount to only a small portion of the useful observations which can be obtained from a map. Consider the following questions:
Where is a given place? What is there at a given place?
What street should I take?—the third on the left!
Where is the village?—on the north side of the river!
What shape does it have?—it is circular!
Is this a winding road?—no, it is straight!
What is the nature of the coastline?—it is jagged!

Useful answers do not always involve a metric evaluation of the distances, but include the possibility of differentiating, ordering, or counting the elements of the information. It only matters that these useful elements be discernible and that the differences, the order, the numbers (within the limits of visually memorizable quantities) constitute a recognizable reference system.

Relational accuracy is the geometric meaning of the arrangement of the signs.

The arrangement of three points as well as their number remain similar, whatever the planar reduction. Consequently:
(1) Any question involving an alternative (≠), an order (**O**), or a small enumeration of elements on the plane can receive an accurate answer. What is read can conform strictly to what should be read.

Relational accuracy can be absolute.
(2) Within the limits of meaning indicated above, relational accuracy is independent of the scale. It can therefore be substituted for dimensional accuracy at the point when the latter becomes insufficient or inefficient. Thus one can receive directions or identify a site with all the efficiency necessary, by counting the number of streets, by observing relationships of arrangement (between the river and the road), angles (before the curve), or structure (boundary between the new suburbs and the old city).
(3) The number of discernible elements is limited for a given area of paper. The number of representable elements thus varies with the square of the scale, and a reduction of the scale will reduce the number of elements which can be represented accurately. However, this will not reduce relational accuracy.

Reduction of the scale involves choosing from the information those elements which must be recorded with absolute relational accuracy. This choice governs the notion of cartographic generalization.

Cartographic generalization

The necessity to "generalize" results from the contrast between the limitations of human perceptual constants and the infinite range of possible reductions of geographic order. Generalization is the spatial equivalent of simplification. In to simplify we categorize.

Generalization thus involves discovering concepts which are applicable to the available signs and which are defined as similar for a certain area, such that this area can be considered as different from neighboring areas.
Generalization also involves regionalization.

The visual *constants* referred to here are:
– the size of the sheet of paper (which, as we have seen, can be considered as relatively constant in contrast to the potentially infinite nature of the information);
– the minimum size required for visible and separable marks;
– the number of different categories which an individual is capable of integrating during the course of observation; this rarely exceeds five.

The *variable* here is the relation between the total number of data elements and the number of elements that can be represented readably.

These constants and the variable must be examined in relation to the three types of planar representations: the point and the line (positions without theoretical area), and the area. Within the range of possible reductions, the ultimate limit of reduction is obviously total elimination.

This excluded, we can state that a point will always remain a point. A finite straight line will become a point. An area will become a point. An open wavy line will become a straight line, then a point. A cluster (group of areas, lines, or points) will become an area, then a point.

TYPES OF GENERALIZATION

Thus, except for point representation, the continuum of reductions includes *critical ratios* which lead to a transformation of the implantation.

Each change in implantation will then be accompanied by a change in the definition of the phenomenon being represented, which means a new level of conceptualization. The legend must also change, posing the problem of defining a new invariant.

For example, the map of "houses and streets" becomes that of a circle representing the "city." The lines representing "canals and rivers" become areas of "marshland" or "polders." The points designating "factories" become "industrial areas." The "tree" gives way to the "forest."

However, there are two possible ways of generalizing.

– One can change the implantation (for example, from a cluster of points to an area), which implies a new level of conceptualization; thus, "mines" become a "coal field." This is "conceptual generalization." It generally involves new information, beyond that being processed.
– One can maintain the level of conceptualization, which implies maintaining the implantation and the planar structure of the phenomenon, but at the same time simplifying the distribution. This is "structural generalization." It can be based the information being processed, provided it is comprehensive.

The critical ratio

The critical ratio (of observed elements to represented elements) depends on both the distribution and the final scale; it can be defined in relation to photographic reduction. For any given information, we can consider three levels of reduction (see A, B, C in figure 1, opposite):
(A) Small reductions. The photograph requires no simplification.
(B) Medium reductions. The reduced photo still shows all the elements of the information, but they exceed the limits of common perception. They are too fine and risk fusing or disappearing. This constitutes a critical ratio, necessitating a choice between the two types of generalization. This choice depends on the nature of the questions being considered.
(C) Large reductions. The reduced photo eliminates part of the information. For these ratios, conceptual generalization is indispensable.

Displacement of the critical ratio

In the range of areas available for expressing given information, *the critical ratio is reached more rapidly . . .*
– when the information is a message representing several different phenomena. In order to support a differential variable, the marks must remain sufficiently large. Note that in a superimposition of implantations (lines over areas, for example) the critical ratio is located on different levels for each implantation.
– when the pertinent questions are on the overall level. This is the case with maps data involving areal data, which are read only on the intermediate and overall levels. A large simplification of the lines in the base map will increase the increase the legibility of the retinal variation, which constitutes the principal focus of the information. Conversely, *the critical ratio is less quickly attained . . .*
– when the information is homogeneous. The distribution of a single phenomenon accommodates very fine marks, approaching the limits of perception (see page 318).
– when the pertinent questions are on the elementary level (position of a site, distance measurements, etc.), the reference elements are chosen especially as a function of their proximity to the new positions being mapped.

In this chapter we will study, in succession, the structural generalization of a wavy line, the structural generalization of a cluster of marks, and the drafting of base maps, which are necessary in current experimental cartography and involve both types of generalization.

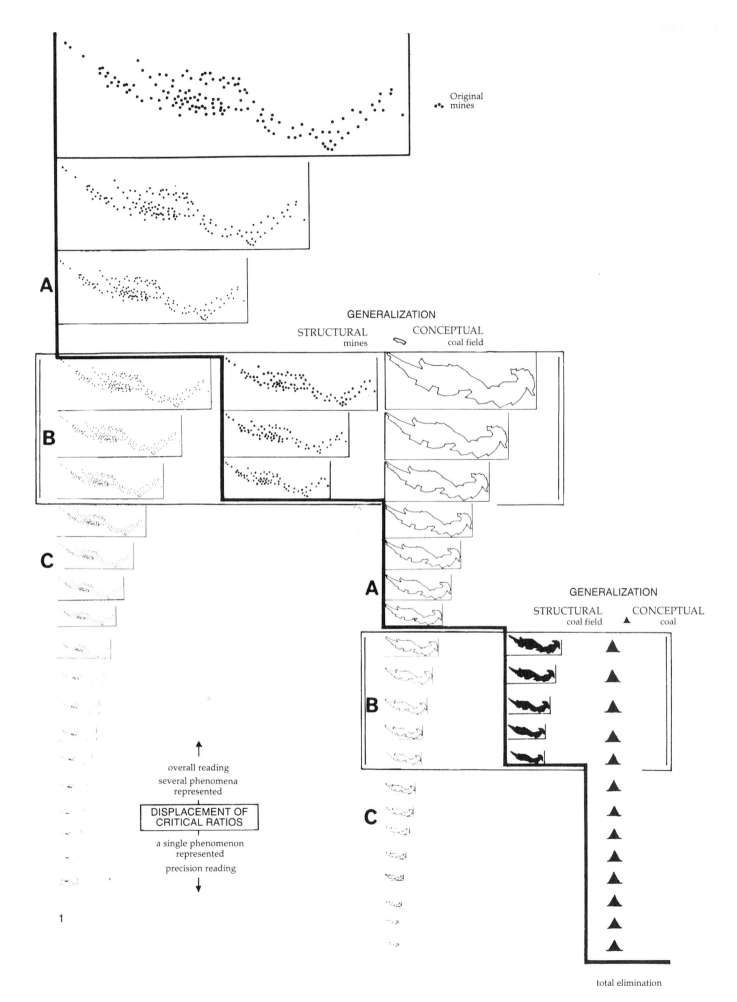

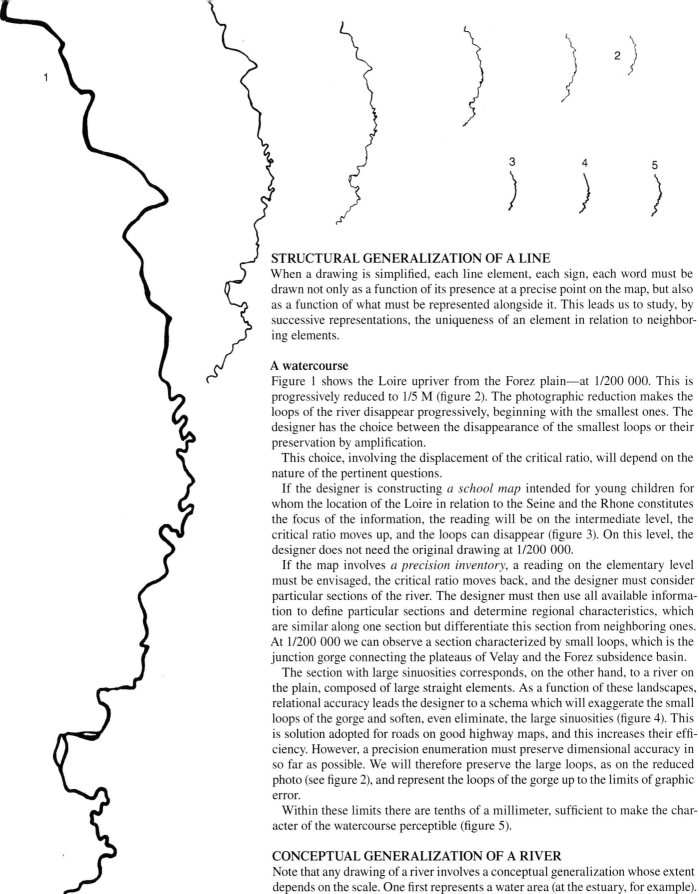

STRUCTURAL GENERALIZATION OF A LINE
When a drawing is simplified, each line element, each sign, each word must be drawn not only as a function of its presence at a precise point on the map, but also as a function of what must be represented alongside it. This leads us to study, by successive representations, the uniqueness of an element in relation to neighboring elements.

A watercourse
Figure 1 shows the Loire upriver from the Forez plain—at 1/200 000. This is progressively reduced to 1/5 M (figure 2). The photographic reduction makes the loops of the river disappear progressively, beginning with the smallest ones. The designer has the choice between the disappearance of the smallest loops or their preservation by amplification.

This choice, involving the displacement of the critical ratio, will depend on the nature of the pertinent questions.

If the designer is constructing *a school map* intended for young children for whom the location of the Loire in relation to the Seine and the Rhone constitutes the focus of the information, the reading will be on the intermediate level, the critical ratio moves up, and the loops can disappear (figure 3). On this level, the designer does not need the original drawing at 1/200 000.

If the map involves *a precision inventory*, a reading on the elementary level must be envisaged, the critical ratio moves back, and the designer must consider particular sections of the river. The designer must then use all available information to define particular sections and determine regional characteristics, which are similar along one section but differentiate this section from neighboring ones. At 1/200 000 we can observe a section characterized by small loops, which is the junction gorge connecting the plateaus of Velay and the Forez subsidence basin.

The section with large sinuosities corresponds, on the other hand, to a river on the plain, composed of large straight elements. As a function of these landscapes, relational accuracy leads the designer to a schema which will exaggerate the small loops of the gorge and soften, even eliminate, the large sinuosities (figure 4). This is solution adopted for roads on good highway maps, and this increases their efficiency. However, a precision enumeration must preserve dimensional accuracy in so far as possible. We will therefore preserve the large loops, as on the reduced photo (see figure 2), and represent the loops of the gorge up to the limits of graphic error.

Within these limits there are tenths of a millimeter, sufficient to make the character of the watercourse perceptible (figure 5).

CONCEPTUAL GENERALIZATION OF A RIVER
Note that any drawing of a river involves a conceptual generalization whose extent depends on the scale. One first represents a water area (at the estuary, for example). The river is represented by two lines or an irregular mark (figure 6). Next, one depicts the axis of the river (i.e., the area becomes a line: figure 7). This line can then be progressively reduced in width to evoke the reduction of real space (this is most often much smaller than the width of the line can suggest).

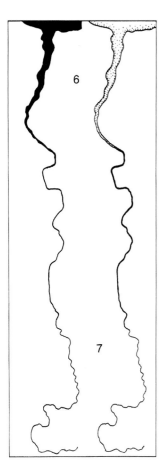

6

7

8
RIVER
ROAD
RAILWAY

9
COAST
LINE

10
CONTOUR
LINE
(ISARITHM)

Rivers and roads; coastlines and contours

In no case can the generalization of coastlines or contours be similar to that of rivers, roads, or railways, etc.; an additional contingency must be introduced. *Rivers and roads* are lines inscribed over a homogeneous area, and there is a good *probability of symmetry* over their entire length (figure 8).

However, *coastlines and contours* (isarithms) are boundaries of areas. On the one side, there is land, on the other, sea; or, for isarithms, differences in population or rainfall. There is a *certainty of dissymmetry*.

For a coastline (figure 9) the angle or small radius of curvature will be turned toward the sea (this is true for all areas, from the Bay of Brittany to the Gulf of Mexico). For contours (figure 10), the small radius of curvature will be turned toward the landform.

USING GRAPHIC ERROR IN A POSITIVE DIRECTION

In cartography, the reader intuitively conceives a correct idea of what is or is not pertinent. For the size of the small loops of a river, the depths of a small bay, the position of a city, it is obvious that 1/4 mm in relational displacement is not an error in terms of dimensional accuracy, *because this error will never be perceived as such*.

No one will attempt to measure precisely a distance from the center of a city. Incidentally what is the center of a city? And what is 1/4 mm? It is often smaller than the total graphic error!

On the other hand, the fact that a city borders a given point on the river (figure 11) or is situated on a plain, at the foot of a mountain, or on the mountain, is relational information of a rigorous, controllable, and practical sort.

All these elements are differentiable within the limits of graphic error, since the designer has been able to represent them without effecting a displacement greater than 1/4 mm (figure 11).

In precision enumerations, the necessary displacements are most often smaller than or equal to average graphic error, and a good designer simply puts the graphic error to use, in the sense that it becomes relational accuracy.

PRINCIPAL PLANAR RELATIONS

Within the limits of graphic error, we can establish a list of the principal geometric relations which we can attempt to maintain (figure 12).

They constitute visible information, and the reader ought to be informed of the differences between A and B. This means that if we draw A, and the information, on a much larger scale, bears evidence of B (or conversely), we have committed a visible relational error, even though we can quite rightly claim not to have committed a dimensional error.

Thus, precision structural generalization is linked to the following basic rule:

We must have at our disposal information on a scale about ten times larger than the definitive drawing and "get into" the map while drawing it. One should never compile a map at the same scale.

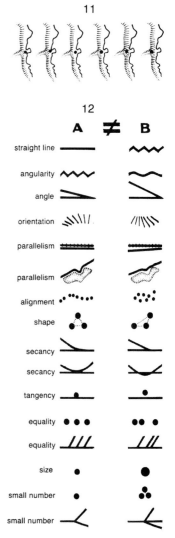

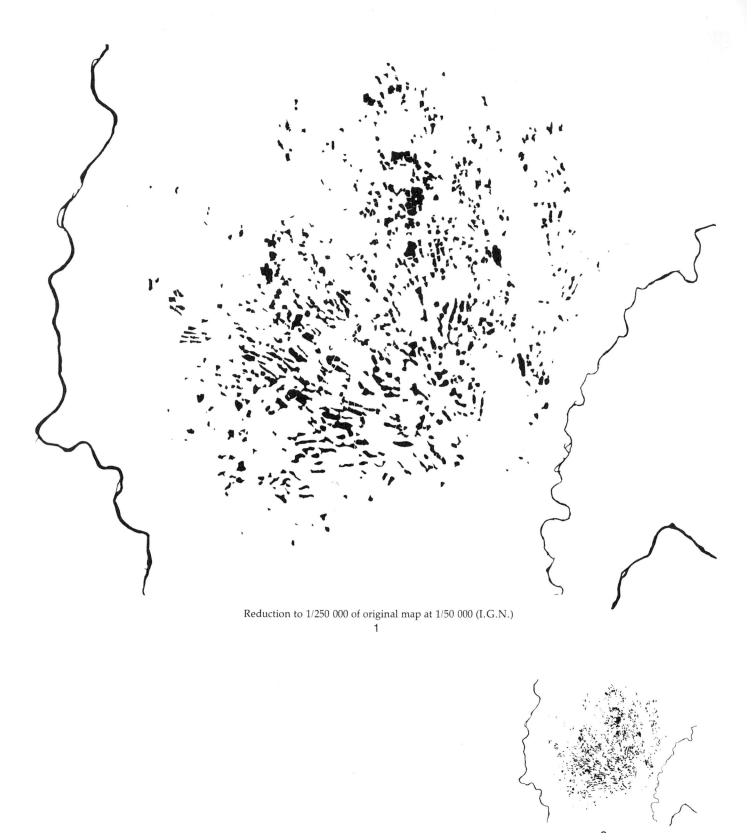

Reduction to 1/250 000 of original map at 1/50 000 (I.G.N.)
1

2

STRUCTURAL GENERALIZATION OF A CLUSTER OF MARKS

The structural generalization of a cluster of marks (islands or lakes, for example) is the most complex of generalization problems, since it involves the notion of a two-dimensional continuum.

The lacustrine area of Dombes, northeast of Lyon, provides a good introduction to this problem, enabling us to define its parameters and establish the elements of a general solution.

Consider a representation of this area at a scale of 1/1 M.

Figures and 1 and 2 are reductions at 1/250 000 and at 1/1 M of the information, that is, of the topographic map at 1/50 000, which includes all the lakes.

Figures 3–9 show maps at 1/1 M, produced by various organizations, such as the *Institut Géographique National*, the *Geographical Section, General Staff* (Standard edition and Army Air Style edition), the *Atlas de France* (hypsometry, morphology, and chorography), and the *Times Atlas of the World*. Which one is best?

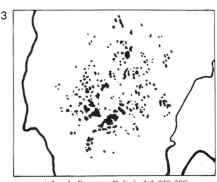
Atlas de France - Relief. 1/1 000 000

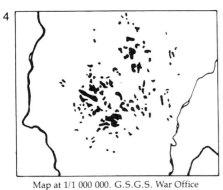
Map at 1/1 000 000. G.S.G.S. War Office

Atlas de France - Geomorphology. 1/1 000 000

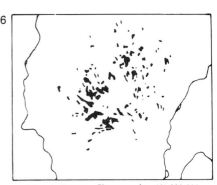
Atlas de France - Chorography. 1/1 000 000

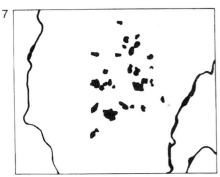
Map at 1/1 000 000. G.S.G.S. Army Air Style

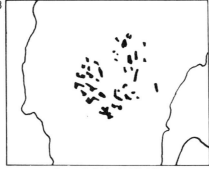
France I.G.N. 1/1 000 000

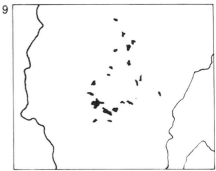
Times Atlas of the World. 1/1 000 000

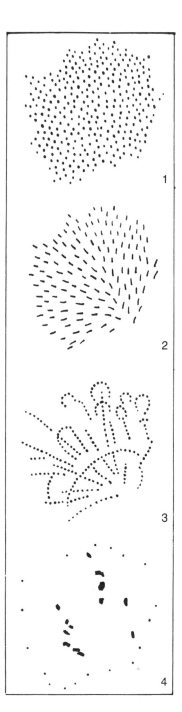

A thorough study of the images in figure 8 shows that absence of method leads to the elimination of certain important lakes, to the invention of other lakes, to the deformation of the lacustrine area, and especially to the disappearance of all the formal elements capable of characterizing the arrangement of the lakes. Now, these are useful pieces information which justify replacing dimensional accuracy by relational accuracy. The notions of parallelism and orientation, for example, indicate the direction in which the reader has the greatest probability of walking without encountering water.

Which drawing should be used?

A cluster of marks is a closed area in which the elements are observed successively on various levels of reading. These observations are translated by differences in shape, orientation, distribution, size, and density.

When, as here, the reduction ratio of the original information is such that a photograph on the definitive scale still permits identifying nearly all the marks, a structural generalization can be carried out in the following manner.

The original is photographically reduced on two scales: at approximately quadruple the final scale (figure 1, page 304); and at double the final scale. The first document is for experimentation. On the second (figure 5) we successively place several pieces of tracing paper and examine the variables separately on different levels.

On the overall level, a simple outline (having a minimum number of concavities) defines an area differentiated from other regions by the presence of lakes (figure 1). This form is accurate as concerns the notion of *lacustrine area*. It is inaccurate in all other aspects.

On the elementary level, the information shows that the lakes are not circular, but generally *elongated*.

On the intermediate level, small visual groupings show that the elongated lakes are often *parallel*. Larger visual groupings suggest a particular type of *orientation* (figure 2). Although superior to figure 1, this figure remains inaccurate.

Still larger visual groupings reveal *alignments* of lakes, often *parallel* to the basic orientation, but sometimes *perpendicular* to it (figure 3). The lakes also display differences in *size,* and the largest ones, along with those defining the boundaries of the area, must not be allowed to disappear under any circumstances (figure 4).

There remains the notion of *density*, which will emerge of itself if one holds to an elementary principle of enumeration at the stage of final interpretation.

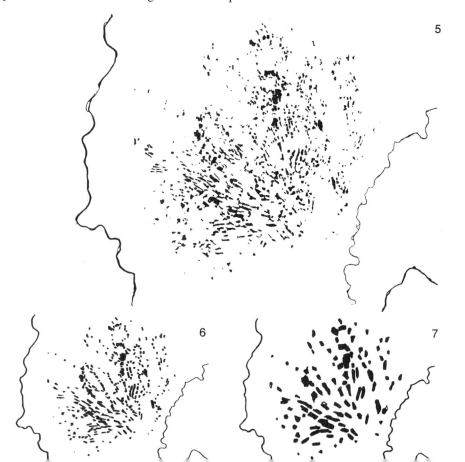

Approximately one mark in two has been retained in figure 6, and one mark in four in figure 7, which causes the final marks to be enlarged proportionately. This interpretation is governed by observations derived from figures 1–4, which are recorded in pencil on the final traced copy.

Figures 6 and 7 approximate, with different degrees of precision, the combination of sensations stemming from the original information (see, for purposes of comparison, figure 1, page 304) and constituting the STRUCTURE of the distribution.

We should note that this type of procedure would not be without interest in the study of aesthetic phenomena.

Note that numerous errors of generalization seem to originate from the I.G.N. map at 1/500 000 (figure 5). The set of maps in figure 8 illustrates the insufficiency of the method of successive generalization without returning to the original source; this also applies to the drawing of rivers (see page 302). Any correct generalization must stem from the original comprehensive document, which must remain under the eyes of the graphic designer.

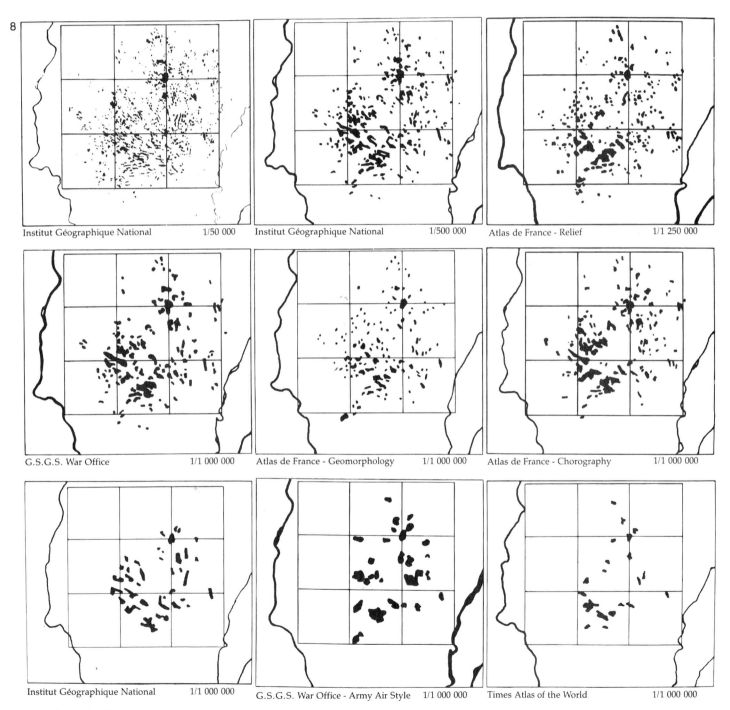

Scale above: 1/750 000.

Base maps

The base map consists of the set of known reference points which are necessary and sufficient for situating the as yet unknown elements of the new information being mapped.

The drawing of a base map always runs up against the following contradictions.

(1) The base map must include all the elements of identification necessary for the construction and reading of the map on all levels.

(2) The base map must be dominated by the new elements being mapped; the reader must be able to select and group them for information on the intermediate and overall levels.

The problem can therefore be divided according to whether pertinent questions will be on the *elementary level*, necessitating a base or background which will permit a precision identification, or on the *intermediate and overall levels*, where a highly simplified reference system will be sufficient.

WHEN PERTINENT QUESTIONS ARE ON THE ELEMENTARY LEVEL

When it is a question of identifying the position of an archaeological site, a factory, a road, or a specific administrative, linguistic, or economic area . . . the information involves the expression of relations in arrangement between the new elements and known neighboring elements.

In other words, knowing that BABYLON is situated at 44° 6′ 48″ longitude east of Greenwich, and at 32° 34′ 36″ latitude north does not constitute complete cartographic information, even though its coordinates permit defining the site down to some 50 m (figure 1).

In the final analysis, the utilization of such information is only possible if one possesses a document on which these coordinates become a point in an imaginable space, identified by reference points which are KNOWN and/or RECOGNIZABLE (figure 2).

If several exceptional cases are excluded (such as those reserved for explorers, topographers, and especially sailors, who have at their disposal no known visible reference point), any positional identification which constitutes new information must be situated on the map in relation to nearby elements which are already known and identified. Practically speaking, the situation of a new position should be recorded on the most complete and precise possible map, given the state of topographic advancement and the scale of the information.

When tracing paper is placed on a topographic map where the background is poorly identified, a loss of information always occurs. Is the site on the right or the left of the road? Does the boundary cross the ridge or its foot? Does the zone encompass this small plain or not? etc.

Numerous archaeological sites in India have been discovered, studied, and described. It has become necessary to organize expeditions in order to rediscover them, because

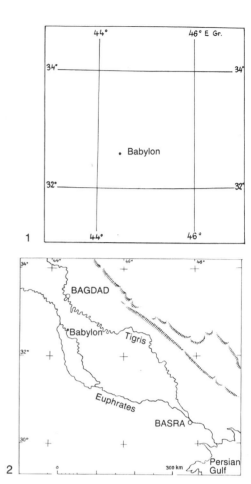

they have not been identified in relation to sufficiently nearby reference points which are known and constant.

With matters involving positional information, whatever the scale, the best document is a complete printed map, as accurate as possible and LOADED with handwritten details, concerning the new element. Two typical objections arise and must be faced.

– "The map is spoiled!" A map which is loaded with new information is a PRICELESS document. The printed map is always available for a sum which cannot be measured against the value of the time spent in discovering and identifying the new information.

– "The map is overloaded, and nothing more can be added!" Unfortunately this objection is often true. The time has not yet come when role of topographic documents has been fully understood. This difficulty can be alleviated by the use of arrows and marginal notations, when the point or area being considered is already loaded with precise notations (figure 3).

Remember that it is imperative to write any proper noun, any specific identification, IN CAPITAL LETTERS or by typewriter.

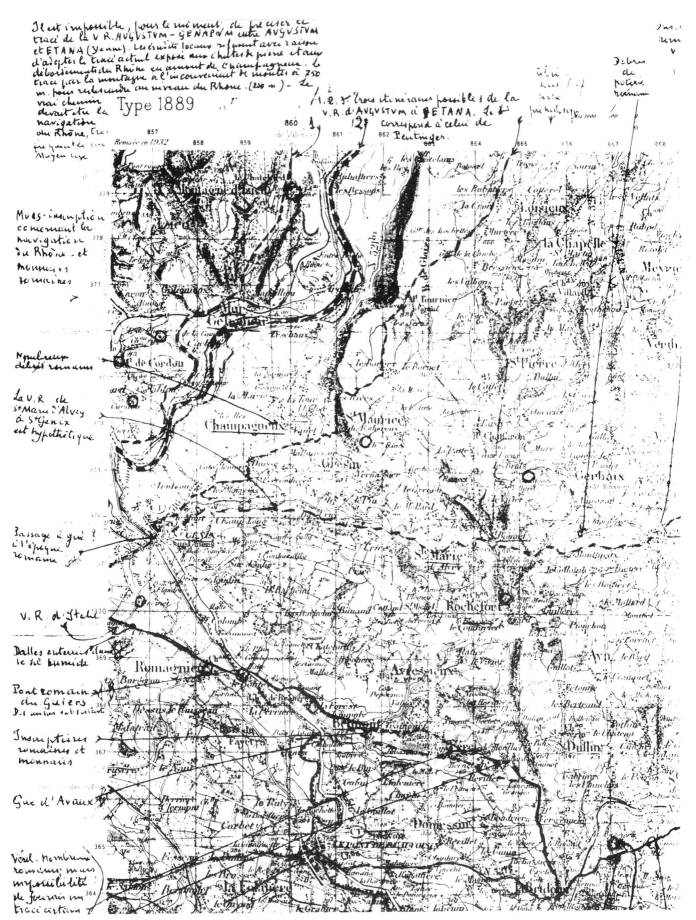

3 PORTION OF MAP OF FRANCE AT 1/50 000 I.G.N.

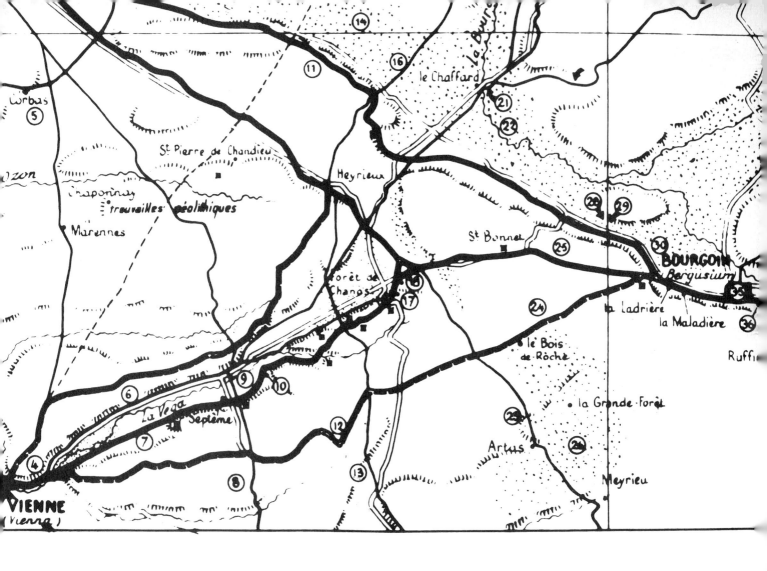

= Present road
— Former road
▬▬■■■ presumed to be Roman

↓ ↙ ↘ Find prehistoric, Roman, post-Roman   ◢ ▲ Milestone out-of-place, in place

A document intended for publication

When the first draft has been completed, how does one proceed toward publication? The best solution is to publish the map which served for identification, printing it in sufficiently light values that the new information stands out in dark, visible marks. But for various reasons, this is not always possible.

We must construct a MAP OF POSITIONAL INFORMATION, displaying the new information on a simplified background, such that the loss of information which this simplification entails will be at a minimum.

Reference points must be chosen according to the following criteria:

(1) *Proximity.* Nearby elements of any kind.

(2) *Stability.* In certain regions, rivers and roads are changing elements. Peaks, steep slopes (cliffs, foothills, etc.) are always more stable.

(3) *Field of knowledge* for the average reader. Positional information will be grasped in relation to the reader's prior knowledge.

Thus an anachronism including former and present roads on the same map, which is easily corrected in the reader's mind, is less to be feared than ignorance of the fact that the two roads are not exactly parallel and that the old road corresponds in certain places to a small present-day road. These are facts which cannot be corrected by further analysis. The same is true for a site, like Ragy, in reality a present-day suburb of Tehran. . . .

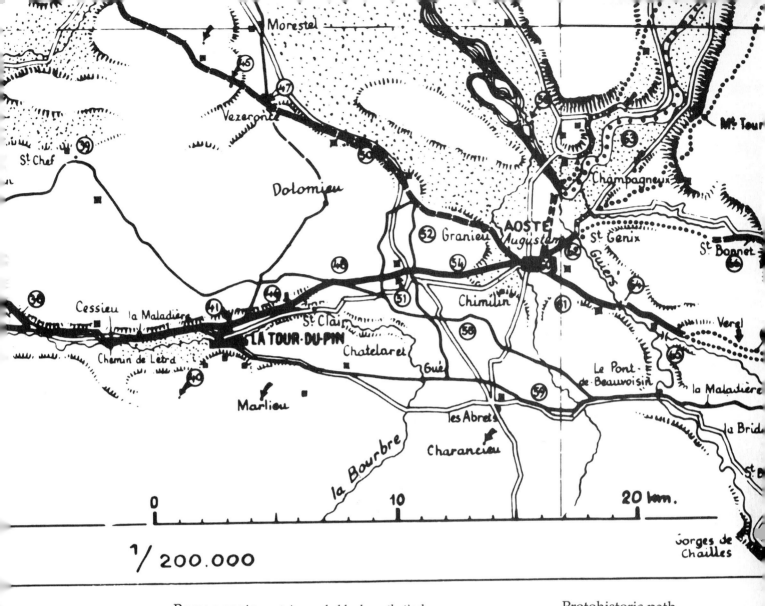

■■■■ •••••••• Roman route certain, probable, hypothetical   Protohistoric path — — —

■ Fortified site   ∴ Ruins   ░░░ Former forest or marshland

Once the decision is made, the reference points chosen can be represented by very simple, light signs; their visibility should never be equal or, even worse, superior to the content of the information.

Figure 1 is the *Atlas des routes protohistoriques et historiques*, Sketch No. 6: from Lyon to Chambéry (compiled by P. Saint Olive, *Bulletin des historiens locaux* [Paris: E.P.H.E.]).

This cartographic sketch at 1/200 000 is prepared from the information contained in a series of maps at 1/50 000 (including the one in figure 3, page 309).

The reference elements are essentially chosen as a function of proximity to the new information. Drafting is thus considerably easier; effort can be concentrated on the relational accuracy the details. Nonetheless, taking the scale into account, it is possible to refer to a topographic map and place all the new elements accurately.

Note that:
(1) it is possible to locate the numerous detailed observations through the use of numbers referring to an index;
(2) everything is drawn by rapid sketching, including the lettering, without any loss of accuracy.

Drafting this map is within the reach of any careful person; it requires no more than sureness of vision and evaluation. The sureness of hand which only a professional draftsman could acquire would not have increased the content of the information nor its legibility.

1

WHEN PERTINENT QUESTIONS ARE ON THE INTERMEDIATE AND OVERALL LEVELS
(Regionalization, external comparisons)

It is not necessary to retain the convolutions of a coastline or a department in order to perceive a departmental density, or a climatic region. For such information there is no cause to utilize the elementary level of reading, and this circumstance enables us to simplify the system of reference points (figure 1), while assuring a better readability on the intermediate and overall levels. The reader who probes such maps for the nature of the coastline, the shape of the boundaries, or the precise position of a site commits an error in reading. These maps are not made to answer such questions, and it would be wrong if they were, since the legibility of the new information would diminish considerably and external comparison, the objective of such maps, would be decreased.

Moving from the precise notation of a geographic site on the base map to the definitive map whose primary purpose is to display areal statistics involves a whole range of intermediate cartographic situations. The plotting of numerous phenomena capable of providing a better knowledge of the Paris area, for example, leads to establishing base maps whose precision is such that no information is lost in the transcription. For example, residential areas, transportation routes, and railway stations are reference points which can serve in numerous plottings.

In these intermediate situations:
an elaboration in two stages most often facilitates the draft work.

(1) Rough sketch on a work sheet
The worksheet should include all the reference points necessary for a precise and rapid identification. It can include the names of departments or communes, the boundaries of urban areas or traffic routes (figure2).

Boundaries can be made less complicated, and areal perception can replace linear perception, by utilizing very fine dotting and areal gradation (figure 2). On an elementary level we can differentiate roads (fine lines) from railways (broken lines), and this also suggests the continuous influence of roads in the economy, in contrast to the discontinuous influence of railways.

The new information, in quantities or qualities per area, is noted in the appropriate place without obliging the reader to search for it in the appended documents. The base map includes all the necessary reference elements.

This preliminary notation must be clear, complete, and unambiguous. At this stage the designer can use all available visual means, particularly color (pencil and ink). The designer will most often record the numbers in ink and represent them graphically by any visual variable which conforms to the laws of image construction.

Thus the definitive draft:
(a) can incorporate the elements of a logical analysis of the problem as well as the elements relative to the nature of the geographic distribution, that is, to the complexity of the image being constructed;
(b) need not be concerned with positional identification;
(c) can be entrusted to a draftsman, once the elements of the legend are determined.

(2) Definitive drawing on a simplified base map
At this stage, the base map no longer has to include the numerous indications necessitated by the rough sketch: names of cantons, boundaries of communes, hydrographic networks, etc. It need only include enough features to permit such elementary identifications as the information may call for (figure 3). It is free to support a striking retinal variable on which the designer can concentrate full attention.

This map can be reduced in scale by the printer, provided this is foreseen and the elements are drawn sufficiently large so that they do not disappear or lose their differential characteristics (figure 4).

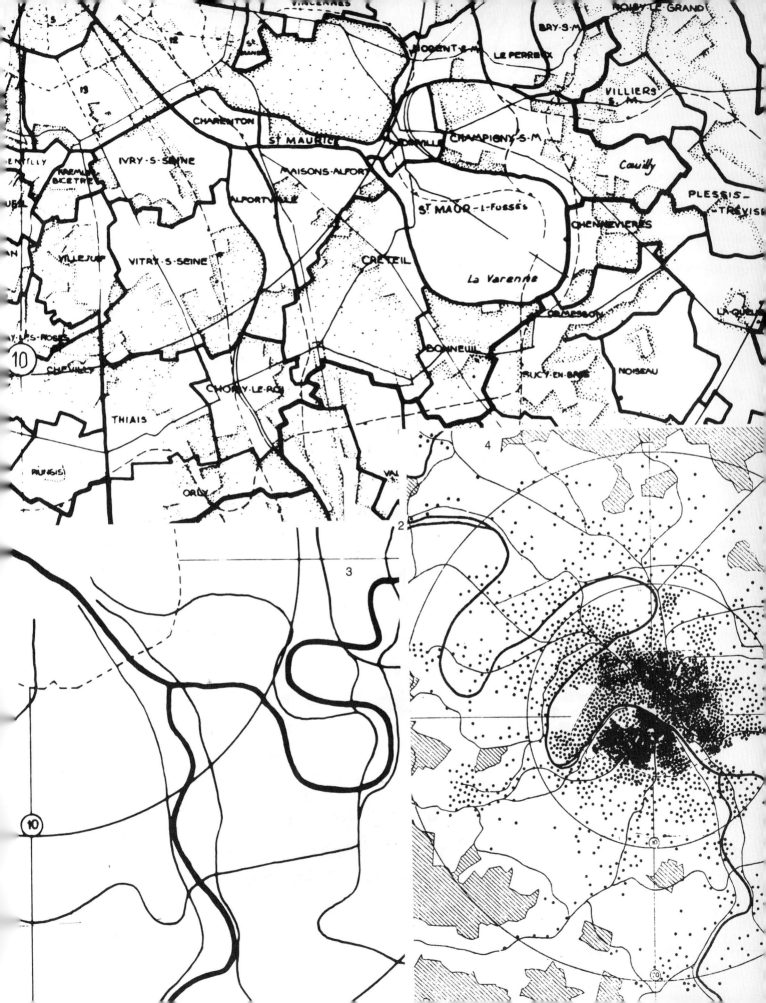

III. — RESIDENTIAL POPULATION AND GEOGRAPHIC SECTORS (1946)

Base map for a city-study

In a city-study, the basic reference points are the streets. A base map including all of them will produce a very precise reference system.

The single line can replace the classic double line, thus simplifying both the image and the draft work. It allows us to draw a somewhat thicker line, capable of sustaining a sizeable reduction, and this avoids the two stages discussed on page 312. A double line can still be used for large arteries, highlighting a fundamental trait of urban structure.

The example in figure 1 is taken from a base map of the twelfth ward in Paris (the rays emanating from the Place de la Nation are readily identifiable). The worksheet at 1/10 000 and easily supports a reduction to 1/25 000 (figure 2); this permits an increase in the number of comparisons and allows the study to be published (P.-H. Chombart de Lauwe et al., *Paris et l'agglomération parisienne* [Paris: Presses Universitaires de France, 1952]).

Note that the segmenting of the circles permits a distribution of the population in a way which, although approximate, is much closer to the truth than that which would result from a single circle per block.

PRECISE POSITIONS ON SMALL SCALES

What must be avoided

Modern research often leads to mapping the distribution of a given phenomenon over a large area. Historical, demographic, ethnic, sociological, pedological, vegetational . . . enumerations, for example, often depict a phenomenon's variation over the whole of large regions, like Europe, the USSR, Africa, the Old or the New World.

In this case, when the scale of the map may vary from 1/1 M to 1/40 M, what are the topographic reference points necessary for internal comparisons, for an efficient reading of the mapped information?

As a general rule, the cartographer retains only the watercourses. This is a solution based on facility, since very often the lines which have the greatest probability of producing a change in the given phenomenon are not the watercourses, but the boundaries of mountain regions, the great monoclinal events, the boundaries of alluvial or swamp zones, forested zones (themselves witness to a characteristic ecological situation). . . .

Along with the watercourses, it is common practice to include several contour lines, generally "master" lines of 200, 500, 1000, 2000 m, etc. This is a very poor solution. Indeed, on these scales the line defined by a given altitude generally loses all meaning. Its crude shape encompasses hills, mountains, and plains within an artificial unity. Furthermore, this system of expression requires tints indicating altitude. It thus covers the entire map with colors which then cannot used to represent the new information.

Authors of modern atlases should remember that their true objective is to show, point by point, the relationships between a given phenomenon and all other phenomena.

It should also be remembered that in many regions of the world "mountains" are low and plains are high, and that numerous coastlines are mountainous and constitute veritable human deserts. Furthermore, the lines of a map should not evoke a plate of spaghetti, but rather the marvelous structures which any designer can now admire during the course of air travel and which in no way resemble contour lines!

The efficient base map

Indeed, such aerial vision should inspire identification of essential geographic characteristics and suggest the form of the base map. One should depict, according to circumstance and scale, the following features:
– the main characteristic surface breaks, boundaries of alluvial plains and hills, foothills, main cuesta lines, principal plateaus and ridges, basins in the mountains, passes; . . .
– the main natural zones, delta zones (internal or boundary), swamp zones (which are often very vast, as in western Siberia), forest zones, boundaries of tundra or permafrost, sandy deserts, rocky or volcanic deserts;
– the structure of rivers, rocky bars, and gorges. When a river cuts through a mountain (which is common) atlases often depict a scene with a large, welcoming valley, whereas, there is nothing more forbidding than a gorge!

But we must recognize that the enumeration of these fundamental geographic characteristics, their precise cartographic notation, and the discussion of their simplification on different scales has not been among the recent preoccupations of geographers.

Consequently, it is not the fault of historians, ethnographers, demographers, . . . and cartographers if, lacking the necessary information, they still commit errors in this regard.

Figure 1. Base map at 1/1 M, Algeria (Frenda region, Géryville).
Mountain regions are in gray.
The horizontal gray bands indicate "less rough" regions.
One can thus distinguish chotts (shallow lakes), sebkhas (smooth flats), and monoclinal reliefs.
Note that high plateaus dominate the djebels (hills) to the north (map from E.P.H.E.).

Figure 2. Worksheet at 1/7.5 M (Pamir region) taken from a "Eurasia" series (E.P.H.E.).

Figure 3. Worksheet at 1/25 M taken from a Eurasia-Africa map (E.P.H.E.).

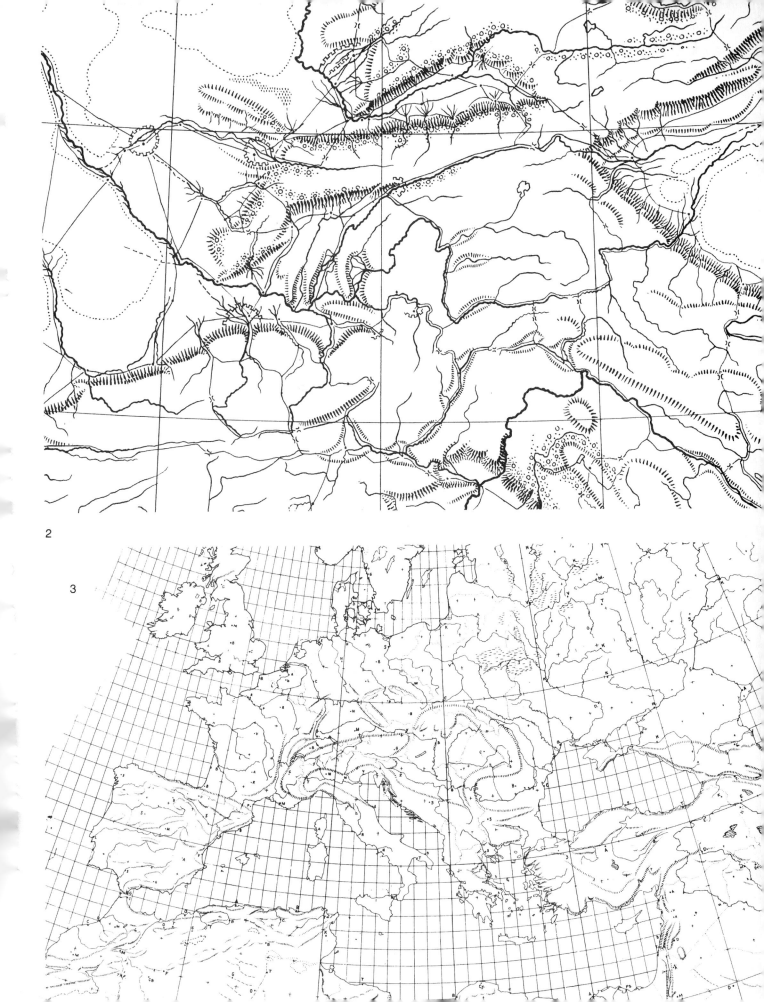

C. Maps involving one component (the geographic component)

GEO (point, line, or area)

In maps involving one geographic component (**GEO**), the points, lines, or areas used to represent that component are uniform in their meaning, which constitutes the invariant.

This uniformity and the precise definition of the invariant must be ensured. Thus "dwellings" should not mean built-up areas; "watercourses" should not include various canals; the term "forest" should not also apply to arboreal thickets or reforested areas.

Such maps can accommodate a large degree of complexity, a great precision in distribution, without presenting difficulties in reading. Indeed, they can be read as a single image, whatever the intended level of reading (see page 176). These maps can also be shown to young children, for whom they create a certain interest, often lively if the invariant is familiar to them (as with dwellings and watercourses). They are easily understood, and useful information can be retained from them. Contrary to general belief, such maps are excellent pedagogical documents.

The three maps opposite (figures 1–3) are reductions at 1/2 M of maps of Poland. These maps were originally drawn at 1/1 M by Professor Franciszek Uhorczak (Warsaw, 1957).
Figure 1: **Buildings** (settlements) were published in red.
Figure 2: **The hydrographic network** was in blue.
Figure 3: **Forest areas** were in green.

The set also includes **cultivated areas** (in yellow) and **prairies** (in light green). The maps are all superimposable, which produces the following combinations:
settlements + cultivated areas;
prairies + cultivated areas;
prairies + hydrographic network;
forests + hydrographic network + prairies; etc.
This set of maps constitutes a remarkable research instrument for problems involving regionalization.

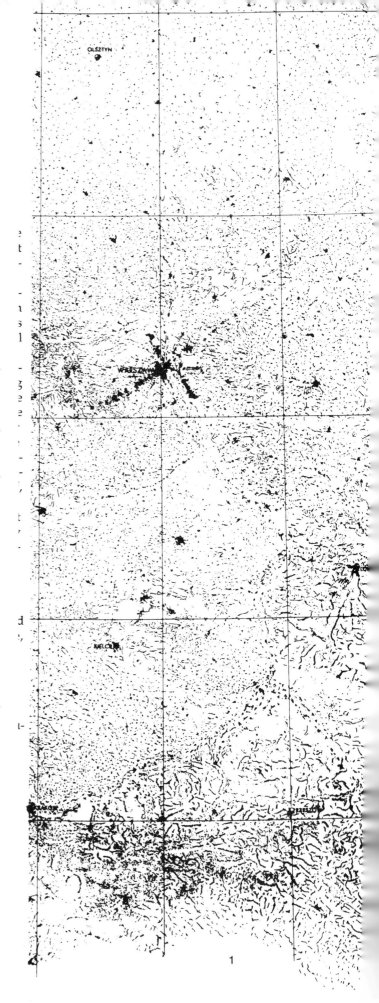

1

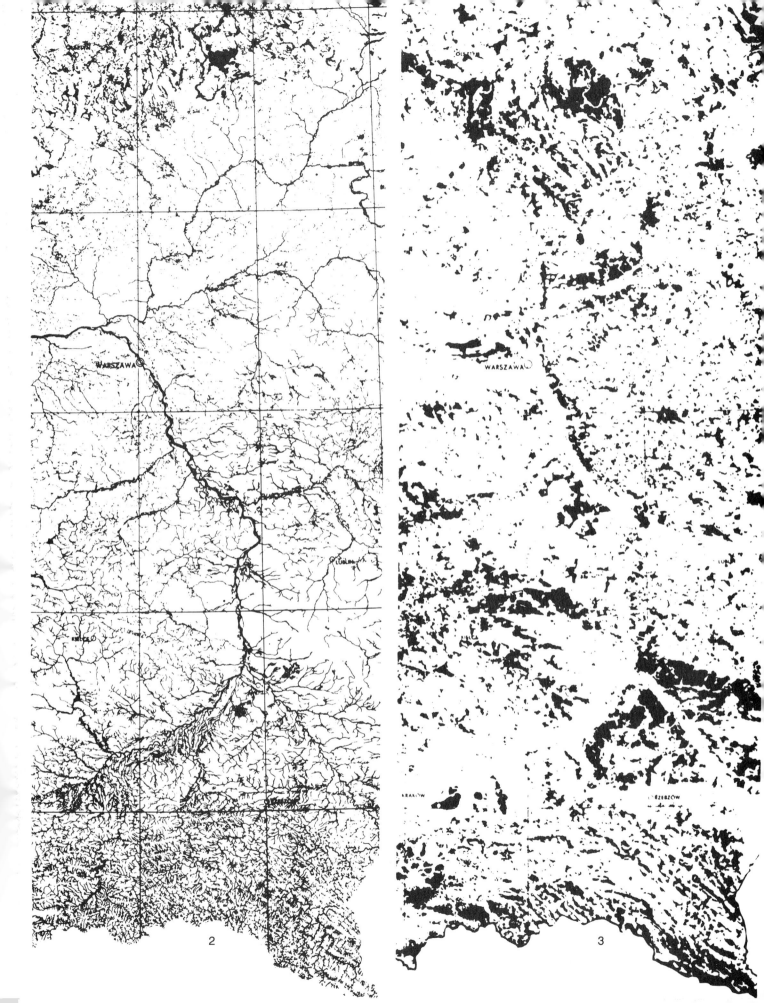

D. Maps involving two components

A cartographic problem involves two components when the information relates a geographic component to a single additional component, which can be:

(1) QUALITATIVE (≠)

Distribution of groups of 200 inhabitants (invariant), according to
GEO—the geographic space
≠ —eleven different ethnic groups

(2) ORDERED (O)

Distribution of travelers' itineraries (invariant), according to
GEO—the geographic space
O —the date of travel

(3) QUANTITATIVE (Q)

Distribution of births (invariant), according to
GEO—French departments
Q —quantities per 100 women between the ages of 18 and 45

We will examine these three cases, discussing for each the three possible representations: point, line, and area.

Even though the geographic component utilizes both of the planar dimensions, it is possible to represent all of these cases as a single image, since an image can accommodate three components. However, the retinal variable must be visually ordered; otherwise, we will be faced with a superimposition of images, which will become all the more difficult to differentiate as the distribution increases in complexity.

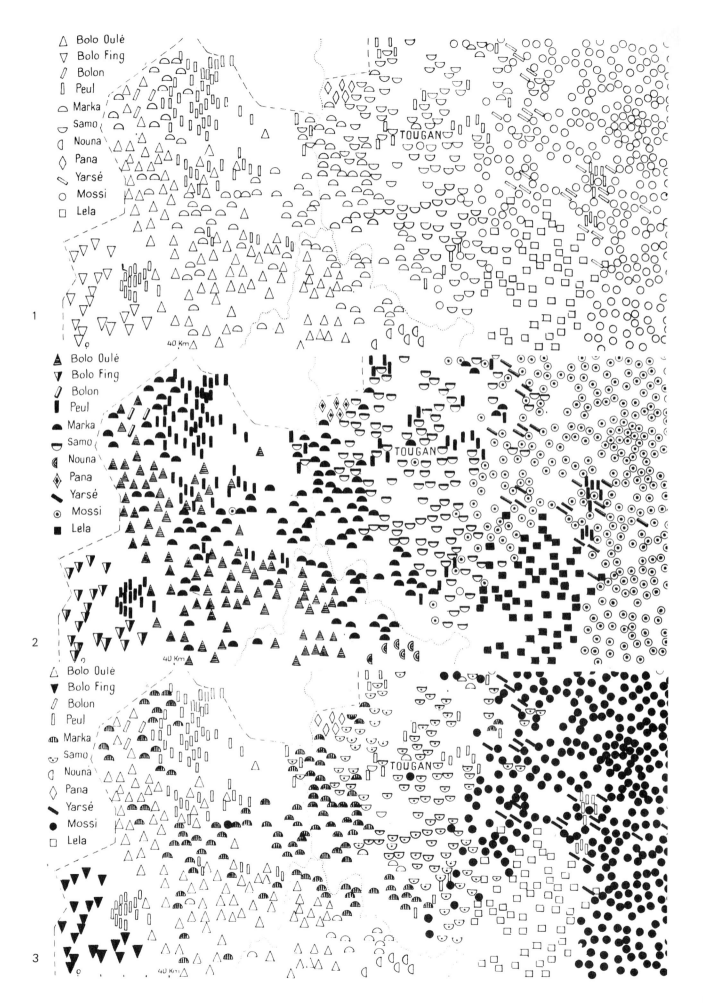

1. Maps GEO ≠ (a geographic component; a qualitative component)

Whatever the implantation, the cartographic representation of any qualitative component (≠) depends primarily on the following question:

Can the ≠ categories of the information be visually ordered?

Example (1)
Distribution of points representing 200 inhabitants according to

GEO—the geographic space (Tougan Region, Upper Volta)
≠ —eleven ethnic categories (figure 1).

Is it possible to give an order (other than the geographic order or that of the quantities) to the different ethnic categories?

A positive response leads to the construction of the maps in figures 2 or 3. Each image depends on the concept chosen for the ordering (age, tribal hierarchy, coloration, height, wealth, etc.).

A negative response leads to the map in figure 4.

From a graphic point of view, visual ordering is always possible, as in figures 2 and 3. It requires the use of size or value variables.

– It greatly increases visual selection, which is indispensable for responding to questions introduced by the ≠ component (Where is a given ethnic group?).
– It facilitates the grouping of categories, since each step in the visual order is perceptible as a set (all the blacks, all the grays, all the whites).
– It permits us to grasp the whole of the information in a single image, which is based on this order and which alone permits overall comparisons with external information.

But it is also necessary to remember that:
– the visual order entails a hierarchic perception, which can be inopportune;
– it excludes perception of the overall density (all categories combined) when it results from size or value variation, since both are dissociative. (Thus the apparent visual density is very different in figures 2 and 3.)
– size variation evokes a quantitative judgment, which is not appropriate in the present problem.

From an informational point of view, ordering can appear quite desirable. But it implies choosing a criterion on which order is based; this criterion may not exist (page 156), may imply a difficult choice, or, as in the example opposite, may seem inopportune. It then becomes a question of achieving the best visual selection among signs with the same visibility, and this means using variables other than size or value.

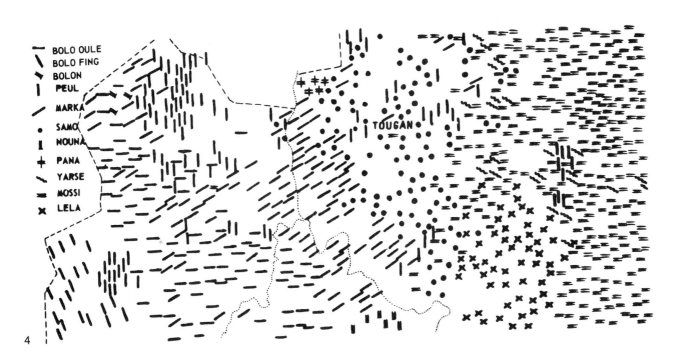

4

GEO ≠ Point

SELECTION OF POINT SIGNS HAVING VARIABLE VISIBILITY

When a concept can order the categories of the component (types of vegetation ordered by vegetal density, types of crops or industries ordered by economic importance), the problem can be treated as ordered: **O** (see page 336).

THE SELECTION OF POINT SIGNS HAVING EQUAL VISIBILITY

When a ≠ component makes the formation of a hierarchy inopportune, or makes density perception (all signs combined) necessary, the problem precludes the use of size or value. Remember that a sign is selective when the reader can, in a single perception, isolate all the points where this sign appears. We can then disregard all the other signs and respond, in one mental operation, to the question "Where is a given category?" The reader need only record the time to identify all the Samo tribe, for example, in figure 1 on page 322, in order to fully appreciate the efficiency of figure 4 on page 323.

Texture, orientation, shape
In point representation (figure 1), with signs having equal visibility (of the amount of "black" per sign), the graphic designer can utilize:
– texture, which has two steps and is selective;
– orientation, which has four steps and is selective;
– shape, which has an infinite number of steps, but is not very selective.

However, if shape variation involves the difference between a point and a line, between a straight line and a broken line, between a line and a cross, shape can provide four selective steps.

The combination of these three variables (figure 1) provides sixteen different signs.

They are relatively selective, and figure 4 on page 323 is based on them. The use of numerous other shapes produces different signs when the reader looks at them one by one. However, these signs are similar (figure 3) on the intermediate and overall levels of reading and prohibit the spontaneous grouping of all the signs of a given shape; there is no selection in the map in figure 1 on page 322. Combinations of signs (figure 2) can be used in simple distributions (grouped geographically).

Facility of drawing
It requires several hours and specialized graphic equipment to draw figure 1 on page 322, which creates a nonselective figuration. It requires one half-hour and a simple pen to draw figure 4 on page 323, which creates a highly selective drawing. One need only attend to the parallelism of the oriented marks, and to the equalization of the amount of black per mark (figure 8).

Note that the signs in figure 5 are differentiated better than those constructed at 45 degrees (figure 4).

Color
As we learned earlier (page 85), a "color variation" implies marks of the same value; since this is the case here, color can be applied to the present problem.

At equal value, for point signs, generally of small size, the maximum number of colors is six: gray, violet, blue, green, brown, and red. The introduction of an additional color (which can only be accomplished by a displacement of all

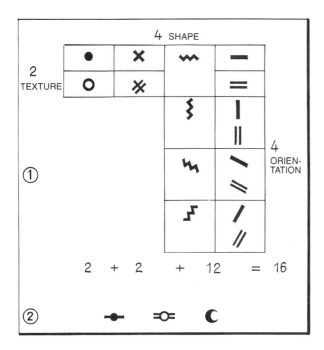

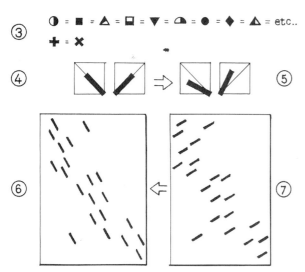

the tones) diminishes the perceptible differentiation among the colors and at the same time their selectivity. On the other hand, the smaller the marks, the greater the loss of discrimination among the colors. As a result, in order to conserve a constant perceptible differentiation, the number of colors must decrease with a reduction in the size of the marks; conversely, this number can increase if one uses larger marks. Furthermore, selectivity diminishes considerably with the complexity of the distribution. In a complete mixture of marks fusion is great and selectivity small.

Remember also that colors produced by traditional pens and inks vary in value. Finally, color requires cumbersome drafting and printing procedures, and its reproduction by most modern means poses practical problems, even if the color variation involves only two steps: for example, black and red.

Efficiency of color

Color variation alone yields only about six selective steps. If we need more than six steps, we return to figure 1. Experience shows that the sixteen steps produced by the combinations in figure 1 are much more selective than the use of sixteen different colors. Furthermore, I am not convinced that six colors are more selective than six combinations obtained from this monochrome table.

On the other hand, if one needs more than sixteen steps, a limit which already considerably exceeds human possibilities of integration, a combination of these sixteen steps with six colors gives 6 × 16 = 96 differentiable categories. However, I do not feel that selectivity is preserved in such constructions.

Visual groupings

In studying figure 4 on page 323, one notices that the points and crosses form a whole which is different from that of the lines (whatever their orientation). Let us imagine that this map also combined two colors (say green and red); then the set of red marks would form a group in relation to the set of green marks, whatever their texture, shape, or orientation.

Thus in the combination of visual variables which ensure selectivity, certain are more "pregnant" and can produce groupings. The designer must take account of this and determine whether the component can usefully *be constructed in groupings of homogeneous categories*. This amounts to imagining the introduction of a new distinction, such as "farmers/shepherds," into the initial series of ethnic categories.

Such an operation occurs in problems involving more than two components. Note, however, that such groupings are only perceptible for two or three different steps (point and line, or violet, green, and red). This phenomenon loses efficiency as the number of groups increases, and becomes useless with six colors or more.

Complexity of the distribution

The mixture of signs plays an important role in selectivity. It is important in all cases, of course, to ensure optimum selectivity by choosing very different signs. However, it is useful to know the distribution before making a definitive choice of signs. There is, therefore, some justification for making *two successive drawings:* a first to discover the distribution, with the help of the temporary signs, and a second which takes account of this distribution in choosing the definitive signs.

It may be preferable, for example, to choose an orientation of signs which emphasizes an oblong distribution (figure 6), rather than an orientation which breaks it up (figure 7), or to utilize the least selective signs, such as those in figure 2, for the most tightly grouped distributions.

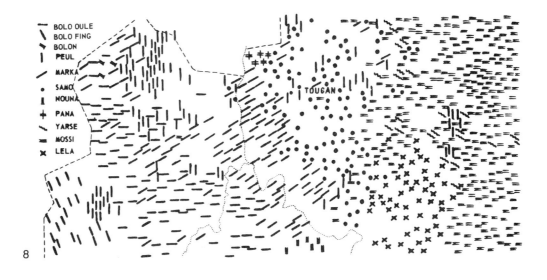

Special case: Several signs on one point

In the preceding example, involving ethnic categories, each point on the plane, i.e., each geographic situation, is characterized by only a single sign. But the information can include several characteristics per geographic point: for example, several industries, several ethnic groups, several administrative organizations per city. This obviously results in a much more complex distribution and an absence of homogeneous groupings on the plane. *Selectivity will be very difficult* to obtain.

Let us recall the general solution to a ≠ problem (page 159).

We construct:

(a) *a series of maps, one per category* of the component, each map capable of providing a response to the question "Where is a given category?" (see page 159).

In fact it is possible to construct one map for two or three ≠ categories (by choosing very different distributions) and to reduce a component with ten or twelve categories to two or three maps.

(b) *a map superimposing all the categories* and capable of responding to the question "What is at a given place?"

Several solutions are possible for drawing this second type of map in point representation.

Figure 1. *An alignment of signs* favors the lettering and reading of the signs, especially if they are always aligned in the same order. However, it disperses geographic groupings and can create positional ambiguity (see page 157).

Figure 2. *A group of signs* favors geographic grouping but, on the other hand, renders reading more arduous.

Figure 3. *A combination of signs* which can be superimposed on a point favors visual grouping, but the number of the combinable signs is limited. When combined, signs become clumsy and risk giving the visual impression of new signs.

Figure 4. *A constant table (chartmap).* One establishes once and for all a table of all the signs (rectangular or polar). It is recorded on each position, and one need only fill in the appropriate boxes. The whole figure is a chartmap (page 119). This system facilitates elementary reading, comparison, and drawing, but the figure is overloaded.

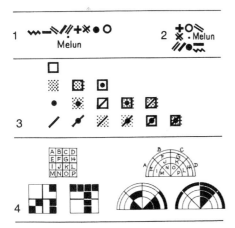

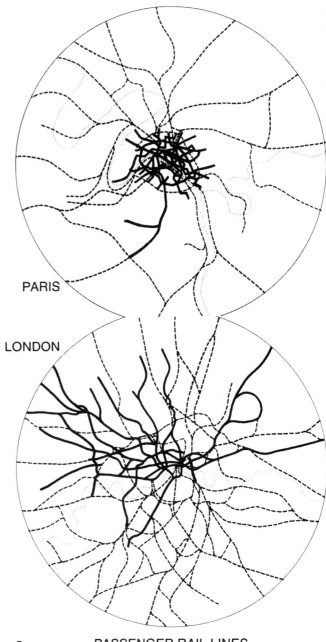

5   PASSENGER RAIL LINES

GEO ≠ Line

CONTRAST BETWEEN BASE MAP AND NEW INFORMATION

The geographical reference elements (**GEO**) constituting the base map most often take the form of lines (meridians, parallels, coasts, rivers, borders, etc.).

When line representation is used for the qualitative component (≠), the first task of the graphic designer is to separate clearly the lines belonging to the base map from those constituting the new information.

A problem **GEO** ≠ (line) thus involves two fundamental visual steps: one, as light as possible, is allocated to the base map; the other, more powerful, is reserved for the new information.

The lighter the first step, the less it will be necessary to enlarge the lines of the second in order to create a sufficient differentiation. Consequently, even though it goes against common practice, the designer should not hesitate to represent the Seine or the Thames by very fine dotting, as in figure 5, in order to increase the legibility of rail passenger transportation in Paris and London (the new information). The curious reader can refer to *The Geographic Review* 49, no. 2 (April 1959): 156–157, and observe the difference in legibility for strictly similar information.

Once this initial problem is resolved, it is within the framework of the new information that one must assess the opportunity for an ordering of the ≠ component. When such ordering is not desirable (when, for example, it is necessary to perceive the overall density of an entire network of lines), this poses the problem of achieving selectivity while maintaining constant visibility.

THE SELECTION OF LINEAR SIGNS HAVING EQUAL VISIBILITY

Except for color, whose contingencies have already been discussed, the retinal variables offer fewer differentiable steps in line representation than in point representation.

At constant visibility, and provided we can use relatively thick lines (1 mm approximately):
– texture provides from four to five selective steps;
– orientation offers two;
– the contrast between edgings or patterns along a line affords two.

Combinations of the above variables can produce approximately nine selective steps in line representation (figure 6).

But in fact, it is tempting also to use a slight variation in visibility. Such is the case with the map in figure 7. Combinations with color can create some forty steps.

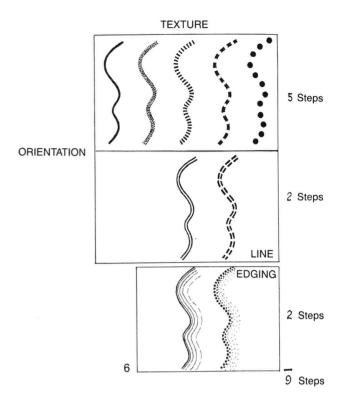

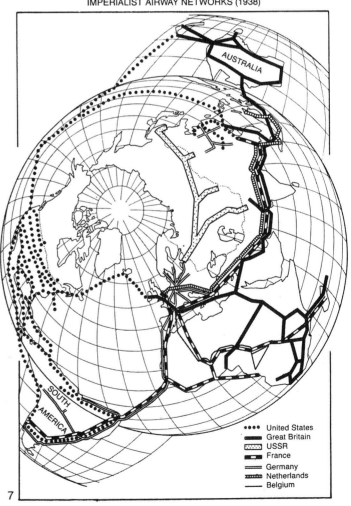

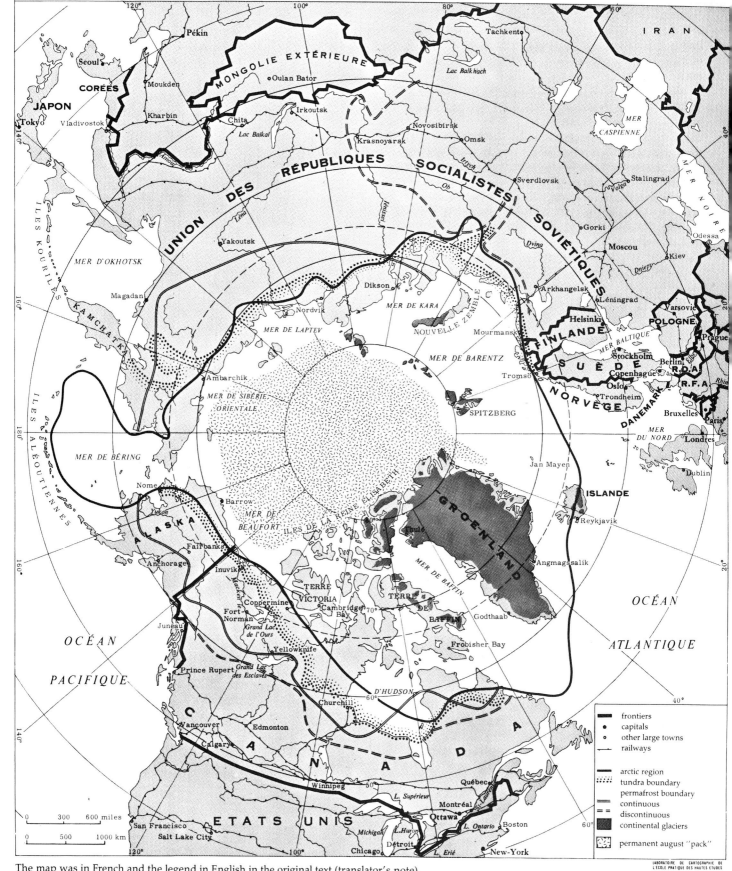

The map was in French and the legend in English in the original text (translator's note).

Nonselective differentiation

In elementary reading, it is necessary to differentiate one line from another, without having to consider the problem of its visual selection on the overall level. For example, we must be able to distinguish rivers, railways, and map coordinates in figure 1.

These differences are based on *texture* variations (which have no selective power when applied to very fine lines) and *shape* variations (which can be added to the lines or take the form of *differences in angularity*).

Note that:
(1) straight lines or very regular geometric lines like meridians and parallels are easy to distinguish from wavy lines,
(2) by the same token, differences in angularity enable us to differentiate several lines which are similar in other respects, provided this angularity is constant along the line. Since the shape differences added to the lines can be of any kind, the total number of distinguishable lines is considerable. Several examples are given in figure 2.

Certain common signs can assume the character of *universal symbols*. However, when this symbolism works against legibility and efficiency, it is generally preferable to sacrifice the symbol. This is the case with the rivers in figure 5 on page 326. Another example is the commonly used system of overlapping crosses which is supposed to represent international boundaries and which certain designers delight in inflicting on the hapless reader.

Note that in order to ensure linear continuity, the dashes must demark the angles, as in figure 2.H, rather than leave them broken, as in figure 2.G.

VARIABLE VISIBILITY

When a *hierarchy* is admissible in the ≠ component, *selectivity improves greatly*, and the number of categories of lines can be increased. The map of different boundaries within the polar region (figure 1) is an example of this.

It is the thickness of the line (size) that most often affords a variation in visibility.

In order to remain selective, *size can be considered to provide three steps*, which, when multiplied by the nine steps defined page 327, gives 3×9, that is, more than twenty selective lines (a loss occurs with small sizes).

Here, size variation is the most selective of the visual variables and can be used to construct *visual groupings* (a grouping of thick lines or medium lines), provided that the steps are clearly marked. In figure 3, groups will form according to 1, 2, and 3, not according to A, B, and C.

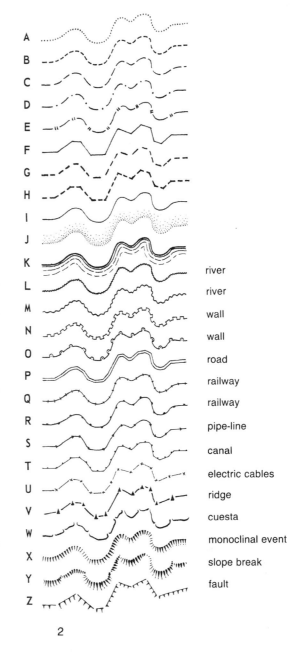

2

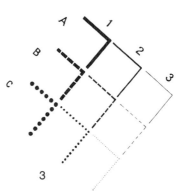

3

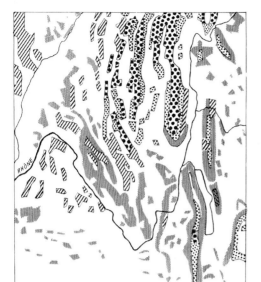

1

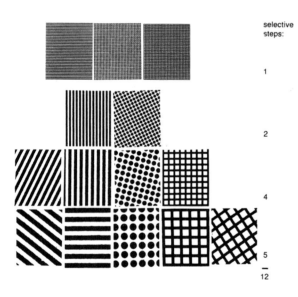

2

GEO ≠ Area

THE SELECTION OF AREAS HAVING EQUAL VISIBILITY

In figure 1, the various types of forest areas do not cover the entire sheet of paper. It can be useful to accomplish the visual sum of the "forests," all types combined. In this case, the categories of the ≠ component (types) are not visually reorderable, and a value variation cannot be used to differentiate them.

At equal value, the selection of area signs (figure 2) should be based on:
– texture, with approximately four useful steps;
– shape, with two to three steps.

The number of perceptible steps depends on the size of the areas; large areas can even lead to the use of differences in orientation (for the coarse textures). Color can also be used to differentiate areas; it has from three to seven steps, depending on the size of the marks and the complexity of their distribution.

Construction of selective signs

Figure 2 gives a total of twelve relatively selective signs. However, they are difficult to draw, and when preprinted patterns are utilized, the choice is not as great as might initially appear. The available patterns are poorly organized, and a logical classing, such as in figure 3, shows that for a given shape and value (for example, a point at 50%), it is often impossible to obtain a series of three different textures. Likewise, a good value progression can only be found for a single texture and two shapes. The patterns using figurative shapes display no visual logic, and the reverse or negative patterns (labeled N in figure 3), only add to the disappointment. In practice, one is obliged to include a slight value variation in the small marks.

The use of preprinted patterns

Preprinted patterns cannot be utilized without taking into account:

(1) *the eventual reduction*. Textures which are too fine do not reduce; they fill in. It is, incidentally, the capacity for reduction which offers the best means of defining a given texture. In photoengraving, a reduction to 50% signifies a reduction of 1000 to 500, that is, a linear reduction by half. A reduction to 75% means 1000 to 750, that is, 4 to 3. In the "texture" column in figure 3 these notations signify that the set of corresponding patterns can support a reduction to 50% (500 0/00), to 75% (750 0/00), or no reduction at all (1000 0/00).

(2) *the drafting material*. A draft on tracing paper or any transparent material can be reproduced by contact printing, without reduction. This procedure affords the best reproduction of patterns. Reduction can be accomplished by using a copy camera with back-lighting, but not all photoengravers have this equipment.

A draft on white paper can only be reproduced photographically, and in this case, the finer the texture, the more the patterns blacken and fill in. A value progression thus risks being destroyed.

PREPRINTED PATTERNS

Functional classing of preprinted patterns

Boxes without numbers denote non-existent patterns

Brands: (1) zentak
(2) Raster
(3) west

Numbers: pattern numbers

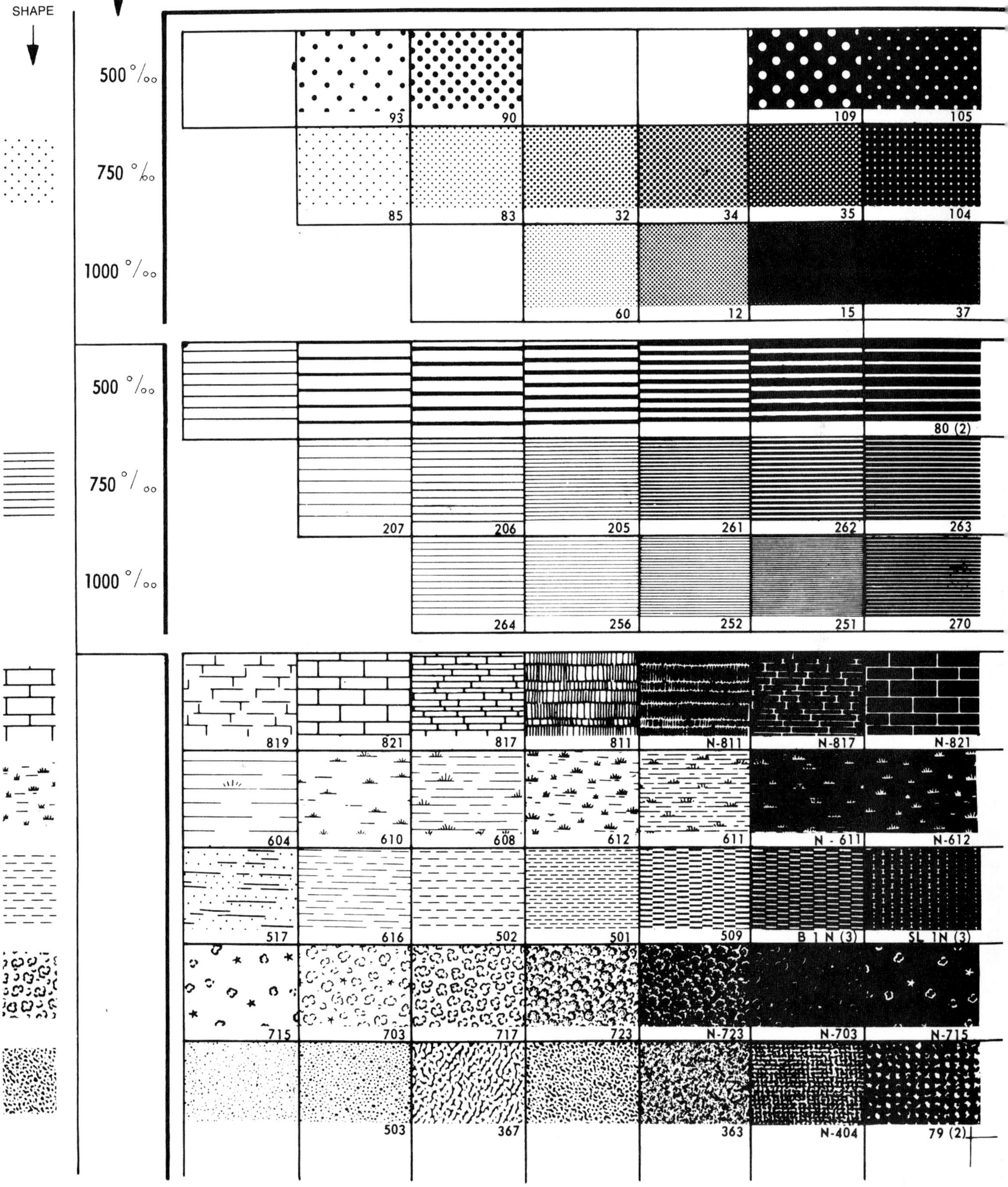

SHAPE	TEXTURE	VALUE →						
dots	500 ‰		93	90			109	105
	750 ‰	85	83	32	34	35	104
	1000 ‰			60	12	15	37
horizontal lines	500 ‰							80 (2)
	750 ‰		207	206	205	261	262	263
	1000 ‰		264	256	252	251	270
bricks		819	821	817	811	N-811	N-817	N-821
grass		604	610	608	612	611	N-611	N-612
dashes		517	616	502	501	509	B 1 N (3)	SL 1 N (3)
stones		715	703	717	723	N-723	N-703	N-715
stipple			503	367		363	N-404	79 (2)

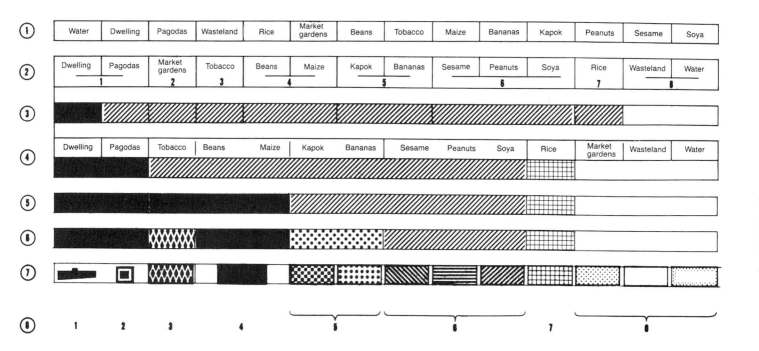

THE SELECTION OF AREAS HAVING VARIABLE VISIBILITY

Whenever the nature of the ≠ component leads to ordering, the number of selective steps increases considerably, and visual groupings can be based on a value variation.

The conception of an ordered legend
Consider the following information:
GEO—*geographic distribution of:*
≠ —*fourteen types of area: water, dwellings, pagodas, wasteland, rice, market gardens, beans, tobacco, corn, bananas, kapok, peanuts, sesame, soya, in the territory surrounding a Cambodian village* (after J. Delvert, *Le Paysan cambodgien* [Paris: Mouton, 1961]).

Interest should be focused on the differences in distribution between the dry season and the wet season. The use of selective signs is thus appropriate to compare a given ≠ category from one map another.

Figure 1. At the top of a large sheet of paper the designer places a list obtained directly from the *raw data*. This list is the point of departure for the ordering procedure, which will use a visual range from black to white.

Figure 2. Immediately below this, the designer orders and groups the various categories, according to a particular criterion. Here, for example, the ordering can be based on the economic or social importance of the various categories. Such distinctions require the introduction of *new elements* into the information. Since the drawing is to be visually ordered, *this order ought to correspond to a meaningful criterion*. The designer will have to determine this through personal research or discussion with the principal investigator or author.

Figure 3. Once the order of importance is established, it seems logical to associate black with the dwellings and white with wastelands and water.

Figure 4. *Only five value steps* can be used to achieve selectivity, which means that some categories will have to be combined. The *pagodas* can be grouped with the dwellings. The *market gardens* appear to be linked to the dwellings, but they do not change place from one season to another (and are also in the hands of non-Cambodians); they are recorded in light values, simply to highlight the black of the dwellings, which they tend to be near. *Rice* generally covers large areas, and the gridding of the rice fields is a suggestive image which can be drawn in a very light value.

Figure 5. The medium gray can be divided in two. After consultation with the author, the designer decided to group tobacco, beans, and corn as "most important." *The five value steps are thus constituted.*

Figure 6. For further diversification, texture is available. It is appropriate to give a textured character to the most "important" element of each group (since texture generally "stands out"); the texture-value combination thus provides *seven selective steps.*

Figure 7. Now we need only to differentiate several remaining categories, for a detailed reading: (a) separation of beans and corn will not be attempted; (b) *shape and orientation* can be used at this stage, since they do not destroy the selectivity which has already been achieved; (c) the particular topography of the pagodas enables us to use a pattern whose effect that is equal to that of black.

Figure 8. This method of progressive analysis produces a range of thirteen categories, distributed over eight selective steps, which are comparable from one map to another.

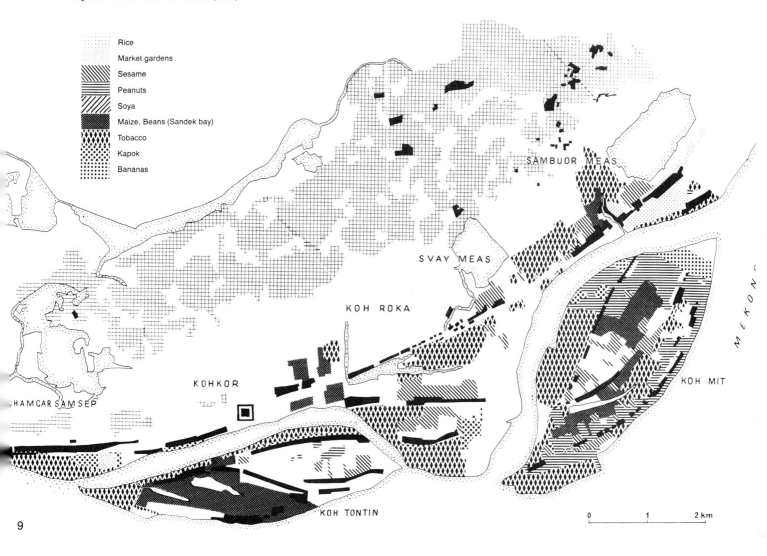

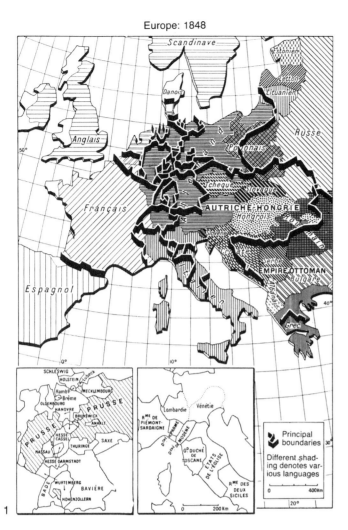

Europe: 1848

SPECIAL CASES

A single area per sign

If we exclude the national borders, the map in figure 1 divides linguistic areas into two groups, differentiated by value. The light value designates linguistic areas which coincide with national borders; the dark value denotes areas not coinciding with borders. Within these groups, an overall selection is useless, since each sign characterizes only a single area. Variations of orientation and shape are therefore sufficient. Their combinations are innumerable and can create, as here, twenty-five signs which will be perceived as different on the level of elementary reading. See page 94 for area signs with equal visibility.

Total superimposition of areas

If one does include the borders, the map in figure 1 becomes the expression of information having three components: **GEO** (area); ≠ languages; ≠ countries. This information raises the problem of the superimposition of areas.

The best solution for this problem, in which the two area systems cover the entire plane, involves the use of a *difference in implantation*. The different languages can be represented by area differences, the different countries by linear signs.

Partial superimpositions

These always create a delicate graphic problem. Consider representing the butter (B), lard (S), and oil (H) consumption in the French kitchen.

Initially we will assume that the boundaries of the areas are clearly defined, as shown in figures 2–5. We should avoid figures such as 2 and 3, which do not suggest a superimposition of butter and lard in the same place, but rather the existence of new categories. These are suggested by the presence of new signs (X), formed by the combination of two or three of the original signs. This ambiguity does not exist in figures 4 and 5, which convincingly demonstrate the selectivity of texture.

Textures of the same power blend and construct new signs. This is what happened in figures 2 and 3. Different textures combined with orientation (figure 4) or with shape and a variation in implantation (figure 5) are distinct and can be superimposed while remaining perfectly separable. This is the graphic solution for the problem.

Gradation

Now we will assume that the boundaries of the areas are blurred. The drawing must suggest a progressive gradation. Points and lines (figure 6) can both suggest a gradation in the textured marks. When the texture is very fine, in the case of flat tints, for example, a staggered arrangement or ribboned bands (figure 7) can be used. As far as possible, the designer should choose an orientation which is perpendicular to the general line of contact between the areas.

The bands are capable of suggesting the variable proportion between the two superimposed categories (figure 7). The width of the bands should be relatively constant at around 3 mm, since too wide bands suggest homogeneous areas within each tone, and too narrow bands pose difficulties of technical realization and visual identification (fusion of the marks).

In geographic arrangements the lettering can sometimes be highly evocative; it can replace the legend while suggesting the general distribution, verbal equivalences, and the real gradation of the phenomenon (figure 8). But it is only utilizable when the distribution on the plane is relatively simple.

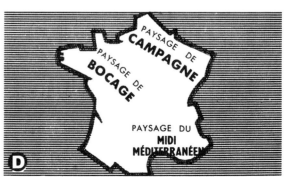

8

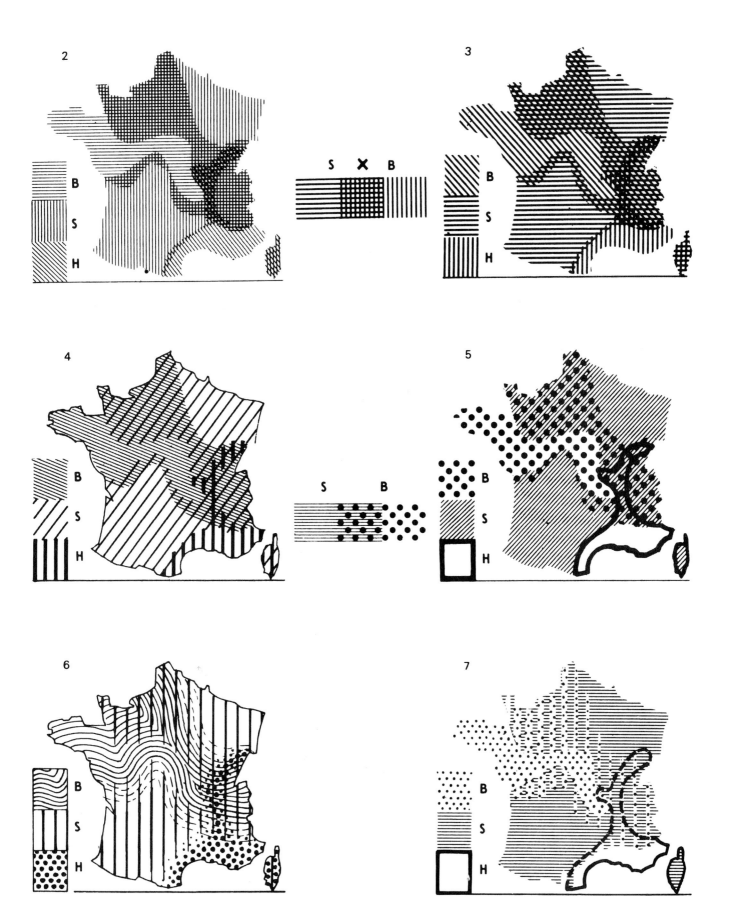

2. Maps GEO O (a geographic component; an ordered component)

A component is ORDERED, and only ordered:
(1) when its categories form a single, universally acknowledged series;
(2) when we accept, a priori, that the same distance exists between each category.

This second characteristic distinguishes an ordered component from a quantitative component. A component is *quantitative* when it manifests both an order and *a variation in quantitative distance* among it's categories.

However, we can use a series of measurements or enumerations as the means of defining categories which are considered as *merely different and ordered*. Although the categories may be characterized by numbers, these are simply ordinal numbers (page 37), and the categories are defined, a priori, as equidistant.

The graphic representation must attempt to maintain this equality of distance or, if one prefers, it must attempt to avoid a priori visual groupings.

Most cases involving order result:
– either from ≠ components to which we attribute a meaningful order, based on some underlying criterion (we have seen several examples of this in section **GEO ≠**, especially on page 332);
– or from a division into ordered categories derived from an interpretation of measurements, dates, or enumerations.

Ordered representations

These are depicted in figure 1 and are derived from the table on page 96, with which the reader should be readily familiar by now.
O   Ordered representations rely on variations in size, value, and texture.
≠ O Most problems will admit variable visibility and can be represented by size or value.
≡ O Some problems require constant visibility (associativity). They will therefore involve the use of texture, which will be limited to some three or four steps.
≠ O Most problems also require selective perception. Size, value, and texture can be combined with other variables, but the number of steps will be limited.

GEO O point

Ordered series of points
Figure 2 shows ordered, selective series which are easy to construct. The different series in the left column are based on value and exclude size, which always evokes a certain proportionality. The associative series using texture will favor perception of the sum of the points, all signs combined; i.e., figure 4 resembles figure 3, not figure 5. However, one can still see an order in the categories of points in figure 4.

Very small points
The use of very small points (or dots) frequently poses the problem of their perceptibility. One solution is a size variation combined with a point-line distinction for the smallest marks (see, for example, figure 6, depicting "Earthquake Epicenters in the North Atlantic, 1910–1956," from C. H. Elmendorf and B. C. Heezen, "Oceanographic Information for Engineering Submarine Cable Systems," *Bell System Technical Journal* 36, no. 5 [Sept. 1957]: 1047–1093.)

This is also an example of the ordered interpretation of quantitative information.

Note that this graphic solution is easily drawn and relatively efficient, even for "freehand" drawing.

Further, with such sign dimensions the largest category is perceived in its proper place in the series, even though it does not involve a greater amount of black than the sign used for the next largest category.

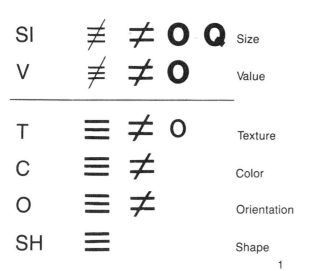

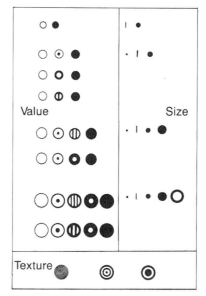

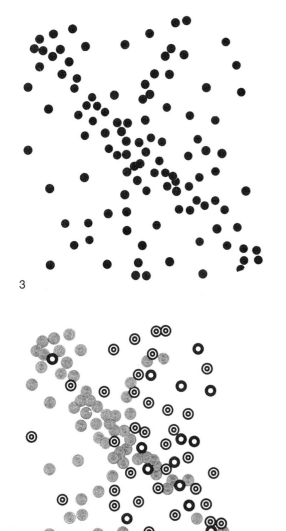

3

4

5

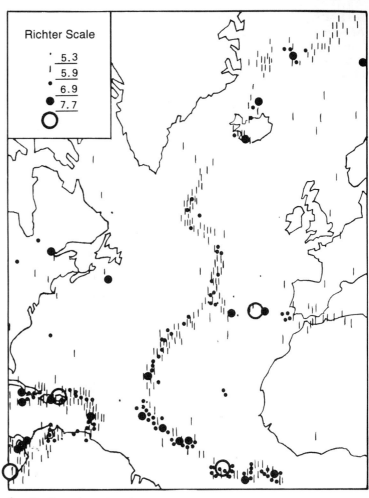

6 Earthquake epicenters
in the North Atlantic 1910–1956

337

Several signs on one point

As with ≠ components, this problem is always delicate.

In the example below, Benedictine abbies at Reims were constructed in the region at various times, often at the same place in several different periods. The series of marks in figures 1, 2, and 3 are not efficient for solving this problem; they are not ordered. Those in figures 4 and 5 are ordered and produce an image (figure 6). A combination with orientation is particularly selective here. Incidentally, with historical problems, one generally associates black with the most ancient period.

GEO O line

Size variation (width of the lines) offers the most convenient and efficient solution for this type of problem.

Ordered and directed lines

Figure 7, which shows itineraries of explorations in the Sahara classed over time, solicits selective perception as well as ordered perception. Furthermore, it is important to indicate a direction for each itinerary. This is achieved by combining size, shape, and texture.

Ordered lines derived from quantities per line

Consider the problem of superimposing eleven administrative systems (economic regions, social security, work inspection, postal regions, military regions, regional planning areas, etc.) that were obtained from a preparatory survey for the determination of administrative regions. The various departmental boundaries can thus include from one to eleven regional boundaries (figure 8).

In this figure, the reader is confronted by eleven steps, irregularly drawn; this number must be reduced in order to arrive at a useful image. The following ordered categories are retained: 1 and 2; 3–6, 7 and 8, 9 and more (figure 9).

These categories closely approximate the boundaries of the regions which were eventually retained.

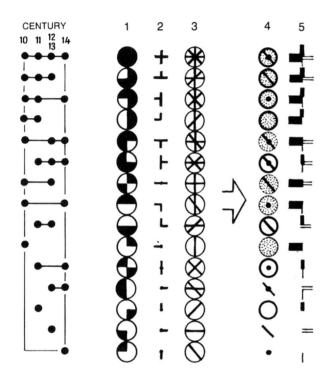

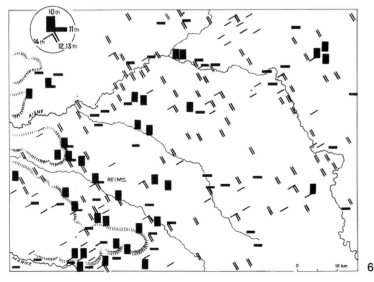

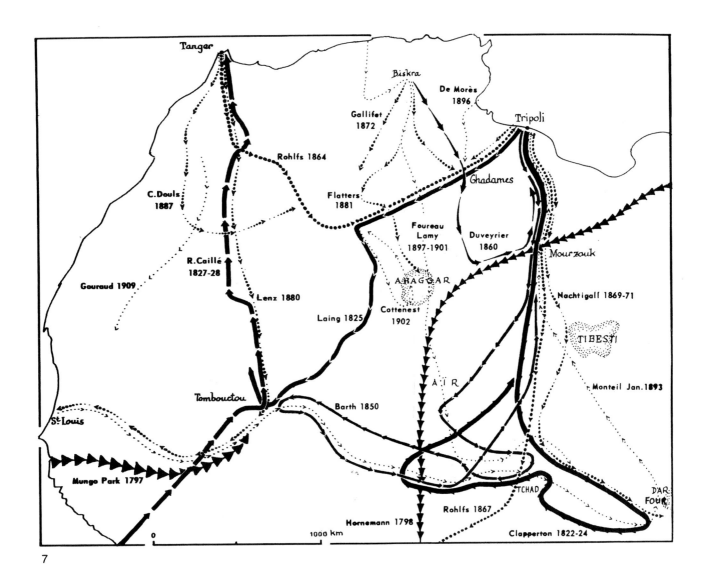

7

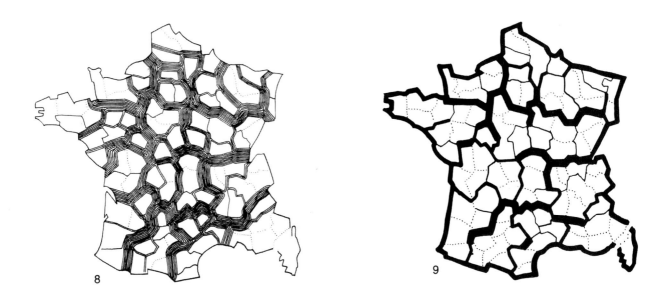

8

9

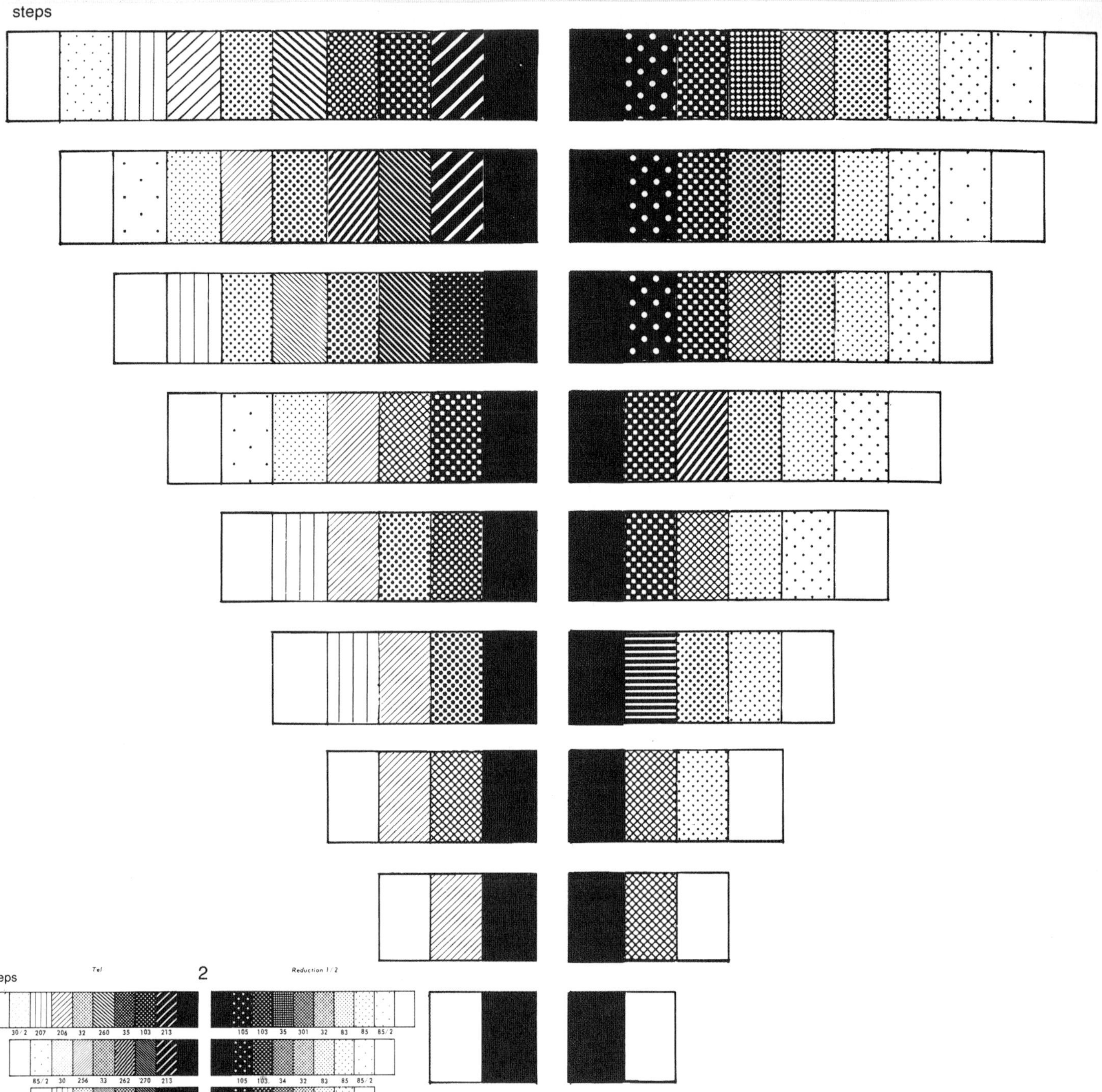

steps

steps

10																		
	30/2	207	206	32	260	35	103	213		105	103	35	301	32	83	85	85/2	
9	85/2	30	256	33	262	270	213			105	103	34	32	83	85	85/2		
8	207	31	252	34	270	37				105	103	301	32	83	85			
7	85/2	30	256	301	103					103	262	32	83	85				
6	207	256	33	25						103	301	83	85					
5	207	256	34							263	32	83						
4	256	301								301	83							
3		256								301								
2																		

Tel   2   Reduction 1/2

1

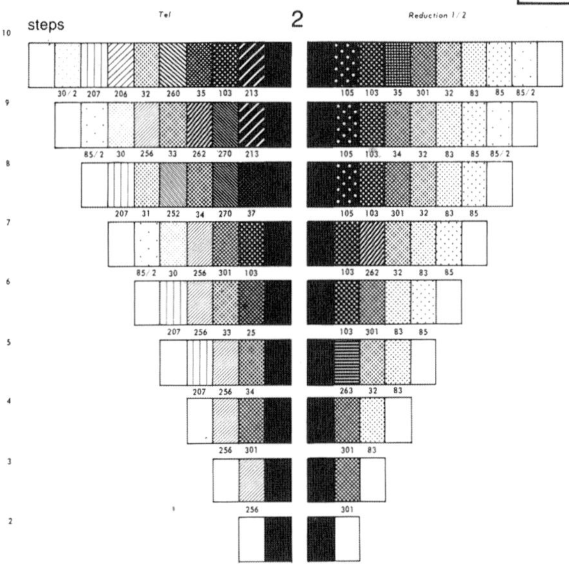

Series of values in combination with texture and shape in preprinted patterns (denoted by numbers)

GEO O area

Value variation, obtained by tones of gray, provides the best solution for problems of this nature.

Order and selectivity
A combination of value with texture, shape, or orientation favors selectivity, as in the example on page 332. Once the order is determined, we can construct a legend which is both ordered and selective.

Useless selectivity
When selectivity is of no practical use, combinations involving tones of gray must be evaluated in terms of photographic reduction. Tones which have too fine a texture do not reduce, and in microfilmed documentation their reconstruction is very poor.

Several series of preprinted grays are given in figure 1. Those series on the left should be avoided whenever large reductions are envisaged, such as that which produced figure 2.

The map in figure 3 is a typical example of an ordered series. The information gives only ordinal numbers. Note that, contrary to the general rules of legibility, black is not used for the first step. This is because so much black here would have weighted down the image too much.

The series of tones is selective, even though the information probably won't generate questions of the type: "Where are the different areas in the third ranking?"

The maps in figure 4, on the other hand, provide an example of an ordered interpretation of a quantitative series (after *The Geographical Review* 49 [1959], figure 11, published by the American Geographical Society of New York). The usefulness of comparing the extent of areas which are *similar* (in quantity) justifies a nonquantitative interpretation here.

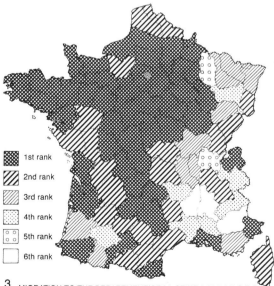

3 MIGRATION TO THE DEPARTMENT OF LA SEINE (1962) I.N.S.E.E.
Each department shows the rank of La Seine in the order of departments migrated to.

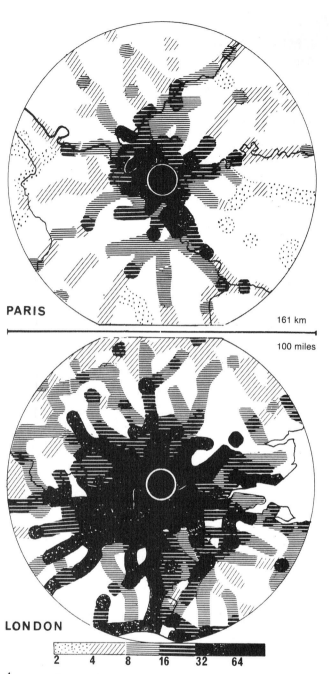

4 Number of passenger trains per day

341

GEO O: The representation of movement on the plane

TIME AND MOVEMENT

Time, like geographic order, can be introduced into any analysis. As a naturally ordered component, it is a universally identifiable concept, on which innumerable comparisons can be based. It can be quantitative, since the day, the year, and recently the atomic second, are recognized as sufficiently stable units.

When the information involves a time component, the correspondences translate a "variation": variation in temperature, price, or size....

Time is linear, and a single dimension of the plane is sufficient to represent it. The other dimension can be used to represent another component. The construction is a "chronological" diagram (figure 1).

When the information involver both TIME and spatial or GEOGRAPHIC ORDER, the correspondences translate a MOVEMENT: movement of the pendulum; migratory, demographic, or social movement.... But when the two planar dimensions are utilized to represent space, no planar dimension remains available to represent the "time" component; this is the basic problem with the representation of movement in cartography. There are three solutions.

– Construct a *series of images* (figure 2). As in the cinema, this solution can be applied to the most complex of movements. But here the number of images is limited by the reading process: with a long series it is difficult to suggest motion.

– Represent *the path and direction of a moving body* (figure 3). This solution can suggest a continuous movement on the plane, i.e., MOTION.

However, we must consider both the nature of the moving body, which can be a point, a line, or an area, and also the complexity of the movement (with or without reversal), which can only be perceived with a simple division of the plane.

– Utilize a *retinal variable* (figure 4). The time component is divided into ordered categories represented by the different steps of an ordered retinal variable. This solution depends on the possibility of defining a small number of categories, since the ordered retinal variables are relatively limited. The figure does not generally suggest movement on the plane. It is a representation of the type **GEO O** (see page 336).

PRINCIPAL TYPES OF MOVEMENT

Movements can trace more or less complex forms on the plane, and the construction of a series of images always permits analyzing them. But the other solutions can be more efficient for representing and animating simple movements.

A continuous movement

The movement of a vehicle, for example, is best expressed by the "trace" of a moving body (figure 5). This *trace corresponds to a change in implantation;* a point traces a line, a line or an area traces an area. But with complex movements (reversals) the area becomes cluttered and indicates direction poorly.

As a result, **only the point produces a moving body capable of suggesting continuous complex movement. It traces a line, which becomes an ARROW when it is directed.**

It is therefore logical in most cases to depict movement by an ARROW (see page 346). The reader is thus asked to accept two conventions: (1) to identify the moving body, whatever it may be, with a point; (2) to consider the trace of this point as the representation of the movement of the body.

A "generation" of points, lines, or areas

The movement of political borders, for example, is a discontinuous movement (figure 6). It can be depicted by a sequence of areas, when the distribution is simple.

This discontinuous representation, like the series of images, can be applied to continuous movements (for example, marine growth), when the title and the nature of the phenomenon preclude ambiguity. Retinal variations of size and value can be applied, often efficiently, to "generations."

Variable speeds

These result from the introduction of a new component: numbers of time units (figure 7). These can be superimposed on lines (length of segment travelled in the time unit) or trace areas (ISOCHRONES).

Systems of relations

These are the result of the accumulation of numerous movements, actual or figurative (figure 8). Their individual direction and motion are generally not elements of interest.

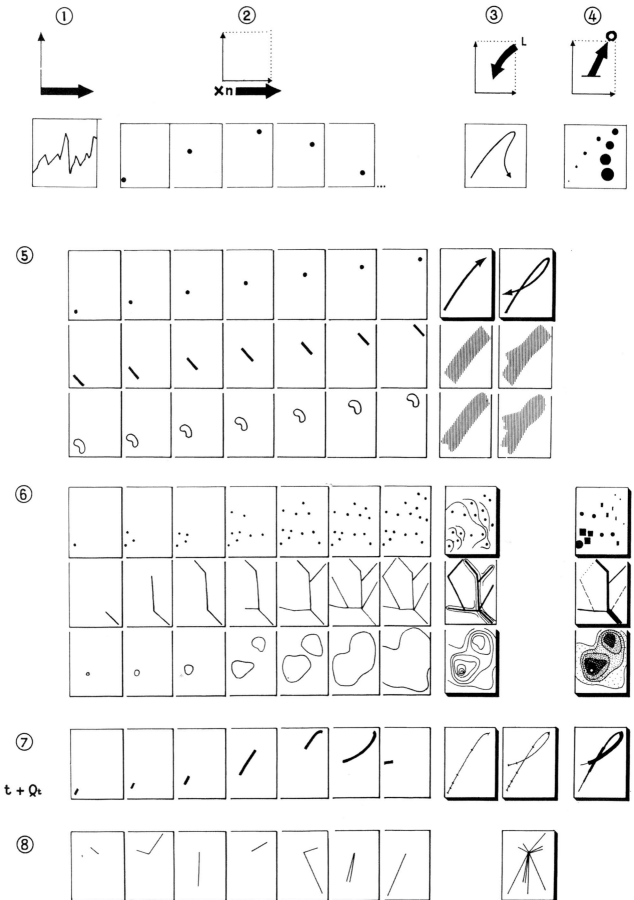

EXAMPLES OF NONDIRECTED RELATIONS

Figure 1. Workers in Casablanca (*Industrialisation de l'Afrique du Nord*, published under the auspices of the Centre d'Etudes de Politique Etrangère). The radiating lines suggest the migratory nature of the phenomenon being represented.

Figure 2. Movement of a person in Paris (P.-H. Chombart de Lauwe, *Paris et l'agglomération parisienne*, cited on page 315). The trace of all the movements of an individual can indicate his or her work, standard of living, and number of relationships.

Figure 3. Mediterranean wheat trade in the sixteenth century (*Mélanges L. Febvre*, vol. II [Paris: A. Colin, 1953]). Tracing the actual itineraries is not sufficient for representing a system of relations. A map of maritime routes, even when weighted, does not show the direction of trade among the centers of activity; it shows the density of the ships at sea. The maritime trade among the cities of Europe and the Mediterranean will only appear in its diversity, weight, and geographic direction, when each connection, even though maritime, is represented by a straight line (figure 4).

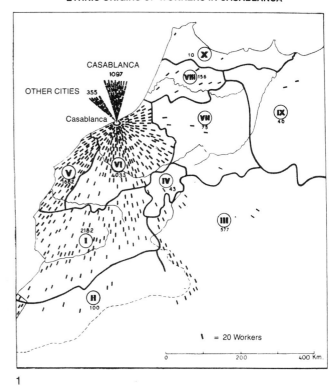

1

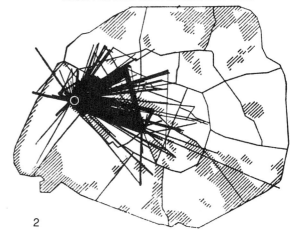

2

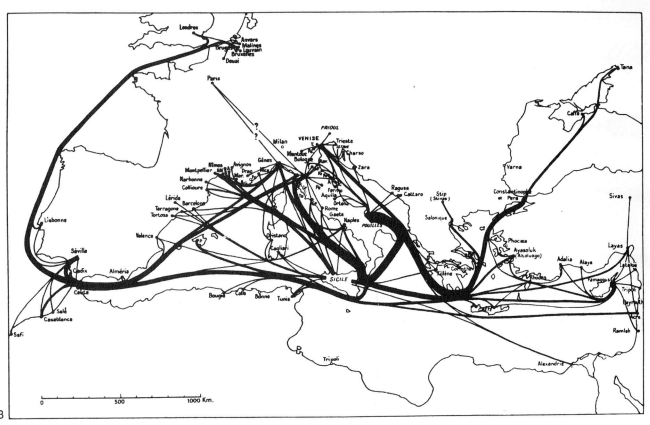

3

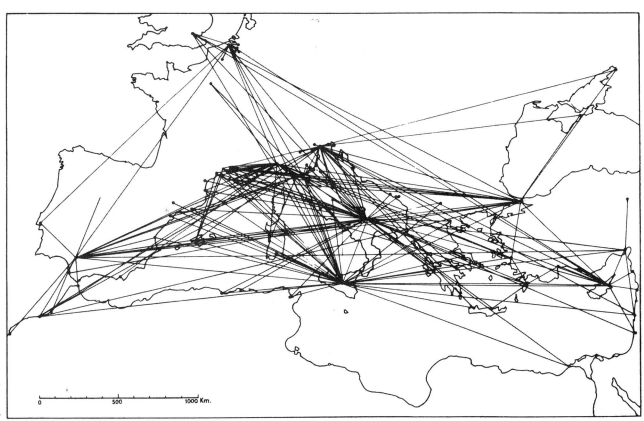

4

345

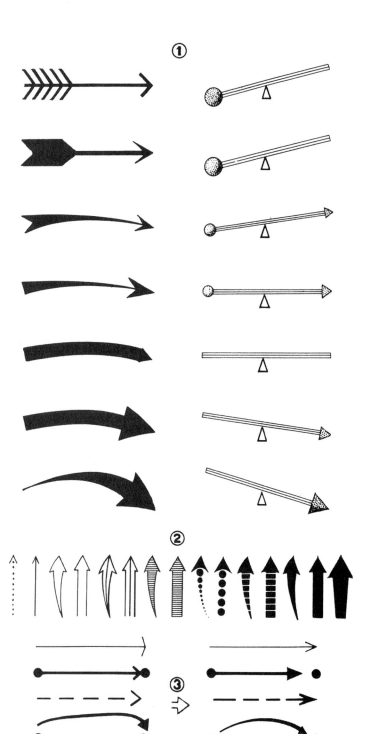

EXAMPLES OF DIRECTED RELATIONS (VECTORS)

The *arrow*, the directed trace of a point, remains the most efficient and often the only formula for representing the complex movement of a point, and, by analogy, that of a line or an area.

The design of the arrow
The use and form of the arrow distinguishes the good designer, one who is conscious of providing a visual response on all levels of reading.

Figure 1. The development of the arrow as a conventional sign (left column) indicates the slow discovery of the role of this symbol (in combination with other visual elements) in the overall perception of the image. The "old-fashioned" signs must be "read" individually, due to errors of visual equilibrium (right column) and the use of "figurative" shapes (arrow or hand).

As these begin to disappear, we come to the "neutral," balanced arrow, an intermediate stage at which important conventional systems (such as the international highway code) have remained. Modern usage and research into visual efficiency indicate that the later forms, barely twenty-five years old, are more efficient.

The "visual weight" attracts the eye in the correct direction. Perception is less ambiguous and much more rapid (figure 5).

Figure 2. Types of arrows varying in size and value.

Figure 3. Here, errors that should be avoided are shown in the column on the left.

346

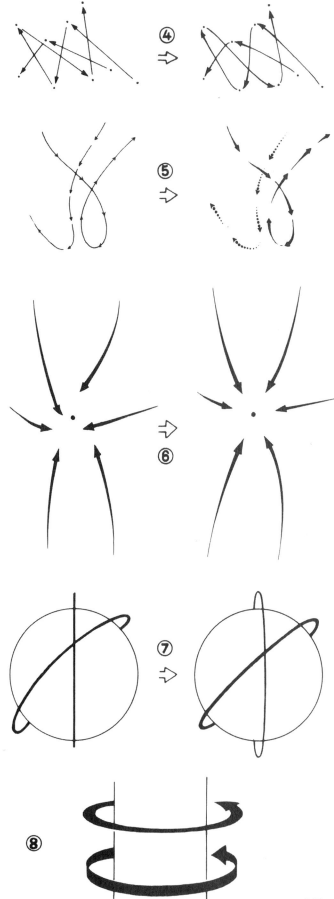

Figure 4. A broken line is a collection of different visual units. A curve, on the other hand, suggests the idea of continuity; the elements are perceived as extended and linked.

By asking the reader to follow a curve, even a rather complicated one, the path of movement appears as a unified whole.

Figure 5. In a complex movement, one must attempt to facilitate the perception of direction. The use of repeated cues, linked to a sequence of detail perceptions, should be avoided. The weight of the arrow, by suggesting the direction of movement, facilitates the reading operation.

Figure 6. When the movements are supposed to be radiating or converging, the axis of the arrow, which the eye unconsciously continues, must pass through the central point. Furthermore, the notion of convergence is strengthened if all the points of the arrows suggest the same circle, whose center is the point convergence.

Figure 7. Any plane secant to a sphere traces a circumference on the terrestrial surface. In perspective it appears as an ellipse. A circle projected on the visual center of the sphere will appear as a straight line in perspective. It should be avoided.

Figure 8. The path of ellipses surrounding a sphere or a cylinder can produce perspective effects which increase the impression of real volume.

The movement of a point

Figure 1. The flow of market clientele at Hili-ba in Chad is a standard type of two-way movement. But the products change with the direction. If one merely uses labels to portray these changes, the map will only produce an overall image of the general movement of trade. To learn more about particular products, each itinerary must be read individually.

The market economy is made much clearer when several smaller maps separate the movements by product, and the geographic characteristics of the Hili-ba center can still be perceived in several images (after M. H. Tubiana, "Le Marché d'Hili-ba," in *Cahiers d'études africaines* [Paris: Mouton, 1961]).

Figure 2. European expansion from the year 1000 (the true beginnings of European civilization) can be suggested by different movements of conquest and reconquest. Note the visual advantage obtained from the use of dark arrows for the regions of movement, while white areas depict the bases of departure (Morazé, Wolff, Bertin, *Manuels d'histoire* [Paris: A. Colin, 1950]).

Figures 3 and 4. Electoral campaigns for the Brazilian presidency in 1950. The candidates' travel itineraries underscore the difference between the complex organization of C. Machado's campaign (figure 3) and the logical simplicity of that of Getulio Vargas (figure 4). The latter was elected.

The numbers indicate the date and place of speeches. The fine dotting groups the places visited on the same day. The form of the arrows differs according to the month (C. Morazé, *Les Trois âges du Brésil* [Paris: A. Colin, 1954]).

Figure 5. Speeds are superimposed on the ships' courses linking Spain to America in the sixteenth century. The slower the speed, the less the space crossed in a day, and the closer the lines used to represent this. Thus slower speeds stand out in darker print (H. and P. Chaunu, *Séville et l'Atlantique* [Paris: S.E.V.P.E.N., 1956]).

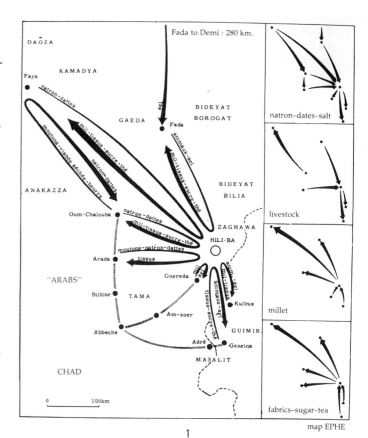

1

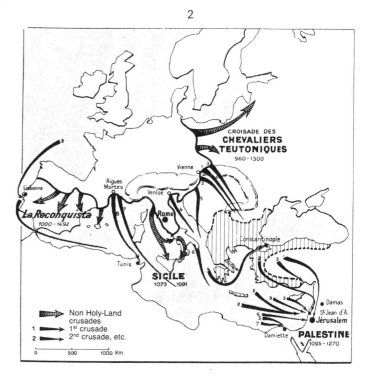

2

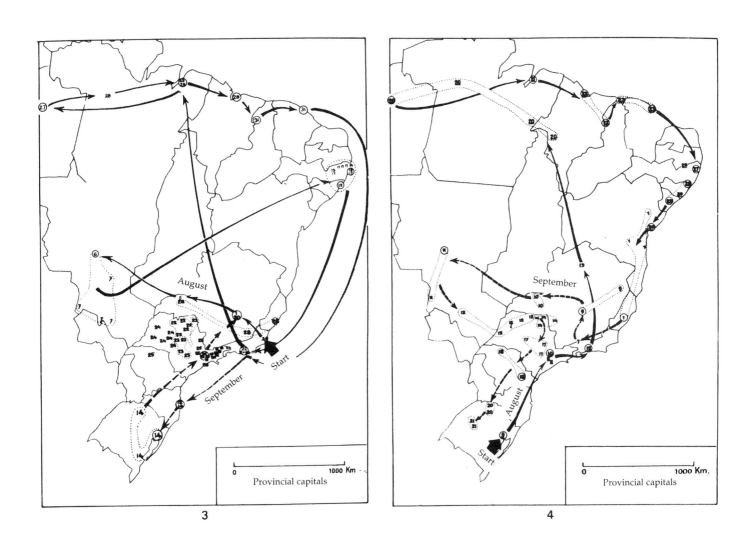

3

4

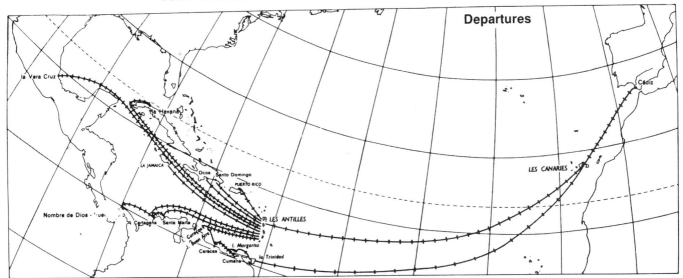

5 **COMPARATIVE SPEEDS FOR PRINCIPAL SHIPS' COURSES**

Departures

The representation of regional migration

This is a classic, highly complex problem.

(1) The observed movements can link any given region to any other region, in either or both directions. The network of the directed movements is therefore very complex in itself.
(2) These movements must also be weighted in terms of absolute quantities.
(3) Finally, the movements must suggest the migratory tendency of a region, expressed by the number of migrants per 100 inhabitants. This tendency is generally inversely proportional to the number of inhabitants.

When it is desirable to represent all these factors on the same map, simplifications are imperative. They can be of various kinds.
(1) Representation of the balance between the two directions will divide the network in half and avoid depicting both directions. But a perception of the total amount of migration disappears.
(2) Not representing movements below a certain quantity will remove a number of details from the information, but these are generally not significant.
(3) Movements below a certain percentage can also be left out. However, a very small percentage can still correspond to an important quantity if the region is highly populated.

Interdepartmental migration in France (1954)

Among the numerous solutions, the map in figure 3 has the advantage of representing the total amount of migration, Paris included, not just the balances.

But two minimums have been set:
(1) All the quantities representing 2% or more of the original population are portrayed.
(2) All the percentages less than 2% are portrayed if they represent more than 10 000 migrants (white triangles).

The area of the arrows is proportional to the absolute quantities.

The length of the arrows (triangle alone) is proportional to the migratory tendency (percentage of emigrants) above 2%.

The arrows relating to the Paris region (Seine, Seine-et-Oise, Seine-et-Marne) are situated in the departments of origin (migration toward Paris) (see figure 2).

The arrows concerning the other regions are situated in the departments of destination (see figure 1).

The composite map (figure 3) obviously highlights the attraction of the Paris region and other large cities, but it also reveals that, with the exception of Paris, the noteworthy movements are almost entirely between neighboring departments. This is a phenomenon of osmosis. (Maps by Serge Bonin, Laboratoire de Cartographie [E.P.H.E.].)

The movements of lines

Here are three examples taken from *Manuels d'Histoire* (cited on page 348).

Figure 4 depicts the battle of Austerlitz. The lines indicate the origin of a given movement. The arrows then trace the direction and development of that movement.

Figures 5 and 6 show Paris on certain days during the revolution of 1848. The progressive advance of the popular masses is represented by successive lines, convex in the direction of the march. On February 23, 1848 (figure 5), these are the only movements. The revolution is succeeding.

In June (figure 6), the government troops repress the insurrection. Arrows of a darker value can be superimposed on the lines to clearly indicate the march of the troops.

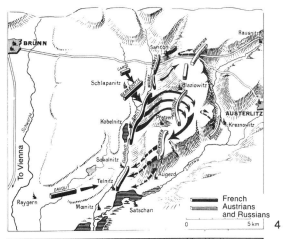

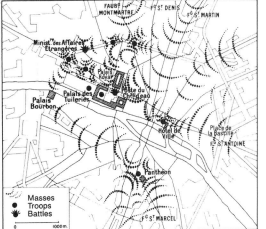

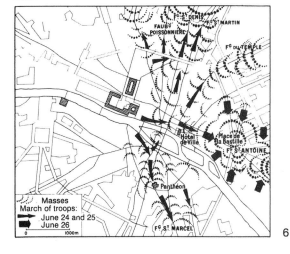

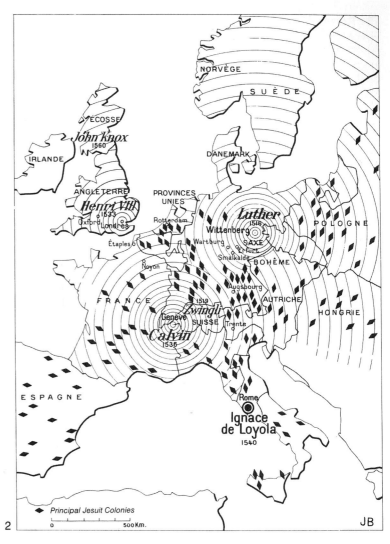

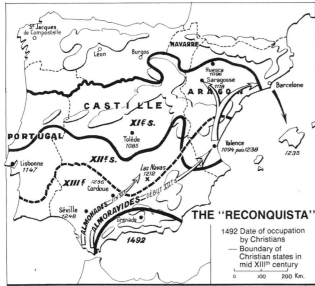

Movements of areas

Let us look at three examples.

Figure 1. The stages of the "Reconquista" involve some backward movement, which means that we must differentiate the lines that show the retreat from the other lines (especially that representing the boundary of Christian states in the thirteenth century). Incidentally, the use of arrows enables us to better understand the momentary reversal of the movement.

Figure 2. The growth of Protestantism provides an example of continuous development in different areas. The concentric circles suggest this progression. Superimposed are the Jesuit colonies, under the centralizing influence of Rome.

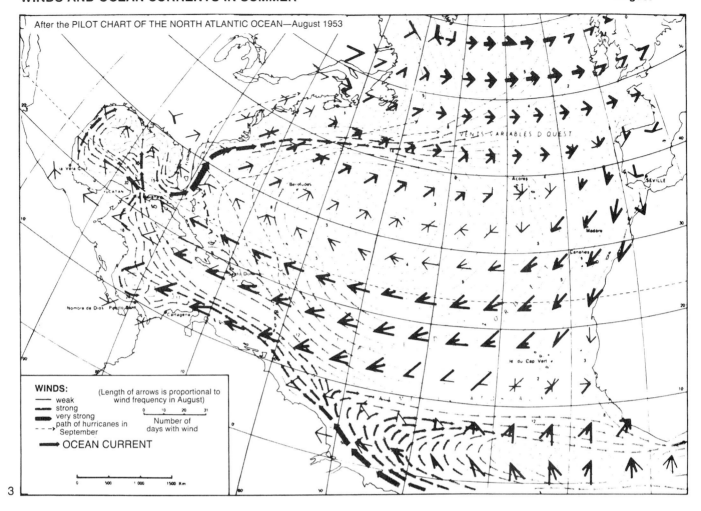

Figure 3. This depicts the direction, force, and duration of area movement (from the Atlas *Séville et l'Atlantique*, cited on page 348).

By using directed signs in a regular manner, it is possible to suggest the general movement of the atmosphere (in black) and to produce a striking image of the winds, based on force (speed) and general direction.

Depicting the movement of ocean currents (figure 4) is simpler, since they only move in one direction. A complete covering of the area by signs permits us to visualize direction and speed in a single overall image.

Note that, in both cases, perception of speed results from the use of a retinal variable: size.

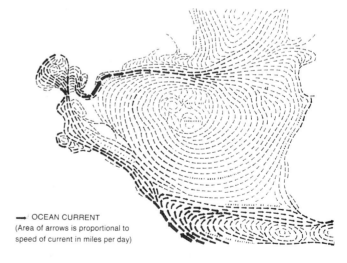

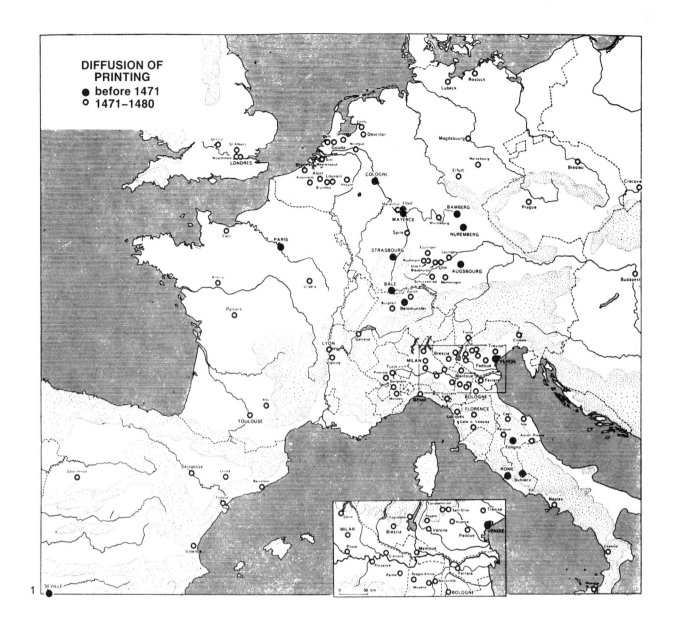

SERIES OF IMAGES

The spread of printing (L. Febvre and H. Martin, *La Diffusion de l'imprimerie* [Paris: Albin Michel, 1958]). The two images in figures 1 and 2 enable us to represent four decennial stages in the spread of printing, provided that the two stages in each image are clearly differentiated (by value here).

Note that the subtle recall of the distribution of the first image (through the use of smaller points) permits a better evaluation of later progress; figure 2 is, in fact, a map of the difference between the two periods.

Figure 3 portrays **the growth of Paris**. For very ancient periods, statistical evaluations are purely hypothetical. On the other hand, archaeological evidence produces a good approximation of the location and extent of built-up areas. Provided that time is depicted uniformly, that is, represented by equal intervals (e.g., centuries), the series of images traces a curve. Here, it highlights the rapid growth of the modern era, dates it, and raises the problem of its

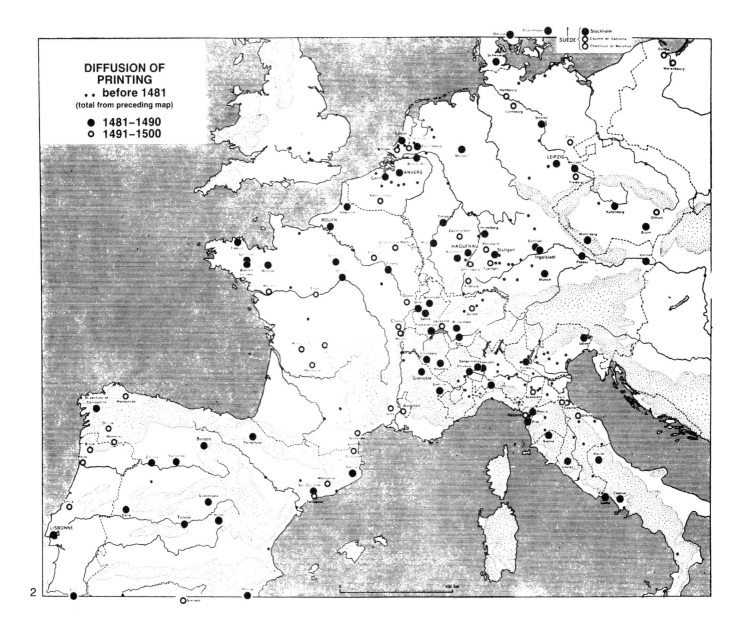

2

nature. Is it a regular progression, in which case an entire page would not be sufficient to represent Paris in 2040? Or, rather, at that time will Paris be not much larger than in 1940, which would produce a drastic change in the overall curve.

With classic historical material, it is customary to choose the dates as a function of main events. Time is not represented uniformly. This is poor utilization of the properties of graphic representation. Since no one would take such a liberty in diagrams, why do it with maps?

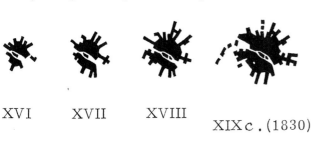

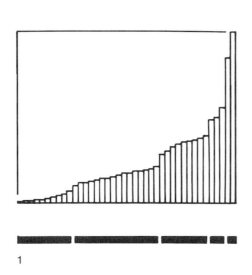
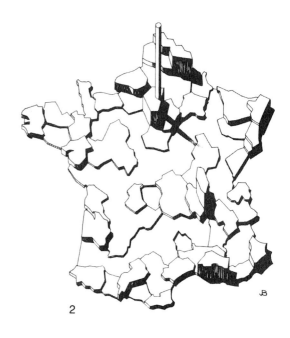

3. Maps GEO Q (a geographic component; a quantitative component)

DEFINITION OF A PROBLEM GEO Q

We will consider a component as QUANTITATIVE when it is agreed that the graphic representation must translate, above all, the VARIATION IN DISTANCE among the categories of the component; this variation is expressed by the quantities.

When the categories are arranged in a line, as in figure 1, the image is a "repartition" histogram (see pages 203–211). The visual groupings, which constitute the steps of this histogram, are independent of the geographic order.

When the categories form a network (geographic or not), the image is a "relief" (three-dimensional histogram), and the visual groupings correspond to the "plateaus" of this relief, as in figure 2.

The map then represents the groupings resulting from the combination of *quantitative distances and geographic distances*. These groupings can be perceived differently from those resulting from the histogram. For example, a step in the latter can be destroyed by the geographic dispersion, whereas ungrouped elements on the diagram can produce geographic areas which are visually more homogeneous (e.g., figure 3, page 375).

Consequently:
(1) A repartition histogram is not sufficient for defining the steps resulting from the complete information **GEO Q**.
(2) Only a three-dimensional histogram (a "relief") is capable of representing such information and constructing all the steps for groupings resulting from the combination **GEO Q**.
(3) Like the diagram, the "relief" must represent all the numbers of the information, AFTER, AND ONLY AFTER, WHICH the observer can define the groupings resulting from the complete information.

Determining the steps is, in fact, the goal of the graphic operation, not its means.
(4) Any choice of steps made prior to the construction of the "relief," even when based on the repartition histogram, will transform the component Q into an interpretation O, whose steps are, a priori, equidistant. (See, for example, the series of values formed by shades of gray on page 77).

In order to make this notion clearer, we can say that **a graphic problem is GEO Q when its goal is to discover the groupings resulting from the *combination of the two components*.**

A graphic problem is GEO O when its goal is to represent the geographic distribution of ordered groups (steps), which have been defined, a priori.

The all-too-frequent interpretations of quantitative areal information according to the latter formula, which unduly transforms the component Q into an equidistant series O, arise principally from failure to use a strictly quantitative form of representation. Visual habit has thus assumed the power of law, even though it does not conform with the fundamental conditions of the problem being posed.
(5) It is particularly important in area representation to construct an ordered and quantitative image whose perception will produce the same possibilities of visual grouping as those obtained with the "relief." The best solution is a series of proportional circles, applied to a regular pattern of points (figure 3).

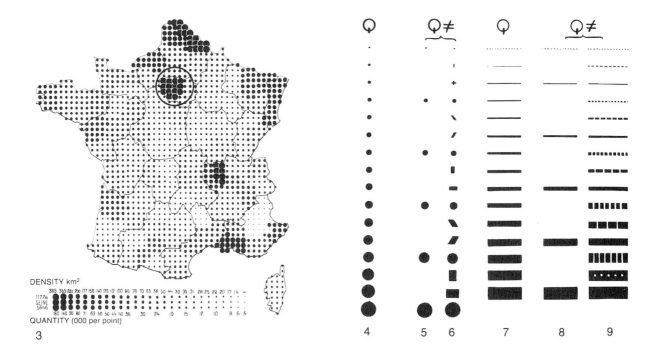

The range of the quantities: extension adjustment

Take a quantitative series the extremes of which are 127 and 11 687. The larger number contains the smaller one 92 times (11 687 / 127 = 92). The series ranges from 1 to 92.

A component Q can range from one to ten million and more (e.g., population). But it can also range from only 1 to 1.2 (e.g., height of individuals, 180 cm / 150 cm = 1.2), and all the intermediate steps are possible. On the other hand, the visual variables have a range which is practically constant for each implantation. One thus encounters three types of problem in quantitative representation.

(1) *The range of Q is greater than that of the visual variables.* We must proceed to a REDUCTION of the quantitative range, either proportionally, or by removal of the "tails" (determination of the extremes).

(2) *The range of Q is of the same order of magnitude as that of the visual variables.* We utilize the STANDARD CORRESPONDENCE: area of the signs = Q. Here the ratios perceived among the signs produce the closest approximation of their numerical equivalents. Number B appears double number A when the area of sign B is double that of sign A.

(3) *The range of Q is less than that of the visual variables.* An image resulting from the application of the standard correspondence will not be differentiated. Two cases can occur:

(a) The lack of differentiation is meaningful. This can happen in series of maps whose legends must be uniform. Certain of these maps will display distributions which are not differentiated.

(b) The variations are meaningful, but the lack of differentiation stems from the nature of the phenomena being measured (height of individuals, yields, etc.). The map is made to be compared with other maps in terms of *phenomena occurring in the same region* this is a FREQUENCY COMPARISON. The variations should be made visible, which means that we must proceed to an EXTENSION of the series Q along the range of the visual variable, until we have a sufficient differentiation.

Density adjustment

In frequency comparisons, the maps must be comparable (equalization of averages). Therefore, the size of the smallest sign must be chosen in such a way as to equalize the total amount of "black" in each map (see page 374).

The level of reading and its implications

The main purpose of a map **GEO Q** is to respond to *intermediate-level questions* (what are the resulting geographic groupings?), and to an *overall question* (which maps are similar in that their concepts have a good probability of geographic correlation?). A series of proportional sizes can be applied, whatever the implantation, as we see in figures 3, 4, and 7.

However, the reader may need to pose *elementary questions*, for which size variation alone is no longer sufficient. When such questions are introduced by the component **GEO** (what is the population of a given city?), they imply that the sign observed on the map is recognizable in the legend.

A redundant combination (see page 187) with shape for points and texture for lines (figure 9) provides the solution.

When the question is introduced by the component **Q** (what is the distribution of a given quantity?), it implies that the quantity being sought is differentiated ($\neq$) on the map. For lines, texture is sufficient (figure 9). For points, it is necessary either to increase the distance between steps (figures 5 and 8), thus reducing their number, or to introduce a redundant combination with orientation (figure 6 here and figure 6, page 377).

GEO Q Point

Example: Distribution of volunteers for the West Indies departing from La Rochelle, from 1634 to 1715, according to **GEO**: place of origin; **Q**: number. Figure 1 shows an E.P.H.E. map (from R. Mandrou, "Les Français hors de France au XVIIe siècle," *Annales ESC* no. 4 [Paris: A. Colin, 1959]).

It is in point representation that the number of available sizes is greatest, since the marks can cover a large area around their center on the plane. It is nearly always possible to adopt the standard correspondence (area of the marks = Q) for a series having a large range.

CONSTRUCTION OF THE STANDARD SERIES

This can be achieved in the four following ways:
The calculation of radii. For area of the circles to be proportional to Q, the radii must be proportional to the square root of Q.

The same is true for the side of a square, the side of a triangle, or any other basis of construction with a figure whose area must vary proportionally with Q.*

Note that calculation of the roots is not useful, since the same result can be obtained in a much more rapid and sure manner by the following means:
The area-radius table-graph (see Appendix) gives the radius of the corresponding circle directly from reading Q.
The preprinted standard series (page 369) gives the preprinted circle corresponding to area Q directly from reading Q.
Calibrated columns (see page 360).

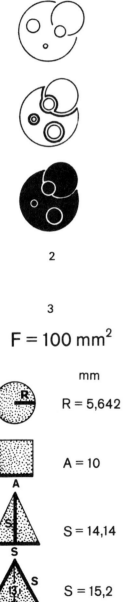

2

3

F = 100 mm²

	mm
R	R = 5,642
A	A = 10
S	S = 14,14
S	S = 15,2

Superimposition of the signs

In point representation, the signs can overlap. In this case the smaller ones must obviously be contained within the larger ones. The best drawing process is illustrated in figure 2.

The legend

The legend must permit an approximate rereading of the Q placed on the map and must indicate the presence of any extensive or reductive scales; but remember that it will never replace a reading of the numbers themselves. The legend should therefore involve:
(1) the smallest sign and its corresponding quantity (visual unit of enumeration);
(2) reference steps;
(3) insofar as is possible, the largest sign and its corresponding quantity.

It is often preferable not to blacken in the signs on the legend (see figure 1), in order to avoid an overall geographic image which is deformed by the visual weight of the legend.

*If one uses signs of different shapes in the same representation, remember that, for an equal area, the principles of construction are different (figure 3). Thanks to the compass the circle will always remain the simplest and most rapid figure to construct.

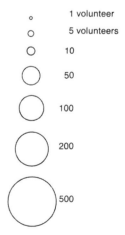

Scale of circles

- 1 volunteer
- 5 volunteers
- 10
- 50
- 100
- 200
- 500

Volunteers for whom we know only province of origin are denoted by white circles

1

Calibrated columns

The map in figure 1 (representing the number of postal checking accounts opened per commune, in the department of Nord, Pas-de-Calais, and Somme) is an example of areas proportional to Q, constructed with a system of calibrated columns. One conceives a series of lines of known graduated width, as, for example, in figure 2. At a constant height, the area of these lines is proportional to the width, and one can make a series of Q correspond to the lines in this same proportion (figure 3). One need only vary the heights arithmetically in order to obtain the areas corresponding to the intermediate Q (figure 4). The lines from figure 2 then combine with the numbers in figure 3 to form the legend for the map in figure 1; it is not necessary to include figure 4 in the legend. With this construction, the base of the column generally corresponds to geographic location.

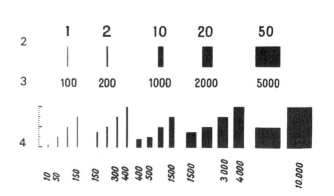

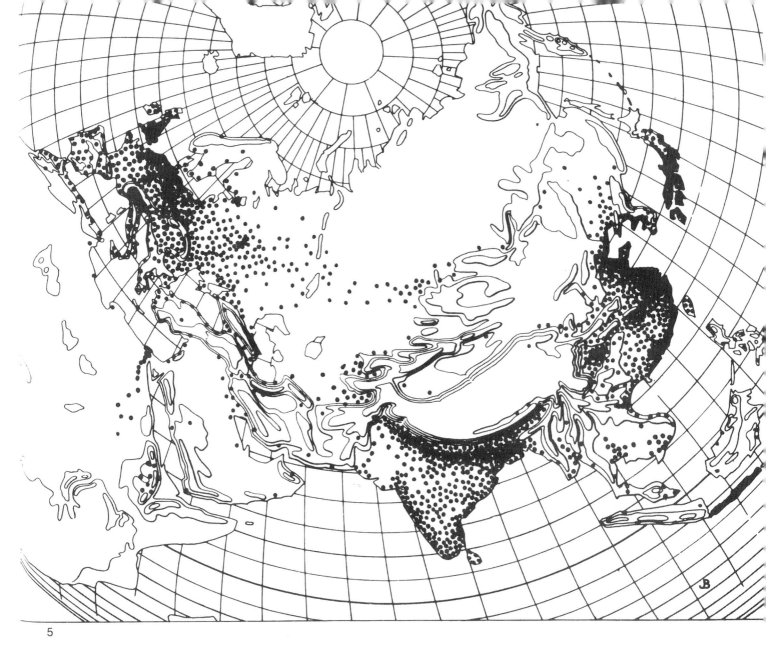

5

A proportional number of equal-sized points (dot map)

Here, it is the number of points which is proportional to Q. This procedure enables us to duplicate a geographic distribution within the enumeration areas. In the map above (figure 5, population of Eurasia in 1936, one point represents 1 M inhabitants) the enumeration areas are the countries, but the internal distribution of the points corresponds to more detailed information

This method works poorly for large concentrations. Extended areas of black should be avoided, and, in the areas of greatest density, the size and quantitative meaning of the points should be chosen so that the eye perceives that the points are juxtaposed, not superimposed.

Common errors

Figure 6. The points are unequal; the boundaries too thick.
Figure 7. The points are too grouped at the center of the area.
Figure 8. The points are too small; the boundaries too thick.
Figure 9. The points are too large and not countable.
The correct construction, avoiding the above errors, is given in figure 10.

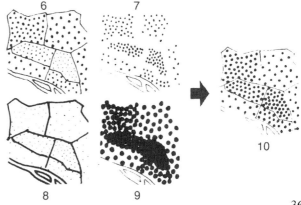

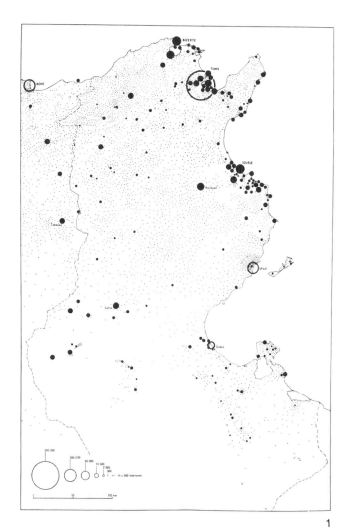

1

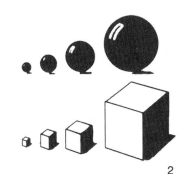

2

QUANTITATIVE SERIES HAVING AN EXTENSIVE RANGE

An extensive range of quantities often occurs with precise population maps, which can involve regions of highly different population densities (see figure 1). Let us recall the fundamental elements of the problem. In Information **GEO Q** each of the two components, quantities on the one hand, geographic space on the other, is by definition homogeneous, and it is a question of discovering:
(1) whether or not the combination of the two components creates a homogeneous structure;
(2) the spatial lines and the quantitative levels which display potential changes in structure.

Information **GEO Q** poses the general problem of aggregation, and the graphic representation must permit us to discover WHERE and for what numbers one can speak of an aggregated structure.

Consequently, when the difficulties of representation lead the designer to a prior definition of, say, two structures (urban and rural) and sometimes two different series, the designer, in fact, inverts the graphic problem, or transforms it into an exposition of conclusions acquired on the basis of other information. It then becomes a problem **GEO Q**, $\neq 2$ (structures), which does not us to discover the characteristics of any changes in structure.

The standard series: S = Q (areas = quantities)

It is always possible to use such a series in point representation (see figure 4). The following factors must be considered:
(a) For maps whose purpose is to discover structures, the choice of the geographic scale depends more on the precision of the questions to be asked than on the eventual publication format.

Incidentally, the homogeneity of the information Q enables us to undertake the study on scales which are sufficiently large to facilitate a precise recording of the data, then to carry out a sizable photographic reduction for publication.
(b) The size of the smallest point can be very small indeed if the base map has a light value.
(c) The quantitative value of the smaller points also has a very great effect. The choice of 50 instead of 1 for the value of the minimum point, for example, will reduce the area of the largest mark fifty times. These larger marks can be replaced by rings (figure 4) whose center must be denoted by a special sign.

Equal-sized points combined with the standard series

Figure 1 shows the population of Tunisia (after R. Lalue, *Annales ESC*, vol. 2 [Paris: A. Colin, 1962]).

This combination avoids the disadvantages of the exclusive use of equal-sized points. Attention must be paid to the visual ambiguity created at the point where the formula changes. If the equal-sized points signify 100 (inhabitants, for example), and the first proportional point signifies 500 (thus having five times the area of point 100), four nearby points of 100 will fuse and be more visible than the point of 500. This can be corrected by slightly increasing the area of the point 500 or by reducing the area of the smaller points.

The series: area = $\sqrt{Q}$

This series involves a reduction in the range of Q (see figure 3). Since the visual series is proportional to the quantities, it permits us to discover the steps in the quantitative series being treated, and it is applicable whenever such a problem is posed.

Representation by volume

The correspondence: area = $Q^{2/3}$ (or radius = $\sqrt[3]{Q}$) can also be stated as: volume = Q, and the quantities are proportional to the volume of the spheres whose diameter would be equal to the diameter of the circles. It is therefore necessary to suggest that the circles are spheres, the squares are cubes, the triangles are pyramids, etc. (see figure 2). However, we must recognize that, in practice, human perception does not evaluate the ratios of volumes (however suggestive they might be) by numbers to corresponding to the real values, that is, to the absolute quantities. Ten percent of the readers will evaluate a volume with only plus or minus 15% error, and whenever the reading involves the intermediate level and, of course, the overall level, only the area of the signs (planar amount of "black") will have a quantitative meaning.

The double series: area-volume (see figure 5)

In a problem which is strictly **GEO Q**, a prior heterogeneity cannot be considered among the possible options of graphic representation. One cannot bend the laws of visual perception, and the homogeneity of the variables is an imperative whose transgression always results in errors of interpretation.

The use of two visual scales (area = Q for dispersed populations; volume = Q for aggregated populations) implies a prior definition of the boundaries between the structures, whereas it is precisely this definition which constitutes the true objective of the representation. It also adds a double visual ambiguity:

(a) at the boundary between the two series, we cannot compare an area and a volume;

(b) in any perception extended over several signs (intermediate and overall levels of reading), the eye perceives a retinal variable as homogeneous, and will implicitly give all the signs a homogeneous meaning, obviously based on the area.

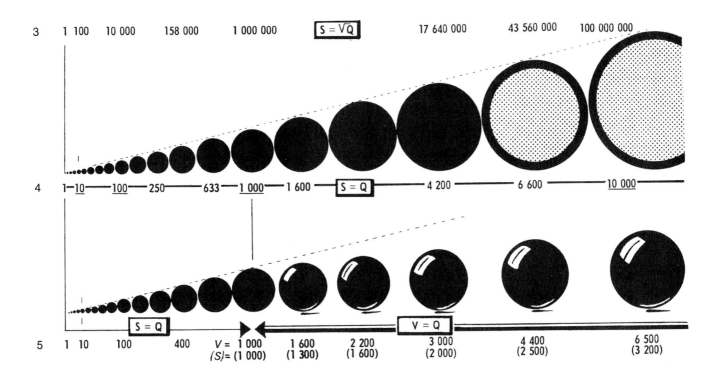

QUANTITIES HAVING A LIMITED RANGE

Figures 1–3 Salary (in deniers) for a harvest day (after G. Duby, "Les Alpes du Sud en 1338," *Etudes rurales* [Paris: Mouton, 1961]).

This is a typical example of a quantitative problem in which the visual addition of several quantities is meaningless. Two simultaneous factors account for this.
(1) The quantities are SINGLE NUMBERS. They define a salary, but do not represent the totality of salaries distributed.
(2) The information is a SAMPLING. All the inhabited places are not included.

Therefore the information only permits us to define the amount of the salaries distributed for relatively homogeneous regions. In this information, Q ranges from 9 to 24, that is, from 1 to 2.6.

The standard correspondence $S = Q$, produces an image which is insufficiently differentiated to reveal steps and define a useful regionalization (figure 1).

The extensive correspondence $S = Q^4$, gives a range of 1 to 46 (figure 2).

The representation is effective if the reader has been trained to read the size of the point as merely a difference in regions. With full knowledge of the facts, the reader can construct a given regional grouping as a function of both the information being treated and whatever other factors our prior knowledge will bring to bear. However, we must take into account the natural tendency toward the interpretation $S = Q$ in point representation. Thus, in the published map it is appropriate to replace the extensive correspondences, which are valuable in the laboratory, by formulae which are merely ordered.

Returning to an ordered variable (see figure 3)
The useful and meaningful steps defined by the preceding representation are transcribed by a qualitative, ordered variation—value (plus shape and texture)—which does not risk leading the reader to an erroneous quantitative interpretation.

GEO Q line

Here the widths of the lines are proportional to Q. The graphic range for representing Q is thereby more limited than in point representation, and we must often set a lower limit of Q, corresponding to the minimum line width, below which proportionality disappears. The maps in figures 4 (passenger flow) and 5 (commodity flow) in France, are examples of overall comparisons. Attention is not focused on quantities, which are always inaccurate and questionable for such phenomena, but on the contrast between the two images, which display noticeably different distributions.

The map of Swedish river flow in figure 6, from M. Lundquist, *The Atlas of Sweden* [Stockholm: Kartografiska Institutet, 1953]), is an excellent example of the representation of linear qualities. Indeed, it portrays a problem involving three components, since the quantities are divided into two categories: medium flow (in black) and maximum flow (in gray).

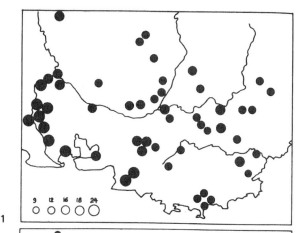

1

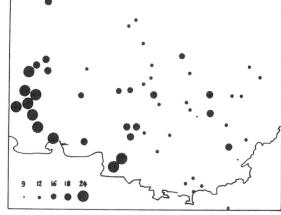

2

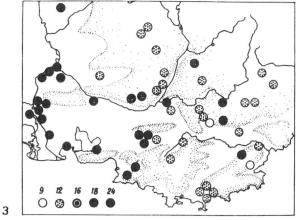

3

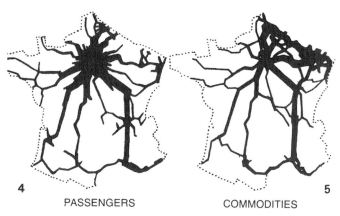

4 PASSENGERS 5 COMMODITIES

RAILWAY TRAFFIC

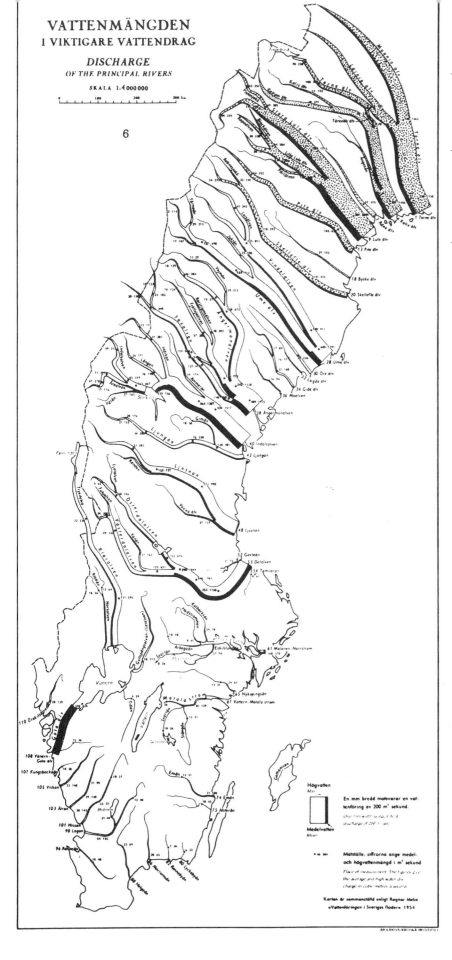

In figure 6, we should note the following:
- the importance of shading in gray and, conversely, the difficulty in perceiving what is not shaded. In these maps, the eye essentially sees "amounts of black," not widths.
- the disadvantage of using too light grays. These have disappeared in the photographic reproduction, as they will disappear from any future microfilmed information (the visible grays have been redrawn).
- the utility of a graphic legend for Q. The legend, "1 mm in width equals 200 m^3 per second," is meaningless in terms of modern microfilmed documentation, which may be reproduced at any scale whatsoever. The largest linear quantities are often inscribed at the convergence points of several lines (highways, railways, etc.).

To avoid the confusion resulting from lines that are too wide, R Bachi proposes a series of signs (figure 7) whose amount of black is readable and measurable (from 1 to 20). The series is applied to a line of constant width (figure 8). However, visual differentiation is obviously less than that obtained by variation of width, and proportionality is difficult to perceive.

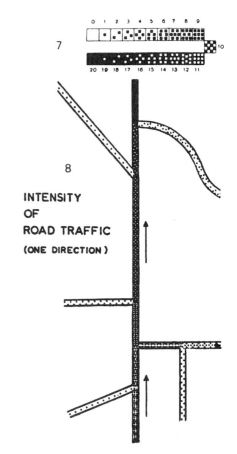

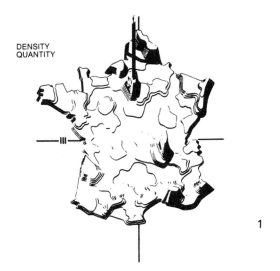

1

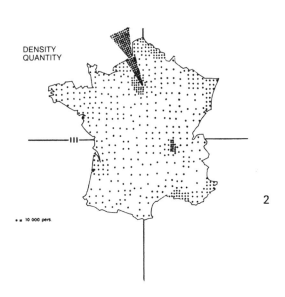

2

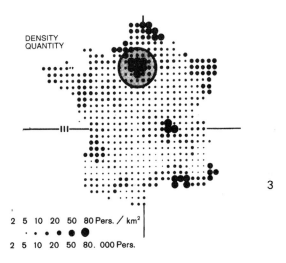

3

4. Maps GEO Q area

Different Q
In area representation, quantities apply, by definition, to the entire enumeration area.
  Consequently:
(1) If we proceed to an internal redistribution, in order to better capture the geographic distribution, we change implantation, and the problem becomes one of point representation.
(2) In any map representing areas of unequal size, *what is seen is Q multiplied by the size the area*. However, Q must remain independent of area size (just like the distribution histogram, in which Q must be reduced to a unitary class).
(3) Any absolute quantity (of persons or objects), which is enumerated by area, must be divided by the size of the area before being geographically applied to this area. In other words, **any QS (see page 38) must be reduced to QS/S, that is, to Q, before being multiplied by S in the image.**

This is true for all types of graphics. In drawing a quantity QS of equal points (figure 2), the designer carries out the division visually, and the result shows the differences in spacing which must be maintained among the points.
(4) A well-constructed image Q displays DENSITIES (which correspond either to the height of a relief representation, or the intensity of value resulting from the distribution of signs in an observed space) and ABSOLUTE QUANTITIES QS (which correspond either to the volume of the observed relief, or to the total amount of "black" in the observed area).
(5) **Any quantity independent of the size of the area can be directly applied to the entire geographic area, without being divided by S.**

Examples of such quantities are: *simple ratios* Qa/Qb (e.g., livestock related to grassland area); *rates and percentages* Qa/Qb × 100 (e.g., automobiles per 100 inhabitants); *densities* Qs/s (e.g., densities per km^2); *samples* S (altitude, price, temperature, etc.) which are applicable to the entire surrounding area which is not sampled. But here the graphic problem consists, in fact, of delimiting the area to which we can attribute the sample value; the solution depends on neighboring values (see page 385).

Graphic solutions

In view of the preceding discussion, we can see that a quantitative problem can only be perceived through the use of a size variation or a variation in the number of elements. Consequently, there are only three possible graphic solutions.

Figure 1. *A size variation suggesting the "third dimension"* (i.e., relief or perspective representation).

Figure 2. *A variation in the number of (equal-sized) points* per unit of area (i.e., a dot map).

Figure 3. *A variation in the size of points* distributed regularly over the area (i.e., a regular pattern of graduated circles).

The last two forms can sometimes be applied to a line, but this reduces the extent of the variation considerably; furthermore, drafting is more difficult and less precise than with a point.

There are other solutions that do not to conform to the definition of the problem.

Figure 4. *A single sign per area.* This formula shows only the QS. Since it does not cover the entire area, it excludes density perception, particularly when the areas are very unequal.

Figure 5. *Isarithms.* When the isarithmic interval is constant, isarithms display the slope (gradient) which links contiguous areas, but they do not permit us to evaluate the height of the steps on the intermediate and overall levels of reading. Thus, overall comparisons are excluded in a series of maps of this kind. The redundant use of value variation only portrays information O.

Figure 6. *Preprinted series of value.* As we have seen, these series translate a prior interpretation and not a component Q whose steps should result from the graphic operation.

The range of Q in area representation

This range is much more limited than in point representation. If it is theoretically possible to construct a column several tens of cm "above" the map (figure 1), this construction must be excluded in practice. Likewise, the number of equal points which remain countable is limited. Once solid black is reached, visual counting is impossible, and, in fact, the means of representation are modified. For example, a "horn" must be used to represent Paris in figure 2. Finally, and exceptionally, points of variable size can form "grape bunches" for a very small area. However, it is always possible to construct a point of considerable size, such as Paris on the map in figure 3.

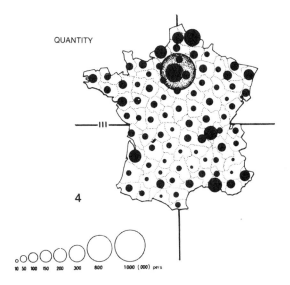

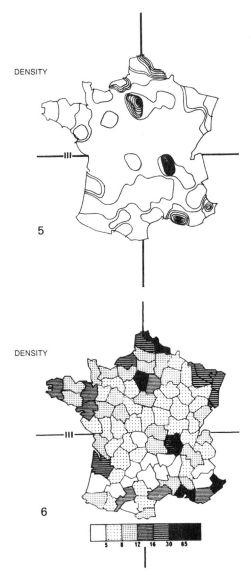

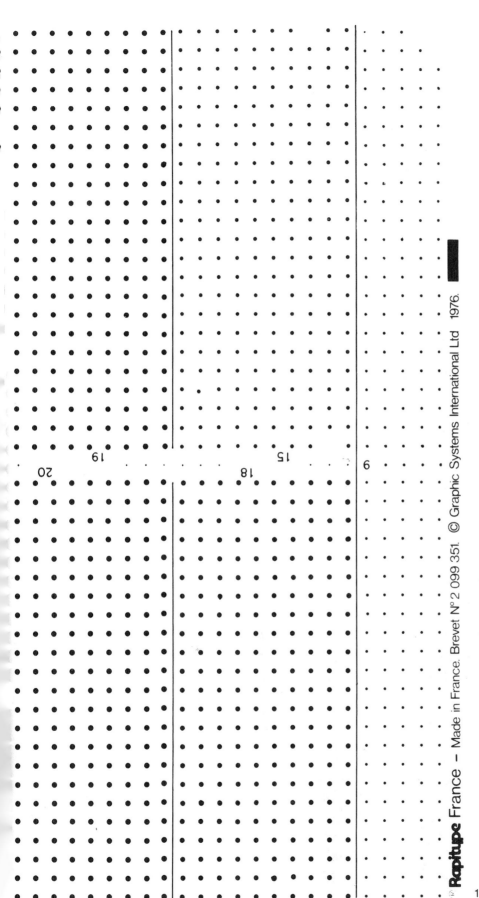

A regular pattern of graduated circles

THE "NATURAL" SERIES OF GRADUATED SIZES

Among the retinal variables, a size variation, applied to a circle, generally constitutes the best representation of a component Q.

This variation is limited in range (size of the extreme circles) and length (number of separable steps) by natural human faculties of perception. We will thus term the series which is necessary and sufficient for graphic representation, the "natural" series of graduated sizes. It is selected from the potentially infinite number of possible circles and corresponds to the limits of visual acuity.

The preprinted sheets shown in figure 1 opposite are based on this series. The blocks of circles are utilized in area representation (size variation of circles in a regular pattern), and they also show identification numbers needed to use the circles in point representation.

CONSTRUCTION OF THE NATURAL SERIES

Range (size of extreme circles)

If the smallest perceptible circle is of the order of 0.2 mm in diameter, the largest circle varies with the dimensions of the figure. However, it will rarely exceed 4 cm in diameter. A series ranging between these two limits thus corresponds to the majority of graphic problems.

Length (number of separable steps)

This is determined by the following:

(1) The minimum perceptible differentiation between two circles, below which supplementary steps would be useless, since they would be imperceptible.

(2) The progression of this differentiation. In a logarithmic progression, the perceptible differences which we perceive among the circles (and not among the grays resulting from density perception) are regular and constant. They follow the natural laws of perception: we see only ratios.

Note that when the circles are very small, the logarithmic differentiation between two signs falls below the threshold of visual acuity and becomes inefficient. In this part of the series, the differentiations become perceptible and useful only among circles 1, 9, 15, 18, and 20, which have been retained, with 19 (see figure 2 opposite).

(3) A practical number of steps in the standard correspondence S = Q. There are twenty divisions between one sign and another whose area is ten times larger, which permits us to retain only one sign in two, one in four, and one in five, while still remaining in a decimal series.

Shape of the sign

Any shape can be used since it is the progression of the areas which is important. However, the circle is the only shape for which the eye can restore the total size, even if only a part of it is seen. Circles can be superimposed, can construct "grape bunches" and can still remain measurable.

PROPERTIES

(1) The natural series produces the quantitative representation which is the most rigorous and efficient of the retinal variables, provided that its progression corresponds to a proportional progression of the numbers expressed.

(2) The natural series determines the necessary and sufficient steps of the statistical series. The problem is to make all the distances expressed by series Q perceptible; the natural series provides the exact value of the perceptible steps. Thus, it is the circles of the natural series which determine the intermediate numbers (steps) of the statistical series, not the other way around, as is often the case.

Any prior determination of the intermediate numbers to which one then applies circles destroys the perceptual properties of the representation. This would correspond to the error that would be committed if the designer of a diagram were to fit the intermediate steps of a linear scale to the numbers to be represented (figure 1). Consequently, the natural series eliminates the problem of the prior choice of a step series.

(3) The natural series also eliminates the necessity of any calculation. The constant progression permits us to move along the series, which allows the following to be accomplished:

– We can establish *tables* such that we can make any statistical series, whatever the numbers, correspond to the circles of the natural series, without calculation.

– We can establish (in addition to the standard correspondence Area = Q) tables which allow us to make a rapid correlation of any quantitative series, by extension or proportional reduction, to the range of the perceptible series (extension adjustment).

– We can choose the extreme circles in view of a sizable photographic reduction (reduction adjustment) or in terms of the total amount of "black" necessary (density adjustment).

– We can introduce this series as a known constant into a data-processing program and make the graphic transcription mechanizable.

(4) In area representation, the natural series produces a representation which allows sizable photographic reductions, without risk of transformation or destruction of the series. Its restoration after microfilming is rigorous (see figure 3, page 375), which is not the case with preprinted shades of "gray."

Tables of the natural series

The tables opposite are the key to using the sheets of preprinted circles shown in figure 1 on page 368. These tables involve the following:

(1) The column of circle numbers (on the left in figure 2). It must be cut out along the line and detached, so that it can be placed alongside the numbers Q of the information, in the appropriate series. Circle number 41 is the tangent circle in the sheets of preprinted points in figure on page 368. For numbers greater than 41, the circles obtained can only be used in point representation. Above number 63, the circles must be drawn according to the radius diagram given at the bottom of figure 1 on page 368. On the right of figure 2 opposite, a second column permits us to obtain the scale $S = Q^6$.

(2) The standard series S = Q (at the center of figure 2). This gives the progression of the areas of the circles.

(3) The principal extensive series (toward the right of figure 2). These permit us to make any limited-range series Q correspond with the range of the natural series between circles 1 and 41, that is, in any problem of area representation.

The scale $S = Q^2$ or radius = Q gives the progression of the radii or diameters, and consequently the corresponding linear ratios which allow prediction of the effects of a given photographic reduction (see page 375).

(4) The two reductive series (toward the left of figure 2). These play the same role as the extensive series but for extended-range series Q. The scale = $Q^{2/3}$ or radius = $\sqrt[3]{Q}$ gives the progression of volumes having the same diameter as the circles.

NUMBERS	CIRCLE	REDUCTION		STANDARD		EXTENSION				VALUE OF AREAS		
		$S=\sqrt{Q}$	$S=Q^{\frac{2}{3}}$	$S=Q$	$S=Q^{\frac{3}{2}}$	$S=Q^2$	$S=Q^4$	$S=Q^6$		N°	BLACK	WHITE
1	·	1 00	1 00	1 00	1 00	1 00	1 000	014 ·	· 1			
		. 26	. 19	. 12 Area	. 090	. 06	. 030	044 ·				
		. 60	. 41	. 26	. 188	. 12	. 059	074 ·				
		2 00	. 70 Volume	. 41	. 292	. 19	. 090	106 ·				
		. 50	2 00	. 58	. 412	. 26 (linear ratios)	. 122	138 ·				
		3 16	. 38	. 78	. 540	. 33	. 155	171 ·	· 9			
		4 00	. 82	2 00	. 678	. 41	. 188	205 ·				
		5 00	3 35	. 24	. 830	. 50	. 223	241 ·				
9	·	6 31	4 00	. 51	. 995	. 58	. 258	277 ·	· 15			
		8 00	. 73	. 82	2 175	. 68 Diameter	. 292	314 ·	· 18			
		10 0	5 62	3 16	. 371	. 78	. 333	353 ·	· 20			
		12 58	6 63	. 55	. 585	. 88	. 372	392 ·	· 22			
		15 84	7 94	4 00	. 818	2 00	. 412	433 ·	· 24			
		19 95	9 44	. 47	3 072	. 11	. 453	475 ·	· 26			
15	·	25 11	11 22	5 01	. 349	. 24	. 496	518 ·	· 28			
	·	31 62	13 33	. 62	. 652	. 37	. 540	562 ·	· 30			
		39 81	15 84	6 31	. 981	. 50	. 584	608 ·	· 32			
18		50 11	18 83	7 08	4 340	. 66	. 631	654 ·	· 34			
19	·	63 10	22 38	8 00	. 731	. 82	. 678	703 ·	· 36			
20		79 43	26 60	. 91	5 158	3 00	. 727	753 ·	· 38			
21	·	100	31 62	10 00	. 623	. 16	. 778	804 ·	· 40			
22		125 8	37 58	11 22	6 130	. 35	. 830	856 ·	· 42			
23		158 4	44 66	12 58	. 633	. 55	. 883	911 ·	· 44			
24		199 5	53 08	14 12	7 286	. 76	. 938	966 ·	· 46			
25	·	251 1	63 10	15 84	. 943	4 00	. 995	2024 ·	· 48			
26		316 2	74 98	17 78	8 660	. 22	2 053	083 ·	· 50			
27		398 1	89 12	19 95	9 441	. 47	. 113	144 ·	· 52			
28		501 1	104 4	22 38	10 03	. 73	. 175	206 ·	· 54			
29	·	631 0	125 8	25 11	11 22	5 00	. 238	271 ·	· 56			
30		794 3	149 6	28 18	12 23	. 31	. 304	337 ·	· 58			
31		1 000	177 8	31 62	13 33	. 62	. 371	405 ·	· 60			
32		1 258	211 3	35 48	14 53	. 96	. 441	476 ·	· 62			
33	·	1 584	251 1	39 81	15 84	6 31	. 511	548 ·	· 64			
34		1 995	298 5	44 66	17 27	. 63	. 585	622 ·	· 66			
35	·	2 511	354 8	50 11	18 83	7 08	. 660	699 ·	· 68			
36		3 162	421 6	56 23	20 53	. 50	. 738	778 ·	· 70			
37		3 981	501 1	63 10	22 38	. 94	. 818	859 ·	· 72			
38		5 011	595 6	70 80	24 41	8 42	. 901	943 ·	· 74			
39	·	6 310	708 0	79 43	26 60	. 91	. 985	3028 ·	· 76			
40		7 943	842 2	89 12	29 01	9 44	3 072	117 ·	· 78			
41	·	10 000	1 000	100 0	31 62	10 00	. 162	208 ·	· 80			
42		12 580	1 188	112 2	34 47	10 59	. 255	302 ·				
43		15 840	1 412	125 8	37 58	11 22	. 349	413 ·				
44		19 950	1 678	141 2	40 97	11 88	. 447	502 ·				
45		25 110	1 995	158 4	44 66	12 58	. 548	594 ·				
46		31 620	2 371	177 8	48 70	13 33	. 652	689 ·				
47		39 810	2 818	199 5	53 08	14 12	. 758	813 ·				
48		50 110	3 349	223 8	57 87	14 96	. 868	924 ·				
49		63 100	3 981	251 1	63 10	15 84	. 981	4039 ·		N°	BLACK	WHITE
50		79 430	4 731	281 8	68 78	16 78	4 097	157 ·		1.	1	99
51		100 000	5 623	316 2	74 98	17 78	. 216	297 ·		9.	2	98
52		125 800	6 663	354 8	81 75	18 83	. 340	403 ·		15.	4	96
53		158 400	7 943	398 1	89 12	19 95	. 466	525 ·		18.	5	95
54		199 500	9 441	446 6	97 16	21 13	. 597	664 ·		19.	6	94
55		251 100	11 220	501 1	105 9	22 38	. 731	800 ·		20.	7	93
56		316 200	13 330	562 3	115 5	23 71	. 870	940 ·		21.	8	92
57		398 100	15 840	631 0	125 8	25 11	5 011	084 ·		22.	9	91
58		501 100	18 830	708 0	137 2	26 60	. 158	233 ·		23.	10	90
59		631 000	22 380	794 3	149 6	28 18	. 308	386 ·		24.	11	89
60		794 300	26 600	891 2	163	29 85	. 463	543 ·		25.	12	88
61		1 M	31 620	1 000	177 8	31 62	. 623	705 ·		26.	14	86
62		1 258	37 580	1 122	193 8	33 49	. 787	871 ·		27.	16	84
63		1 584	44 660	1 258	211 3	35 48	. 956	6043 ·		28.	18	82
64		1 995	53 080	1 412	230 4	37 58	6 130	219 ·		29.	20	80
65		2 511	63 100	1 584	251 1	39 81	. 310	401 ·		30.	22	78
66		3 162	74 980	1 778	273 8	42 16	. 494	587 ·		31.	25	75
67		3 981	89 120	1 995	298 5	44 66	. 633	780 ·		32.	28	72
68		5 011	104 400	2 238	325 5	47 31	. 878	978 ·		33.	32	68
69		6 310	125 800	2 511	354 8	50 11	7 080	182 ·		34.	35	65
70		7 943	149 600	2 818	386 8	53 08	. 286	392 ·		35.	40	60
71		10 M	177 800	3 162	421 6	56 23	. 498	608 ·		36.	45	55
72		11 220	211 300	3 548	459 7	59 56	. 718	830 ·		37.	50	50
73		15 840	251 100	3 981	501 1	63 10	. 943	8058 ·		38.	57	43
74		19 950	298 500	4 466	546 3	66 33	8 175	293 ·		39.	63	37
75		25 110	354 800	5 011	595 6	70 80	. 424	536 ·		40.	71	29
76		31 620	421 600	5 623	649 4	74 98	. 660	785 ·		41.	80	20
77		39 810	501 100	6 310	708 0	79 43	. 912	9042 ·		18+41.	85	15
78		50 110	595 600	7 080	771 8	84 24	9 173	305 ·		✦+41.	90	10
79		63 100	708 000	7 943	842 4	89 12	. 441	587 ·		26+41.	95	5
80		79 430	842 400	8 912	917 3	94 41	. 716	856 ·				
81		100 M	1 M	10 000	1 000	100	10					
Progressions:		1,258	1,188	1,122	1,090	1,059	1,03	1,014				

J. BERTIN

(5) The series of values of the blocks of circles, given in percentages of black (bottom right of page 371). This is obtained from the standard series and gives a value series for the circles, such that absolute black, in area representation, corresponds to 100%. Block 41 would be 80% black and 20% white. For values darker than 80%, it is sufficient to add circle number 18 (5%) to this white, which gives 85%; to half-shade this white, which gives 90%; or to shade it by 3/4 or add circle number 26, which gives 95%.

THE NATURAL SERIES IN AREA REPRESENTATION

Properties of a regular pattern of graduated circles

Length of the variable. The perceptible lines of a screen always have an area much greater than the smallest perceptible dot. A regular pattern of circles utilizes more fully the entire visual range and, in fact, constitutes the longest available size variable.

Reading. When the natural series is applied to a regular pattern of circles it produces a quantitative perception, whatever the level of reading. The reader who is content with elementary perception and who focuses on only one circle will nonetheless be informed of an absolute quantity and a density. The same is not true for a representation based on the number of equal-sized points, nor, of course, for a representation based on steps of value.

At the intermediate level, the reader is informed of a density and an absolute quantity (if applicable) by visually summing the circles.

On the overall level the reader can construct an image either with two steps (dark, light), or three (dark, medium, light), or more, and compare it to any other image. In all cases the reader remains informed of the numerical value of the steps which are retained and of those which are excluded.

Problems GEO Q area: Standard correspondence

Let us consider the problem of representing the quantities of population per department (figure 1).

(1) *The base map.* In area representation, the natural series is applied to a regular pattern of points. The circles must be visible. It is the perception of their size which matters. Furthermore, the spacing between their centers must not, after reduction, be smaller than 1.5 mm.

Knowing this, and that the preprinted transfer sheets (page 368) have a spacing of 5 mm or 2.5 mm, determines the scale of the drawing. If the scale of a map is such that each point corresponds to 1000 km^2, or to 100 . . . , the same numbers (with a difference of 000 or 00) express both densities and quantities per point (see, for example, page 137).

In a departmental map of France, there will be from 10 to 40 points per department. In the case of a canton map (which involves some 3000 cantons), a regular pattern of points must be prepared in advance, such that there is at least one point per canton.

(2) *Calculation of densities.* When the information is in absolute Q per area, we must calculate the densities per km^2. If the number of points per area is known, it is sufficient to divide Q by that number.

(3) *The scale of the circles.* This involves the choice of the correspondence between the circle numbers (from figure 2, page 371) and the enumeration area.

Quantities of population, and in general all raw numbers, should be represented, insofar as is possible, according to the standard correspondence (S = Q).

We proceed in the following way:

– We first study the extremes, the "tails," of the statistical series.
– The department of the Seine (a very important extreme) must be excluded and drawn later.
– The three largest departments can constitute overlapping points (bunches of grapes) and make up the "black" parts of the map.
– We slide the column indicating circle numbers along the column S = Q (as shown on the left of figure 1, opposite) and place circle 41 next to the numerical value of the fourth largest department (282). We must be careful that secant signs (above circle 41) cover only smaller and separate areas.

(4) *Translation of the series into circle numbers.* Once the column is fixed in position, it is sufficient to read and record—in nonphotographic blue pencil—the circle numbers corresponding to the numerical value of each department on the map.

(5) *Drawing or transferring of circles.* If only the tables of the natural series (and not transfer sheets) are at one's disposal, the circles can be drawn with a compass in sequence (all circles having number 20, then all the 21s . . .). The transfer of preprinted sheets is obviously more precise and faster.

(6) *Drawn circles.* Circles larger than number 41, which must never cover more than 1–3% of the plane, are drawn in clusters, according to the radius of the circle of the corresponding number. Thus: Department of Rhone, radius of circle number 43; Seine-et-Oise, number 45; Seine, radius of circle number 73. This latter radius is taken from the radius diagram 60–81 of the preprinted sheets (see bottom of figure 1, p. 368).

(7) *Case of the "extremes" (Department of the Seine).* The scale of the pattern of points on the maps in figure 1 opposite uses one point per 250 km^2. Thus the department of the Seine has two points. In the case of very large quantities, it is preferable to construct only a single circle, equal to the total population, that is, at double density. On this level the eye perceives ratios of quantity more than ratios of density. A similar solution occurs when we use one point per 1000 km^2. We construct the circles on the basis of total population, and, since the Seine has only 600 km^2, we reduce that circle in the proportion 600/1000.

In both cases it is sufficient to report the exact numbers in the legend.

Note that an open circle can be used to replace a large solid circle.

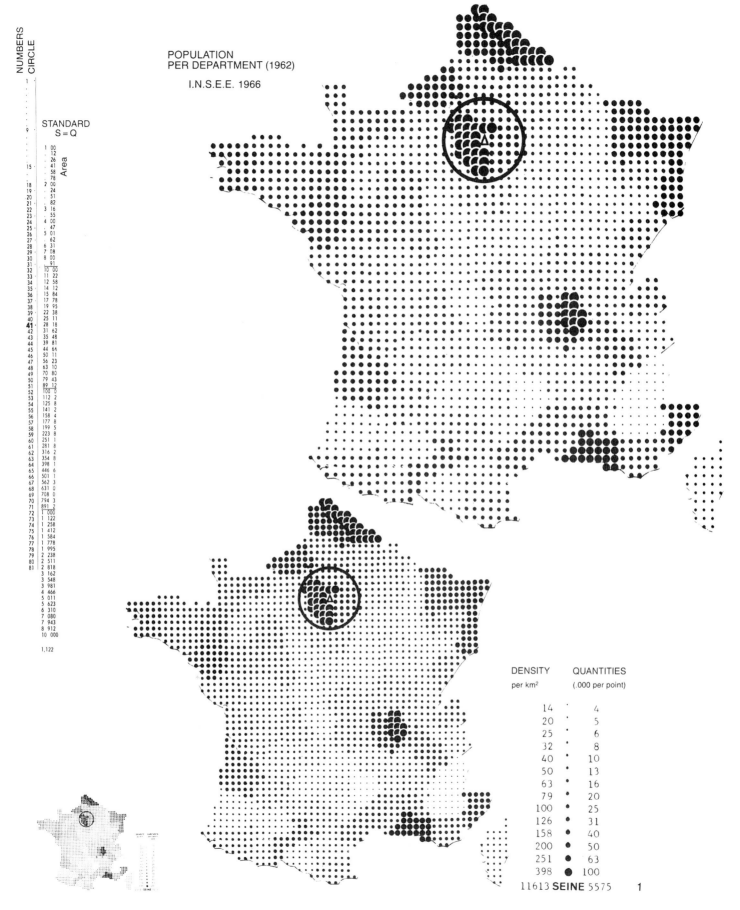

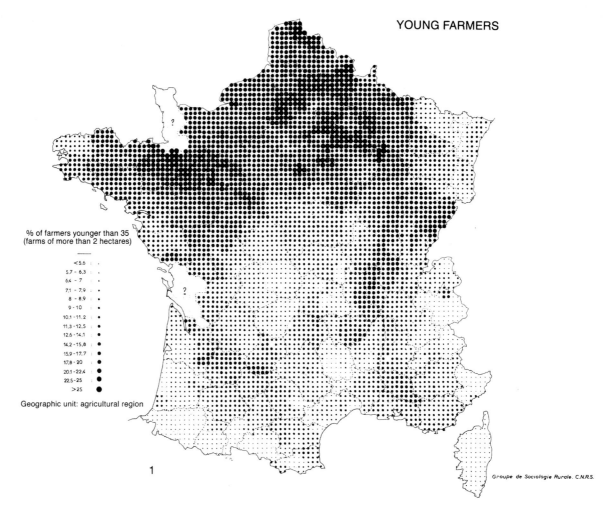

YOUNG FARMERS

% of farmers younger than 35
(farms of more than 2 hectares)

<5.6	·
5.7 – 6.3	·
6.4 – 7	·
7.1 – 7.9	·
8 – 8.9	•
9 – 10	•
10.1 – 11.2	•
11.3 – 12.5	•
12.6 – 14.1	•
14.2 – 15.8	•
15.9 – 17.7	•
17.8 – 20	•
20.1 – 22.4	●
22.5 – 25	●
>25	●

Geographic unit: agricultural region

1

Groupe de Sociologie Rurale, C.N.R.S.

Problem GEO Q area: Limited range in quantities (extension adjustment)

Let us consider the problem of representing the quantitative series produced by the percentage of farmers younger than thirty-five years of age (figure 1). The range of the series is approximately 1 to 5. The procedure outlined on page 372 is modified as follows:

(1) Calculation of densities is inapplicable with a percentage (as here), a rate, a ratio, or a sample;

(2) The correspondence between the series Q and the series of circles must be adjusted to suit this problem.

When quantitative series whose range is less than 10 (from 1 to 9, from 1 to 5 . . .) are translated into the standard correspondence $S = Q$, they produce an image which is not very legible. We must use an *extension adjustment*, that is, adopt an extended series $S = Q^2$, or Q^4, or even Q^6. We choose the series that allows Q to range from circle 41 to circle 9 or its neighbors; here the series $S = Q^2$.

Density adjustment

Just as distribution or repartition diagrams are only comparable if their averages are equalized, different distribution maps can only be compared if the total amount of "black" per map is perceptibly the same from one image to another.

The map of the total population in figure 4 does not permit us to perceive that its distribution is similar to those of population II (industry) and III (tertiary sector), in the same figure. In figure 5, where the total amount of black is more or less similar for all four maps, the distributions can be compared.

The calculation of the amount of "black" resulting from a given correspondence can be furnished immediately by a computer, and a density adjustment can be carried out by a different fitting of circles (thus the necessity of choosing a range which is slightly less than that of the natural series).

In manual procedures, this adjustment can result either from a repartition histogram (provided it is weighted by area) or from a comparison of the totals of each series, which we should try to equalize.

The margin of photographic reduction

This is quite large for a regular pattern of points. The canton maps in figures 2 and 3 involve some 3000 enumeration areas and about 6000 points. Their reading on the intermediate and overall levels remains efficient, even with a reduction where twelve maps are included within a 21 × 27 cm space. The comparison and classing of multiple quantitative series, the primary objective in modern statistical research, often entails such reductions, and a regular pattern of circles provides the means of accomplishing this goal.

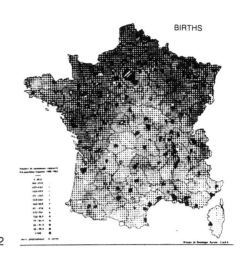

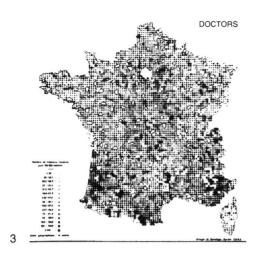

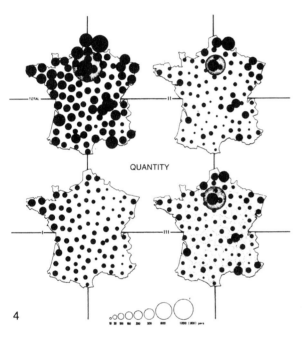

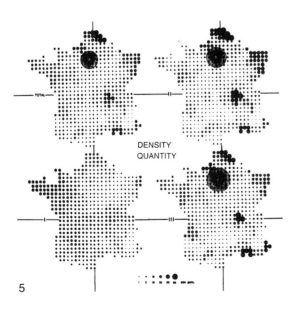

THE NATURAL SERIES IN POINT REPRESENTATION

GEO Q point: Standard series

Let us consider how to represent the population of Tunisia (page 362).

(1) *A calculation of densities* is not useful in point representation.

(2) *A correspondence between the quantitative series and the standard of circles* is almost always possible with very extensive ranges. However, although the scale of circles is usually determined by correspondence with circle number 41 in area representation, here the scale depends on the size of the smallest circle. Consequently, we must take into account the photographic reduction envisaged and introduce a *photographic reduction adjustment*. A linear reduction by half corresponds to a shift of twelve circle numbers; circle 56 has a diameter double that of circle 44, and any circle N has a diameter double that of circle number N–12.

The table $S = Q^2$ (figure 2, page 371) gives the sequence of numbers proportional to the diameters of the circles, and, thus, the reduction corresponding to each shift. A reduction of 200 to 100 (2 to 1) does indeed correspond to a shift in twelve numbers. The principal reductions correspond to the following shifts:

6 to 5: 3 numbers	2 to 3: 7 numbers
4 to 5: 4 numbers	3 to 5: 9 numbers
3 to 4: 5 numbers	2 to 1: 12 numbers
5 to 7: 6 numbers	5 to 2: 16 numbers

We can thus define the final series as it will be printed and published, and, by knowing the ratio of photographic reduction, determine the exact series which must be drawn.

GEO Q point: Limited range in Q

Extensive series are used in point representation to investigate regional groupings resulting from limited-range statistical series. Such is the case with the map in figure 2 on page 364, which utilizes the correspondence $S = Q^4$.

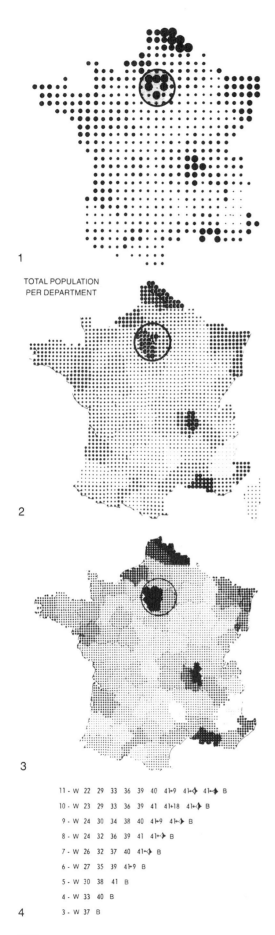

1

TOTAL POPULATION
PER DEPARTMENT

2

3

4

DENSITY OF THE POINTS

The regular pattern of graduated circles is not independent of the density of the points or, consequently, of the reduction and size of the final image. What varies among the three maps (figures 1, 2, and 3) is the spacing of the original pattern. In figure 1, there is one point every 1000 km², in figure 2, every 250 km², and in figure 3, every 100 km². In all three cases the standard correspondence S = Q has been applied. However, the visual impression is very different for the three figures. In figure 1, the eye perceives each point. It gauges the "weight of black"; it is sensitive to the *area of the point*. The natural series of graduated sizes and its tables, page 371, have been developed to correspond to this type of perception.

On the other hand, in figure 3, the eye no longer perceives the points as such. The image is based on the *value differences* resulting from the different tones of "gray," and it is the shape of the lines separating the steps which naturally attracts attention. Here, the frame of reference which the eye uses to evaluate a given distribution is the equidistance of the value steps (which is expressed by the equal visibility of the line separating two successive steps, at any level of value).

We saw on page 75 that equidistant steps are obtained by using both the progression of the amount of "black" and the inverse progression of the amount of "white." The natural series of graduated sizes (and its tables) is not applicable to very fine patterns, which only involve a perception of the values. However, we can derive from the table (figure 2, page 371) the percentages necessary to construct up to eleven equidistant steps (figure 4 gives the numbers of the points to be used).

Beyond eleven steps, a new series must be constructed. A procedure is now being worked out for up to twenty-four steps.

The division of series Q of the information is then arithmetic, and, for example, a series ranging from 12 to 42 represented in ten steps gives 42 − 12 / 10 = 3 as an interval, and thus the steps are:

12–15–18–21–24–27–30–33–36–39

In figure 2, the situation is intermediary. The eye perceives *both the size of the points and the value*, resulting from the ratio of the amount of black to the amount of white per area. At the present stage of experimentation, the natural series and its tables can be used in these intermediate cases, provided the three smallest circles (numbers 1, 9, and 15) are given places 15, 16, and 17 respectively in the column of circle numbers.

DRAFTING BY MACHINE

A typewriter or lineprinter which is keyed with the series of circles and whose lining is equal to its spacing permits machine typing of the circles; the typist simply refers to a document indicating the correct circle number for each point of the grid.

In area representation, when the areas are quite numerous (as with canton maps), typing saves considerable time.

Automation

A computer can be programmed with the following instructions:
(1) the "address" (number of department, canton, commune) corresponding to each point of the grid (the x and y coordinates on the sheet of paper);
(2) the natural series of circles (column of circle numbers or N);
(3) the statistical series being processed (component Q);
(4) the correspondence between Q and N (that is, the position of the column of circle numbers beside the series of Q).

When this is done it requires only several thousandths of a second for the computer to integrate these various instructions and provide the typewriter with the instructions (in the form of a series of punch-cards) for an entirely automatic typing. When one considers that the x and y positions of departments and cantons are common to thousands of statistical data, that the natural series is common to all data, and that modern statistical series are already on punch-cards in numerous cases, one can see that all that is required for each representation is to provide the correspondence QN, that is, to create one or two punch-cards. The rest is automatic. For the map in figure 5, which was the first ever to be drafted automatically using this formula, the typing of the legend and the title, which were not then automated, required as much time as the drafting of the entire map.

The machine mapping of quantities per area permits us to introduce into the memory bank of an electronic calculator *only the new series*, generally in absolute quantities. The computer then produces the geographic map expressing either these quantities or any transformation of them which the researcher may call for.

The typewriter and lineprinter are fast being replaced by the photocomposer and even the cathode-ray tube. In line representation, a photopositioner programmed for the natural series can trace lines that have a width proportional to Q.

In point representation, the photopositioner permits the plotting of the proportional sign at each point, and to the extent that x and y can be usefully programmed, it can operate automatically.

THE NATURAL DIFFERENTIAL SERIES

To answer questions of an elementary level, such as, "Where is a given step?" or "What is the value of a given sign on the map?" an element of differentiation must be added to the natural series. The most efficient one, as we have seen, is orientation. The differential series, part of which is shown in figure 6, permits the identification of each sign while the progression and properties of the natural series are retained.

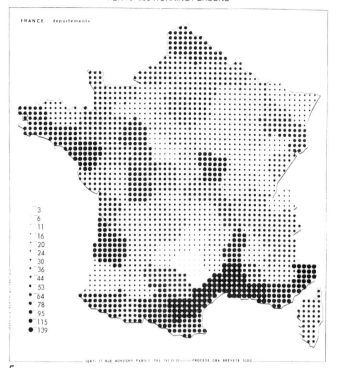

5

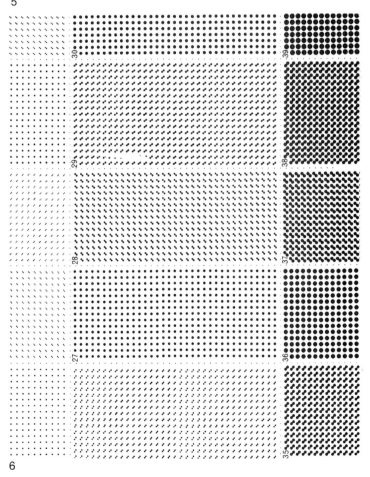

6

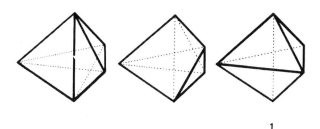

1

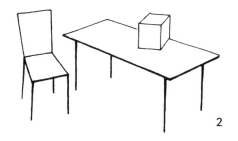

2

3

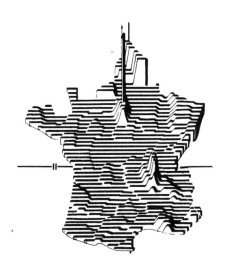

4

Perspective representation

The visual illusion of three-dimensional space results from the associations acquired through previous perceptual experience. The observer will translate perceptual deformations into the differences in distance which generally produce such deformations. The size, value, texture, color, orientation, and shape of an object change with distance, and it is by carefully "deforming" these visual variables that the painter or decorator can create the most convincing "illusion" of depth.

In graphics, representation of space on the plane can be obtained even more easily. The observer must merely be capable of interpreting several visual sensations (and sometimes only a single one) as the meaningful and pertinent deformation of a known characteristic.

A variation in the thickness of the lines of a network (figure 1) is sufficient for suggesting a volume. It evokes the total shape of the object as well as the transformation of value and size which usually accompanies differences in distance. But this variation must be meant to signify such differences. This is achieved by seeing to it that any other meaning is excluded. Thus a figure formed by two thicknesses of lines, signifying two quantitative levels, cannot be put into perspective by a simple variation in thickness of lines.

This variation in line thickness must also be appropriate; that is, none of the spatial positions suggested by the volume should be inverted. What is at the front is thicker; what is at the back is thinner or lighter.

Shape variation
This is the easiest and most efficient type of planar deformation.

However, no sense of deformation can exist without prior knowledge of the nondeformed phenomenon.

A cube, a chair, or a table can appear to us "in perspective" and suggest a three-dimensional space (figure 2). However, the volume of a pebble cannot be suggested by a perspective deformation, since its initial shape is unknown (figure 3). Therefore, it is useless to deform elements whose initial shape is not familiar. It is not the perspective drawing of a hydrographic network or a topographic site which creates

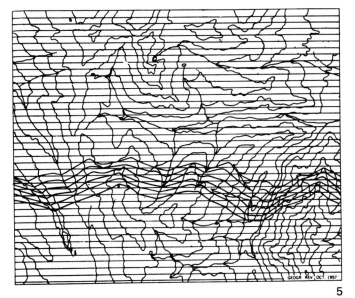

5

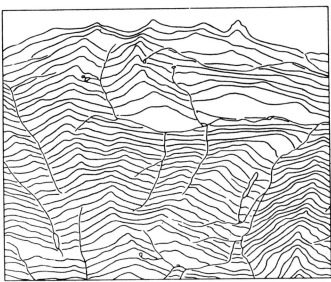

6

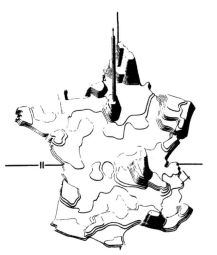

7

the sensation of space in a "block-diagram," but the deformation of the recognizable elements alone—that is, of the characteristics which are conceived as belonging to any volume, whether a landform or any given construction.

The universal characteristic which can be applied to the representation of any volume is REGULARITY. One can easily conceive that the regular sectioning of a regular volume yields regular lines. *This rule holds* for visible and meaningful irregularities which translate irregularities of volume. This is the basis for any representation in figurative relief. However, irregularities represented by lines are not ordered, and there is often confusion concerning the "sense" (up or down) of the irregularities (a confusion between valleys and peaks). SHADING, whose principles are familiar, adds sense to differences expressed by lines.

FORMS OF REPRESENTATION

Any ordered component can be constructed in apparent relief. This construction is not the prerogative of topographic relief. Four different forms can be used:

Regular vertical sections (figure 4)
This concept is particularly applicable to complex variations (perspective diagrams, page 253) and to topographic relief.

Inclined sections (figure 6)
In this representation (set forth by Arthur H. Robinson and Norman J. W. Thrower, "A New Method of Terrain Representation," *The Geographical Review* 47, no. 4 [October 1957]: 507–520), the inclined sections are derived from a contour map by application of a grid (figure 5) corresponding to the equidistant spacing of the contours, at an appropriate scale. Each intersection of a contour and its corresponding horizontal line on the grid is connected to the next higher or lower intersection. In the final drawing only the inclined sections are retained (figure 6).

Regular horizontal sections (figure 7)
Here, one merely draws each contour with its corresponding elevation and does not draw those that are "hidden" behind elevations.

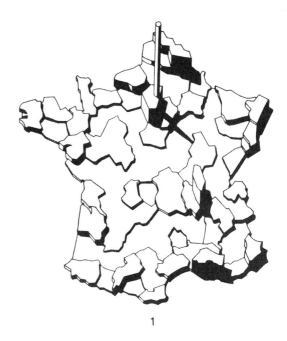

1

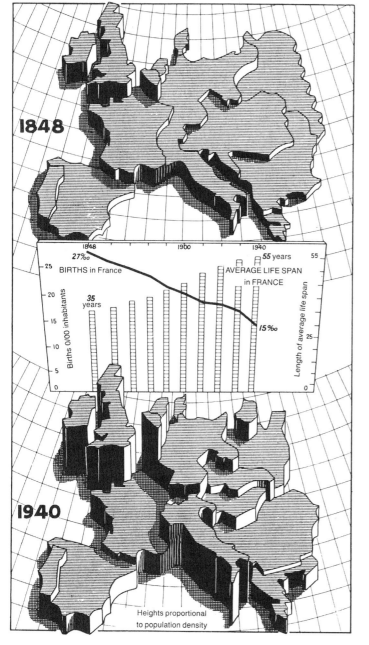

2
DEMOGRAPHIC EVOLUTION (1848–1940)

The outline of a volume (figure 1)

The combination of the preceding techniques enables us to draw the outline or silhouette of a volume. But when the regularity of the sections is no longer represented, other regular elements must be used to replace them. In figure 2, it is sufficient to elevate the plane shape of the countries by an amount established by the scale of heights. The height and placement of the vertical walls suggest the deformed regular system, and thus the volume. However, the "hidden" parts must be minimal, and representation in relief does not allow the use of high walls in the foreground.

TOPOGRAPHIC RELIEF AND THE BLOCK DIAGRAM

Representation in relief enables us to utilize the principal geomorphologic shapes and the natural elements which constitute our environment. Various characteristics, such as features of rocks, nature of the soil, vegetation, or crops, along with human settlements, can be inscribed on the topography and depict the landform as it is seen, with all its classic positional correlations. In any regional description, this should be the first graphic constructed.

An overall perspective is useless

In regional studies, an overall perspective is useless and often regrettable, since the deformation prohibits comparison of the landform with other factors, particularly those obtained through statistical cartography. Furthermore, by replacing this nonsuggestive (since the shapes are not familiar to us) deformation by a perspective of the details (perspective "as seen") the designer can save time and construct the relief on a piece of tracing paper (called a transparent "flap" or "overlay") placed directly on the map. We thus retain precise control of the new, pertinent information at each site.

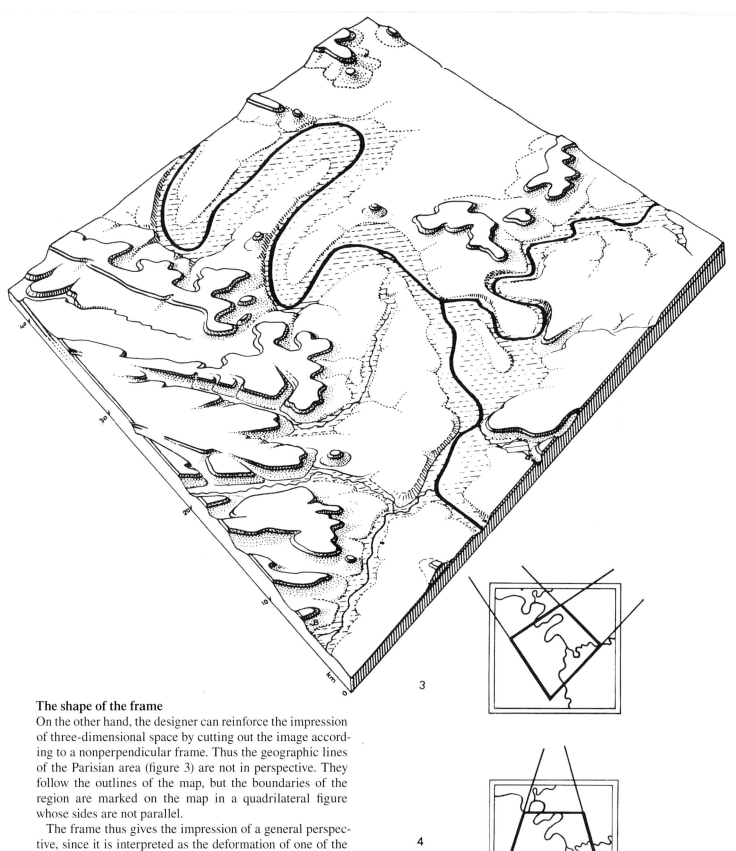

The shape of the frame

On the other hand, the designer can reinforce the impression of three-dimensional space by cutting out the image according to a nonperpendicular frame. Thus the geographic lines of the Parisian area (figure 3) are not in perspective. They follow the outlines of the map, but the boundaries of the region are marked on the map in a quadrilateral figure whose sides are not parallel.

The frame thus gives the impression of a general perspective, since it is interpreted as the deformation of one of the most familiar rules: perpendicularity.

A trapezoid produces the same result (figure 4). It merely requires that the convergence point of the nonparallel lines be situated toward the top of figure 3 or figure 4.

Perspective of the details

It is the details which generally constitute the known regular elements whose deformation suggests relief. The following are the most common:
– meanderings of watercourses, which flatten out horizontally (figure 1);
– topographic indentations, which tend toward the horizontal in flat regions (figure 2), which follow the slope of the terrain in rough regions (figure 3), or which flatten out vertically for ridges (figure 4);
– valleys and large watercourses, which change regularly in width with orientation (figure 5) and flatten out as a function of their proximity to the horizontal axis of the image;
– side views, which hide part of the landform and permit drawing vertical silhouettes.

From these elements one can derive the principal geomorphologic shapes in homogeneous terrain (figure 6) and with a stratified lithology (figure 7). It must be stressed that a block diagram is a schema, which extracts and emphasizes the principal characteristics of the landform. Before drawing it, a precise goal must be determined. Accordingly, an efficient drawing cannot be constructed without having traced all the following characteristics on the map:
– lines of ridges and peaks . . .
– breaks in general terrain, terraces . . .
– boundaries of alluvial zones, flat bottoms of valleys . . .
– troughs, notches, basins . . . .

In the final analysis, the block diagram is merely the translation of a good morphologic map into signs that suggest perspective. One should not hesitate to reserve sufficient space for the representation of a characteristic feature, such as the flat bottom of a valley, by encroaching on the surrounding plateaus. Whatever the complexity of the landform, it should always be expressed in the minimum number of characteristics. The outline must define the shape. Only shading, achieved by dashes, points, a wash, or reinforcement of the characteristic lines, can be used to "enhance" the figure, discreetly of course.

When apparent perspective is adopted, vertical displacements are practically useless. In the map in figure 1 on page 359 they are nonexistent. The hydrographic network is in place, which does not hinder the perception of the general elevation of the masses of mountains, the Cévennes and the Causses. One need only use topographic indentations to suggest the general slope of the terrain, which can be brought out further by the overall shading.

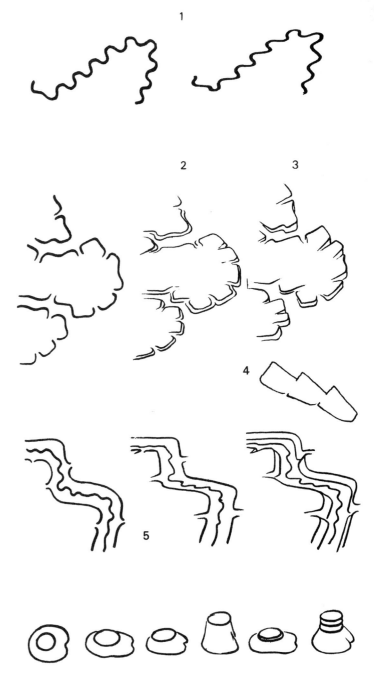

In the map in figure 1 on page 359 the shading diminishes progressively and is used to highlight an outline drawing. On the map above (figure 1), the relief results from the shape of the light and dark features drawn against a gray background, corresponding to the horizontal areas.

Overloading and superimposing deprives such representations of their efficiency as base maps, except with simple, homogeneous distributions (page 359) or for particular phenomena which conform closely to the relief.

Scale 1 : 5000   Contour interval 20 m

2

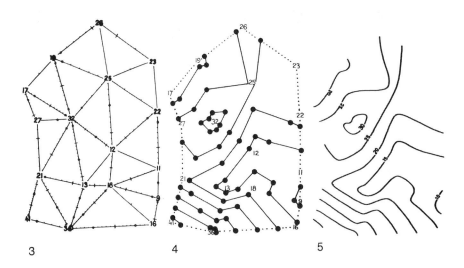

3 4 5

Isarithms

Isarithms are applicable to any quantitative component. They are only conceivable as equidistant; if not, they merely show the boundaries of areas, which must be interpreted in succession. The efficiency of isarithms is very limited.

The map in figure 2, which depicts the Salt Mountain of Djelfa (Algeria), surveyed by E. F. Gauthier and published in the *Annales de géographie* (Paris: A. Colin, 1914), shows that isarithms are an excellent means of recording and displaying SLOPES, but that their properties stop there. They do not permit us to perceive in a single image the total elevation, the actual volume, the altitude, nor even the "sense" (up or down) of the slope, for which we must return to the elementary level of reading (counting the contour lines, reading the elevation numbers).

Isarithms are therefore used
– to represent slopes or breaks in the distribution of a component Q;
– to discover, at the laboratory stage, the main lines of a distribution;
– to superimpose on the same drawing several components, depicted in various implantations, with the goal of discovering precise positional relationships, in elementary reading.

Isarithms do not permit us
– to carry out overall quantitative comparisons;
– to represent a component QS, that is, absolute quantities calculated for variable areas (the densities must be calculated);
– to represent a sparse sample, that is, information involving unknowns whose numerical value cannot be inferred from the known points. Only a retinal variable can correctly express such a sample (see page 364).

Construction of isarithms
Let us look at a series of numbered points (figure 3).

By definition, the information includes all the numerical values, such that the value of any point in the space can be estimated.

Whenever one decides to employ isarithms, this postulate, which alone permits us to interpolate from the known points, is implicit.* For precise construction, it is necessary:
(1) *To draw all the possible triangles among the points* (figure 3), without intersection. The shorter of two alternative lines will be selected (for example, 13–32 rather than 21–12).
(2) *To divide each triangle side into equal parts*, as a function of the numerical values and the unit chosen. Here, the unit is 1, with principal values at intervals of 5. The line (or trace) is "smoothed" by observing three or more values. Thus, between 12 and 25, it is possible to take account of the flattening revealed by line 25–26, and confirmed by 25–23 and 25–22. The spaces would then be narrower near 12 and wider near 25.
(3) *To link the equal values thus discovered*, which replace the original sample values (figure 4). It is not possible to depart from the network of triangles; the information does not exist outside of it. Attempting to do so will lead to ambiguity.
(4) *To "smooth" the information by drawing the isarithms* (figure 5), which is reasonable if one considers that angularity, that is, discontinuity on the plane, is highly improbable on these conceptual levels.

---

*This postulate also determines the "positioning" of the value within an area. Either the position of the "center of gravity" of the phenomenon and its value are known—which may deviate from the mean value and produce a supplementary value—or else this value is unknown, and the mean value cannot be placed elsewhere than at the geographical center. However, within the boundaries of the area, it can be moved slightly as a function of the neighboring values (smoothing of the surface profiles).

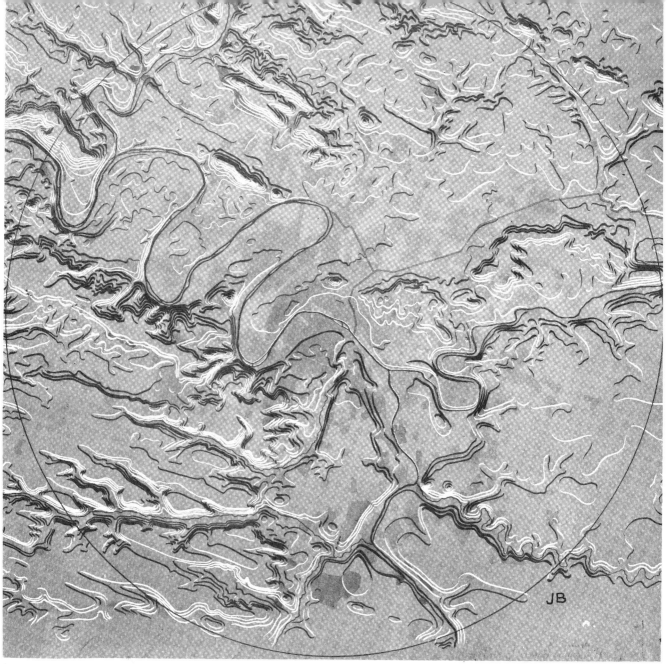

1

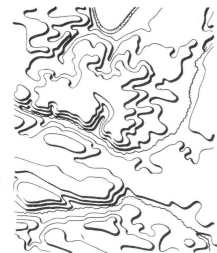

2

Sense of the slope and shading

Isarithms do not indicate the sense of the slope (see figure 2, page 385), which will only be apparent in very simple and familiar distributions. However, in meteorological sketches, for example, it must be inferred from appropriate signs. Thus, the sense is generally suggested by a series of tints (redundant combinations with value or color, which obviously overload the image), or by shading. The simplest way is to shade the contours by darkening the lines. One either imagines a light coming from the top left (figure 2)—which is familiar, probably due to "right-handedness" and to resulting lighting habits—or a light coming from the top of the image. In this case, the shading combines with a variation in

thickness, which is added to the perception and avoids relief inversion (figure 3).

On a gray background, the white and black lines reconstruct the lighting of actual relief (figure 1). The visual distance between gray and white must be perceptually equal to the distance between gray and black. Shading alone can express the main classic geomorphologic shapes, especially if medium gray is utilized to represent the absence of slopes (figure 4). The photograph of a relief model, in plaster, now constitutes the safest means for evoking geographic relief at a large scale, provided the plaster is as vigorous as the landscape it is meant to represent.

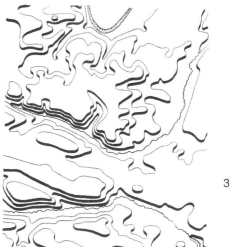

E. Cartographic problems involving more than two components

A cartographic problem involves more than two components when the information expresses the relationship between a geographic component and two or more additional components.

The graphic representation must then utilize four components or more. It cannot be realized in a single image, and the totality of relationships cannot be perceived instantaneously.

The designer will therefore have to choose among the following:
– *an inventory map*, which is comprehensive but must be read point by point;
– *"processing" maps* (collections of images), which represent the comprehensive information on separate maps which are comparable and classable;
– *a cartographic message* or synthetic schema, which superimposes the essential aspects of the information in several simplified images.

Any inventory can produce a message, and, in fact, any message results from an inventory followed by processing. By the judicious use of cartographic methods, we can avoid useless inventories and directly construct the processing drawings from which we then derive the message.

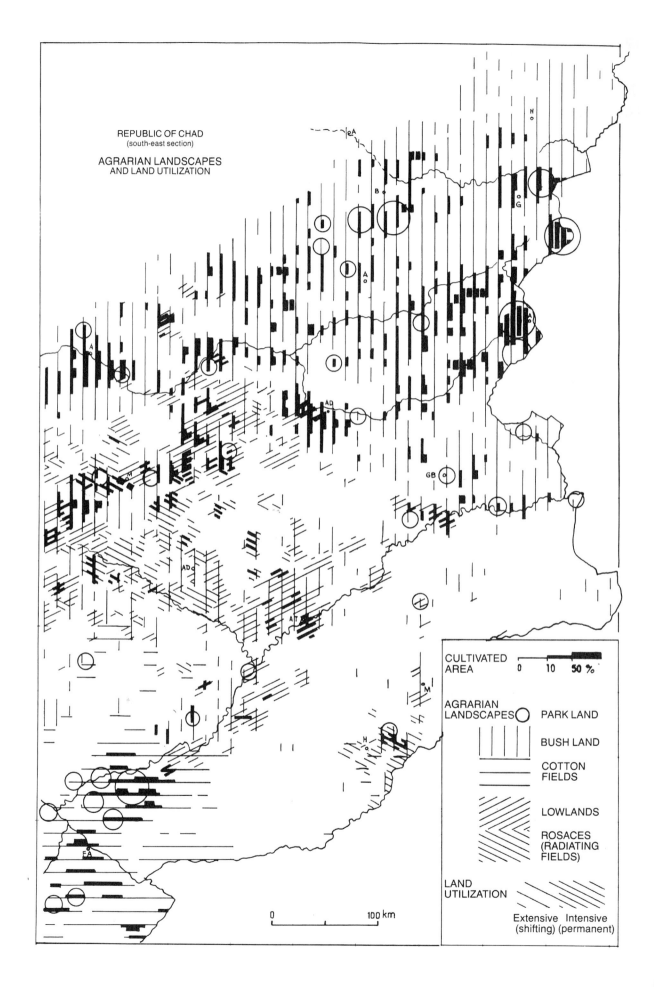

1. Inventory maps (comprehensive figurations)

We utilize a cartographic inventory questions involving the geographic component are the most pertinent ones.

Finding a geographic location is a common problem. The topographic map, the city plan, and the subway diagram constitute collections of reference points which permit us to orient ourselves, to know in advance what we are looking for, or to identify what we are looking at. With these types of document, the elementary level of reading is the normal level: "What is there at a given place?"

The intermediate and overall levels are virtually excluded; for example, the study of relief developed only when geographers had at their disposal the relief map alone (orohydrographic edition), separate from the overall topographic map, even when the various components superimposed on the overall map were differentiated by all available means of visual selection.

One also constructs an inventory at the point of a first graphic recording of complex information. Such is the case with the map by J. Letarte (figure 1, page 150: Agrarian Landscapes in Chad), which is merely the translation in black and white of the original, which was done with colored pencils directly on the topographic map at 1 M. This map constitutes the preparatory stage for graphic processing.

We also examined (figure 1, page 165) the maps which had to be established next in order to "process the information" and perceive the totality of the information in only a few images. The very simplified message which could be obtained from the whole of the information was given in figure 2 on page 163.

But there is an infinite number of levels of simplification, which can appear to muddle the distinction between inventory, processing, and message, if precise definitions are not used (see page 160). Indeed, certain simplifications can be utilized as inventories and also as messages. Such is the case with the map in figure 1, which is no longer a comprehensive inventory (the geographic shapes are simpler than in the given information), although it does display the useful totality of the categories of the nongeographic components. It can be considered as a research instrument, even though it is not strictly speaking "comprehensive."

Note that the graphic is both efficient and easy to draw. It can serve as a standard for numerous problems and can be analyzed as in the table in figure 2.

The overall image is based on the amount of land utilized. Furthermore, it is possible to visually select each of the subsets defined by a category of the thematic components. For example, we can see only the park, the bush areas, or the "rosaces" (see page 163), etc., and can compare these different distributions.

Thus, we can observe that the "rosaces" and lowlands are always intensive (permanent), that they almost never occupy more than 50% of the land, and that they are often combined with the bush areas but practically never with cotton.

Finally, any elementary question, such as, "At a given place . . . ," can be answered immediately and completely. But this representation could only be conceived after an attentive study of the distributions drawn in figure 1 on page 165.

2

INFORMATION			GRAPHIC CONSTRUCTION
GEO (20 000) aerial photos			simplified
O (4)	Cultivated area:	desert 0–10% 10–50% + 50%	size variation, of lines
≠ (5)	Agrarian landscapes	parkland bushland cotton fields lowlands radiating fields	variation of orientation and shape (parkland) groups of oblique lines (lowlands and radiating fields) in contrast with orthogonal lines
≠ (3)	Land utilization:	intensive (permanent) intensive-extensive extensive (shifting)	spacing of lines (reduction to two categories: intensive and other)

Useful inventories

Two additional components can produce a useful inventory, provided that their combination is meaningful. The overall image is based on the result of this combination. One example is the map of Spain in figure 1, page 188, where population combined with tax rate produce a single image based on the total amount of tax revenue; another is the wind map in figure 3, page 353. This is also the case with the map of agricultural population in the United States, by Roberto Bachi, given in figure 1 here.

The problem can be analyzed as follows:

GEO —American states
Q   —total population (number of large squares)
Q% —agricultural population (represented by the arithmetic scale [figure 2] perfected by R. Bachi, which permits graphically reading, on the elementary level, a quantity of 1 to 100 down to the single unit)

Thus in elementary and intermediate reading, one can focus either on the total amount of population or on the percentages of agricultural population. Furthermore, the overall image involves the absolute quantity of agricultural population, since: Q total × Q% = Q partial. However, this overall image is very different from the image which would be produced either by the percentage of agricultural population, or by the total quantities.

These last two elements are indispensable in the processing of a large data set, where the classing and grouping of factors is imperative; as, for example, with the factors of regionalization in the national budget (Paris: Imprimerie Nationale, 1965, 400 maps).

It is possible to load a map to the limits of elementary legibility, but such figurations are most often inefficient and thus useless. Such is the case with "chartmaps," constructions where diagrams having 2, 3, n, components are "strewn" over a geographic background.

The chartmap is hardly ever efficient, except when the third component is very limited or when it can be represented with three variables; two examples of this are shown in figures 3 and 4.

Figure 3 shows grain supplies in Paris (after M. Baulant and J. Meuvret, *Prix des cereals, extraits de la mercuriale de Paris* [Paris: S.E.V.P.E.N., 1960]).

GEO —cities
≠   —two types of quantity: grains and shipments
Q   —the quantities

Note that the ratio, Q of grains transported/Q of shipments,

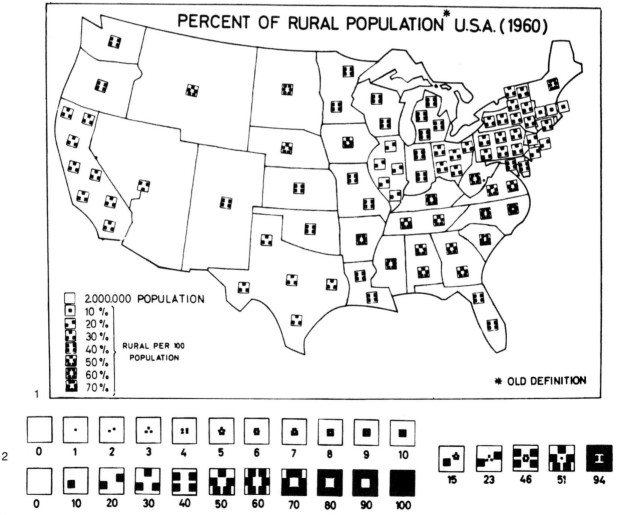

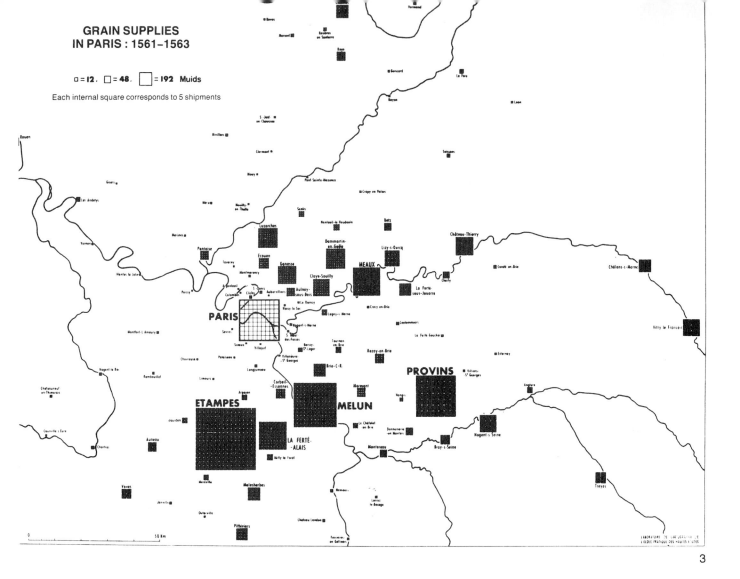

gives the average size of the shipment (and at the same time suggests the probable means of transportation: boats, wagons, animals).

The overall image is based on the quantity of grain. Internal division produces an efficient perception of the size of the shipment.

Figure 4 shows the devaluation of money (after Fernand Braudel and Frank Spooner, "Prices in Europe from 1450 to 1750," in *The Cambridge Economic History of Europe*, vol. 4, ed. E. E. Rich and C. H. Wilson [Cambridge University Press, 1967], pp. 374–486).

GEO —sixteen cities
Q   —silver per monetary unit in 1750 (based on 100% in 1450)
≠(2) —silver remaining (in black), silver lost (in white)

The ≠ component has a length of 2: it is a binary component, and the overall image is based on the quantity of silver remaining (black). The precise measurement of each percentage can then be read, image by image, according to the angle at the center.

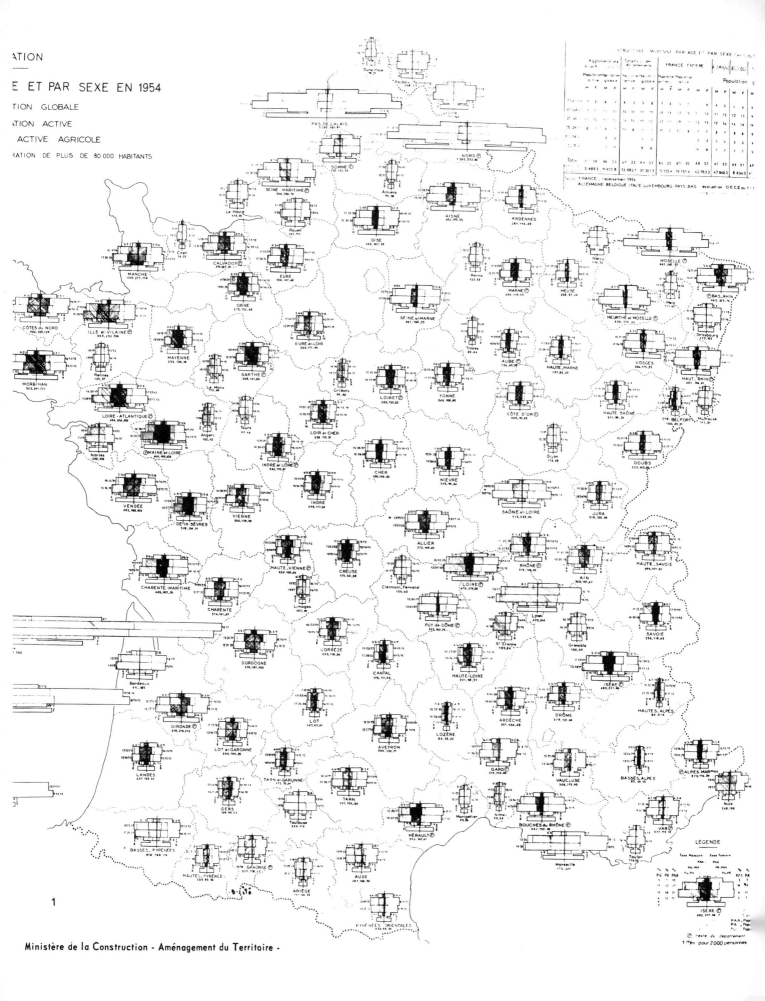

Useless inventories

A "chartmap" is practically useless when the length of the ≠ component is extensive: an example of an inefficient inventory is the distribution of the work force, by sector, age, and sex, as shown in figure 1. The original map is at 1 M and consequently measures 1 m × 1 m (Paris: Ministère de la Construction, 1959).

INVARIANT   —*work force*
COMPONENTS   —**Q** *according to*
        GEO (120)–*ninety departments plus thirty cities larger than 80 000 inhabitants*
        ≠ *three main employment sectors (I, II, III)*
        O *six age-classes*
        ≠ *two sexes*

The problem has five components and thus requires six visual variables for its representation.

To grasp the total information given in figure 1, we would have to memorize 120 × 3 × 6 × 2 = 4320 successive images!

This "figuration" (see page 151) cannot be compared with other data, nor does it permit regionalization.

Such information should not be graphically represented by a "chartmap." It is a useless drawing, an ill-conceived dictionary which cannot even serve as an initial airing of the information. Indeed, in constructing the appropriate images, one realizes that the original statistical tables are much more practical and, obviously, more accurate.

Here, a graphic representation is justifiable only after processing, and it must lead to a limited number of images. Two types of images can be constructed:

(1) images **GEO Q**:
× six age-classes
× three sectors
× two sexes ... that is, thirty-six images
(There is one image per age-class for one sector and one sex.)

We can obviously eliminate one or another of the three components and construct eighteen, twelve, six, three, or two images.

(2) images **GEO O** (O denoting the order of the types resulting from the components **Q** and **O**6):
× three sectors
× two sexes .... that is, six images
(There is one image per sector and per sex.)

We can also eliminate one or another of the two components and construct only three or two images. The process outlined below is of the second type. It combines all the sectors and treats only the masculine sex. (However, the approach is the same for a given sector.) The information is as follows:

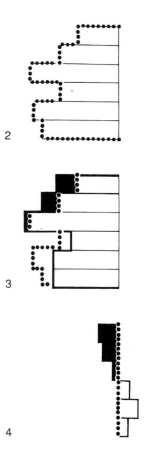

INVARIANT   —*total male work force*
COMPONENTS   —**Q** *according to*
        GEO *(90) (only the departments have been considered)*
        O *six age-classes*

This information has three components, thus requiring four variables.

It must therefore be reduced again in order to construct it as a single image. This reduction is obtained by diagonalizing the image formed by the component **GEO**, considered as linear and reorderable, and the component **Q**. We can proceed, as on pages 231 and 245, by constructing one linear card per department (this is the best method) or as below by constructing one diagram per department.

Take the diagram of all of France (national mean) (figure 2). The departmental diagrams are compared to it (figure 3). The differences positive (in black) or negative (in white). The mean is brought to a straight line (figure 4).

The diagram of deviations from the mean is thus constructed for each department.

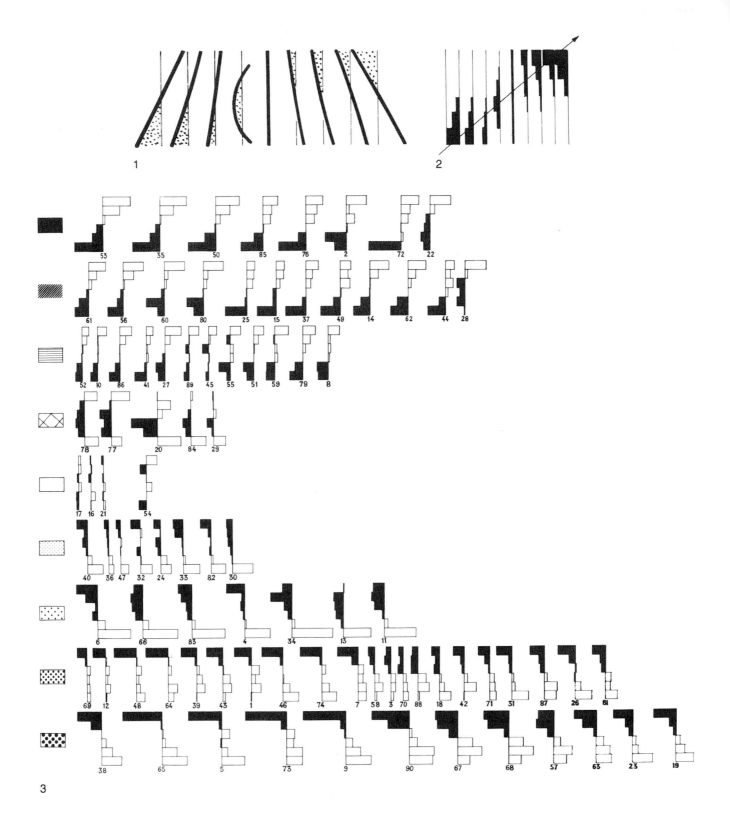

DOMINANT AGE CLASSES

Male Population in 1954 According to
Age-Classes
Departments

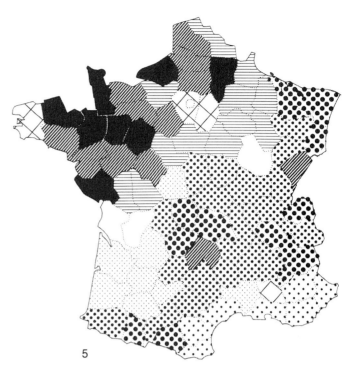

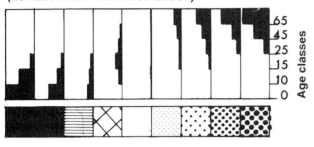

Types of distribution
(deviation from mean distribution)

4

5

The departmental diagrams must be made classable. They are drawn on tracing paper; one or two copies are made; then these are cut up.

By successive comparison and classing we discover similar types of distribution (figure 1), which are ordered according to the principle of diagonalization (figure 2), and we construct a display (figure 3), in which all the departments are ranked.

All that is required is to adopt an ordered retinal variable and project this display onto the map (figure 5), which forms an image, for which figure 4 is the legend.

A comparable image can be constructed for the feminine sex and for each sector. As a result, the whole of the information contained in figure 1 on page 394 could be perceived in six images.

The relative efficiency of one construction over another can be expressed by the ratio of the number of images necessary for their overall perception, that is:

$$90 \times 3 \times 6 \times 2/6 = 3240/6$$

Positive-negative series

A *texture variation* permits us to separate the series of departments into two parts: predominance of youth (values in fine textures); predominance of elderly (values in coarse textures). See also figure 5 on page 231. The fine textures are generally obtained by lines, the coarse textures by points.

In the absence of color, texture is the only retinal variable available for efficiently separating two ordered series. This is the best graphic solution for the problem of separating positive and negative quantities.

2. Processing maps* (collection of comprehensive images)

The general solution for any geographic problem involving more than two components is a collection of images, each one comprehensive.

FIRST EXAMPLE: ECONOMIC CENSUSES IN SPAIN

(From G. Da Silva, *En Espagne: Développement économique* [Paris. Mouton, 1966]).

Historical censuses are numerous in Spain, but they are often subject to caveats. Is this a reason not to utilize them? The rigorous recording of the data, according to the formula of "the regular pattern of proportional circles" (see page 369) permits us:

(1) to take some 1500 statistical data points involving ten components, make them graspable in thirty-six comparable images, and represent them on a double page 42 × 27 cm (see following pages, 398 and 399);

(2) to approach the problem of critically evaluating the census, based on the numbers themselves and not on non-quantitative external information;

(3) to carry out all the desirable comparisons among these components and propose a general reduction of the information to several images.

*See also *G.I.P.*, pp. 161–167 (translator's note).

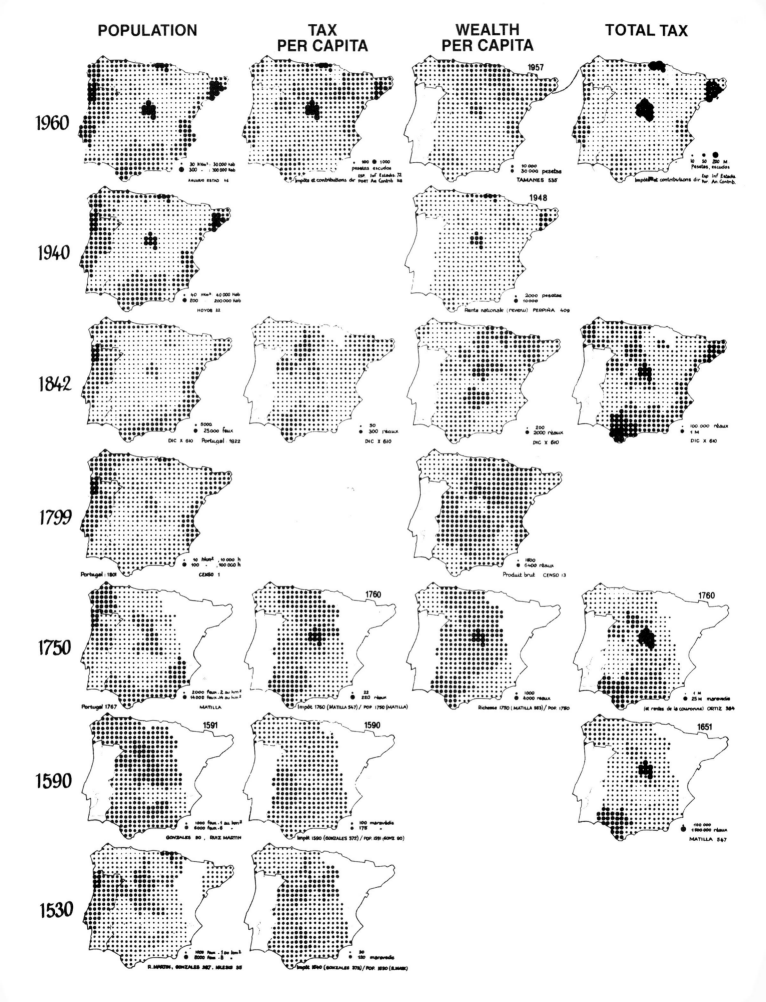

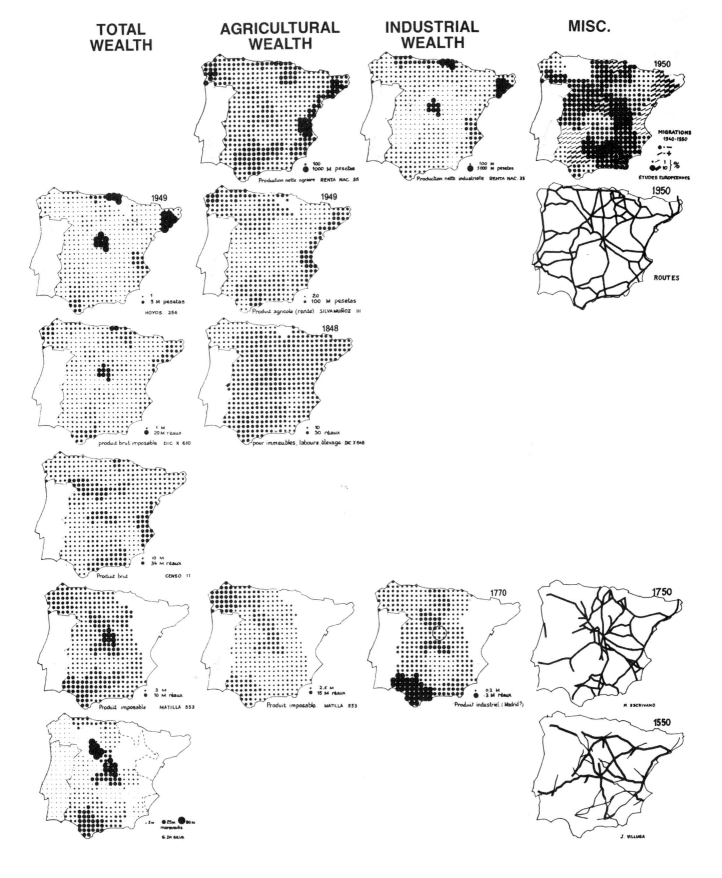

The graphic problem

The task of the graphic designer is to make the statistical data easily graspable, while ensuring comprehensivity. The designer must avoid a priori simplification of the data, a reduction of the quantitative series to a small number of predefined steps. The regular pattern of graduated circles provides the means of accomplishing this. It is also important for all the maps to be comparable, whatever the nature of the phenomenon and whatever the time period involved.

It is clear that the only conceivable unit is the total of each particular quantitative series. Otherwise, we could never compare population and currencies, or even different currencies (pesetas and maravedis, for example, are units whose values vary over time and whose bases are still to be determined). For the same reasons, it is inadvisable to represent the variation in total population from one period to another; it would dominate and overwhelm regional variation and make any other comparison impossible.

We must focus only on the regional distribution of each phenomenon, independent of the total differences, which are not usually comparable between different periods. In other words, the eye compares *geographic frequencies*, that is, the regional percentages of a total constant volume.

A "density adjustment" permits us to relate all the representations to each other.

Criticism of statistical documents

The properties of graphic expression also enable us to resolve the delicate problem of the credibility which should be accorded to given information.

There is no perfect information, but only degrees of error, which oscillate between the acceptable and the unacceptable. In the present case, one generally evaluates the level of acceptability according to a historical criticism of the sources, which involves determining the conditions of the original elaboration of the numbers, along with the competence of the researchers and the authors.

This criticism is based on comparisons with inventories which are similar in nature, region, and time period. The tendency is obviously to reject given information *en masse* when several samples have yielded bad results. Since cartography is considered to involve only "accurate" representation (but what is accuracy?), such information is often abandoned. However, the graphic representation of information is truly the best possible instrument of control and evaluation, when the graphic system allows us to represent the information integrally. Instead of relying on samples, it paints a total picture of the potential error. The graphic representation of erroneous data is not to be feared during the course of research.

On the contrary, it always yields useful information, either concerning the content—to the degree that, for example, even if all the numbers are false, the general tendency can still remain valuable and confirm a relationship discovered elsewhere—or concerning the extent of the error, or its principal characteristics. It allows us to approach the problem of historical error from the error itself, not from the potential causes.

Consequently, the statistics of 1799, reputedly erroneous, have been retained in the maps on pages 398 and 399. Each person can evaluate them and discover certain probable aspects, even in a series of flagrant abnormalities. Likewise, we have preserved information from the sixteenth century whose accuracy is questionable, but which in general underscores tendencies which can be considered as probable.

An attempt at overall summary

Faced with such a collection of comparable data, it is tempting to seek a formula capable of producing an overall summary.

Justification of the categories retained

(1) It is known that a single statistical number means nothing and that only the comparison of several numbers is meaningful.
(2) No number is strictly accurate, but the *tendency* resulting from the comparison of two numbers *can* be meaningful (if one limits oneself to the tendency).
(3) This tendency has an even greater probability of being meaningful, if it is found again in more observations of different nature and origin. The probability of meaning obviously results from the observation of the greatest possible number of data points.

In the present study, it is precisely the graphic system which permits the maximum number of comparisons. It also affords the assurance that visual perception does not result from prior transformations, which reduce the quantitative variations to several arbitrary classes. It is therefore legitimate to define at least the two extreme tendencies, one toward the minimum (predominance of white on a map), the other toward the maximum (black), and it is indispensable to reserve an intermediate category for the unconfirmed or doubtful tendencies (gray).

These are the three terms which have been set out in each phenomenon and for the seven characteristic regions, which emerge from the collection of maps (pages 398 and 399).

Utilization of categories: R *m* P

Furthermore, it seems legitimate to add similar tendencies, if one takes into account the fact that in each phenomenon, black represents a "rich" tendency (R) and white a "poor" tendency (P). We will define gray as "medium" *(m)*. An addition has been attempted here on an experimental basis.

By accomplishing, within the framework of a region and a period, the total of the phenomena observed in each tendency, then by making the total of the observations constant across regions and periods, we can transcribe the *frequency* per period of each one of the three terms R *m* P, and we obtain the final figure (figure 1). It is indeed a "chartmap," but it involves only seven images. It is therefore easily understandable.

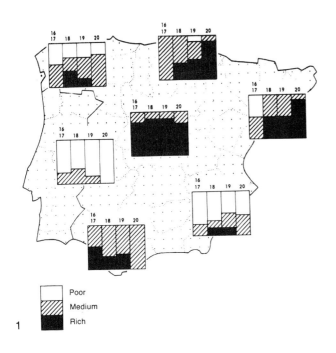

SECOND EXAMPLE: POLITICAL GEOGRAPHY OF THE DEPARTMENT OF ARDECHE

Here is another example of graphic information-processing. In 1949, André Siegfried published his *Géographie électorale de l'Ardèche sous la IIIe République* (in *Cahiers de la Fondation Nationale des Sciences Politiques* [Paris: A. Colin, 1949]), a small book, rich in substance, and now a classic. Siegfried's book is loaded with maps illustrating the different political factors, and indeed the essential points of his demonstration are derived from geographic similarities. The map, far from being a ritual "illustration," constitutes the basic, indispensable material for his analysis.

His study is presented in the form of a geographic description, where physical, human, and economic factors are discussed successively. It ends with the analysis of political behavior. In the conclusion, attention is focused on the distributional similarities of several phenomena, and these similarities permit an explanation of political attitudes to be sketched out. In expressing his thoughts, Siegfried relies mainly on language, where he excels, and his approach is intuitive. If it had been deductive, he would not have let a cartographic arrangement as improbable as that in figure 2, be published. He would have required as much logic of the graphic demonstration as of his verbal argument, and he would no doubt have developed the latter otherwise. In fact, the work contains the elements of a concise and rational demonstration, which can be reconstructed simply by using the published documents and the very phrases of the author.

401

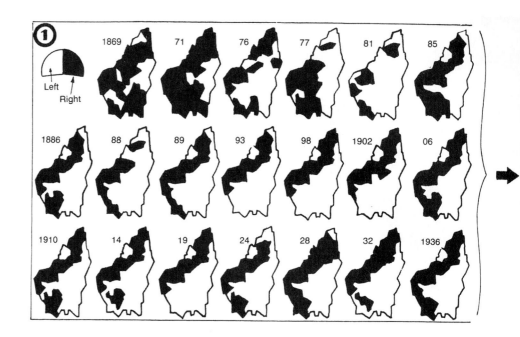

1st confirmation: More detailed maps yield the same distribution

2nd confirmation: Stability of the period

3rd confirmation: Vote of representatives on significant issues

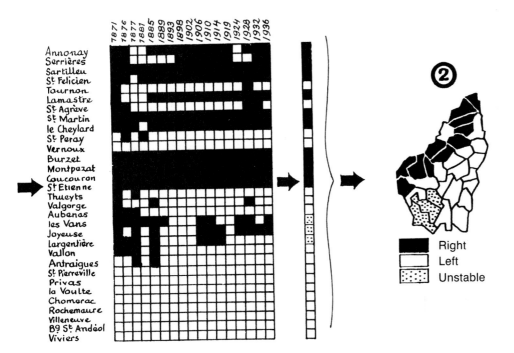

In his book, Siegfried proposes investigating the factors of political behavior in the department of Ardèche in the following manner.

First question: What is the political physiognomy of the department of Ardèche?

An answer is obtained by mapping all the known votes (figure 1). Each map is discussed and defended (pages 75–106). A first confirmation is achieved by creating a more detailed distribution between the right and the left, through the introduction of intermediate steps.

Looking for a possible evolution is imperative when confronting documentation extended over more than a half century. The series of maps in figure 1 already reveals a great deal of stability, which is confirmed by a graphic (second confirmation).

The votes of the parliamentary delegates in particularly significant ballots confirm the first results (third confirmation).

Such stability leads to the construction of a table (which could have been diagonalized [figure 3], thus permitting the inclusion of the canton of Serrières—the most northern—among the unstable cantons and leading to a further confirmation of the main thesis). Then the author drew a map (figure 2), which he titled "Summary of Political Opinion in Ardèche." This map constitutes the detailed answer to the first question.

Series of comparative factors

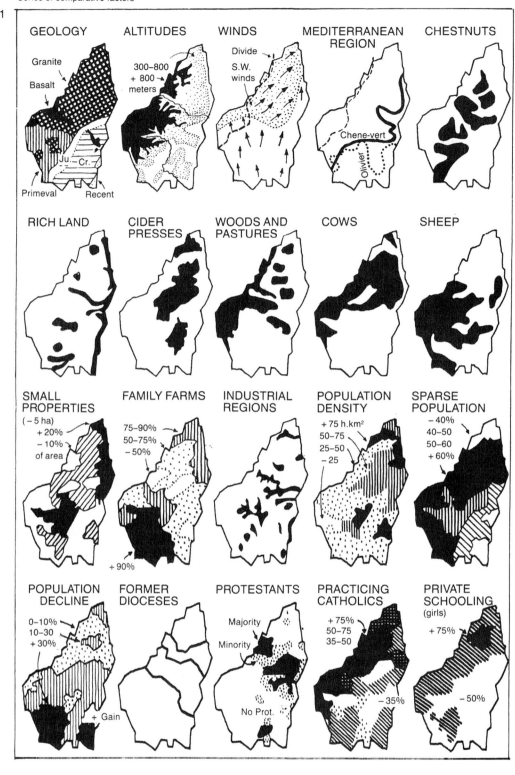

404

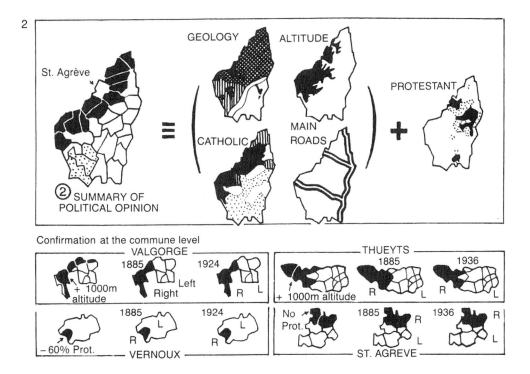

Second question: What are the fundamental causes of the political physiognomy of Ardèche?

To answer this question involves increasing the elements of comparison (figure 1). They are obtained by the mapping of all the factors, obvious or remote, which may relate to public opinion. Each map is discussed, since categories are defined not only as a function of the phenomenon, but also of a potential resemblance with the map in figure 2 on page 403. Three factors display a similarity in distribution with that map: geology; altitude; Catholicism. Siegfried also suggests the importance of main roads, which we have added in figure 2 here. He next shows the interference of the distribution of Protestants, which explains the anomaly of the St.-Agrève canton.

This is what is expressed by the graphic equation in figure 2.

In parentheses are the factors whose distribution is close to that of the main schema. The sign + introduces a factor which is necessary to complete the analogy.

"One need only consider the map of religious affiliation to realize that the cantons . . . which are the most Catholic, are at the same time the most oriented toward the right . . . . At this point the geologic map and especially the altitude map are to be retained as giving the true explanation . . . However, it is a singular exception which points us in the direction of new explanations . . . the St. Agrève Canton, although located at an altitude of 1000 m, votes on the left. The reason is simple: it is a Protestant canton . . ." (Siegfried, pages 112 and 113). A confirmation on the commune level affirms that the correspondence observed in figure 2 applies not only to the general tendency but also to a precise reality on the most detailed level of observation.

Thus two pertinent questions, two graphic processes, and two answers achieve the essential purpose of this remarkable study, the reduction by comparison and classing of fifty-three main images corresponding to four components (**GEO**, **Q**, **O** of time, ≠ evidence).

One can see here that the essential task of the researcher is to class the available records, the maps, and to define groupings and layouts which highlight previously hidden relationships. To arrange and group the maps in the best way is the fundamental problem, and, in graphic processing, it is not the duty nor even the province of the "layout editor."

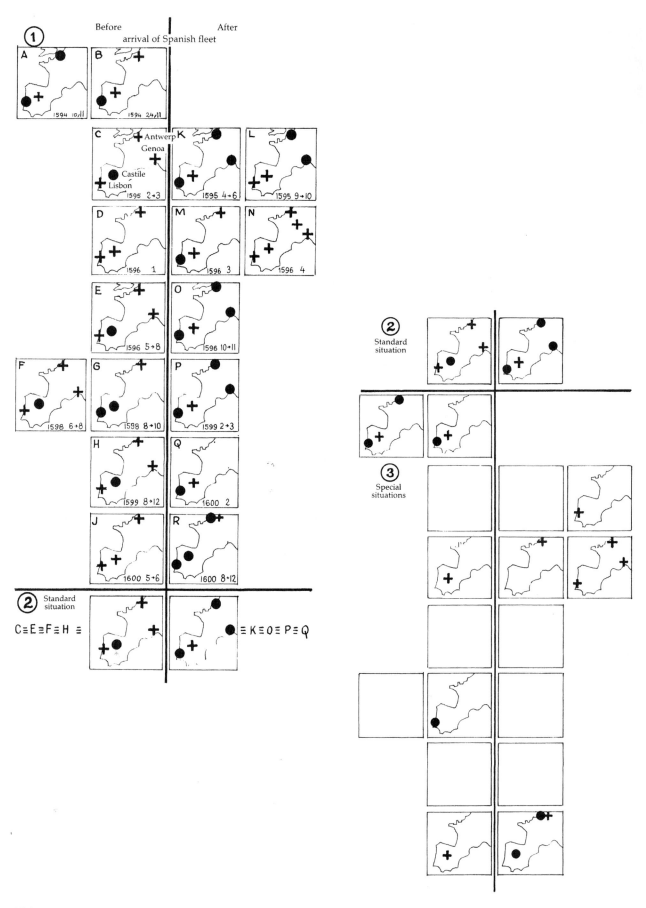

THIRD EXAMPLE: FINANCIAL FLUCTUATION AT MARKET PLACES

(From G. Da Silva, *Stratégie des affaires à Lisbonne entre 1595 et 1607* [Paris: A. Colin, 1956].)

Here is a simpler example of graphic information-processing. The arrival of the Spanish fleet loaded with precious metal has a repercussion on the financial marketplaces. Metal coin, on the one hand, bills of exchange (an instrument for both transfer and credit), on the other, are the two terms in an exchange where merchants take account, either of the "largeness" or abundancy of the metal (or numerical value), or, on the contrary, of its "smallness" or rarity, at a marketplace and at a given moment.

The data show this alternative over time and for the principal marketplaces. The situation is always very different before and after the arrival (always unpredictable) of galleons from the Indies.

First question: Is there a general situation, repeated frequently, which could be considered as a standard situation?
This involves indicating:
– each particular market situation:
  Largeness +
  Smallness •
– the period "before" or "after" the arrival of the fleet.
  Figure 1 attempts to represent this.

A comparative study within the "before" series, then within the "after" series, enables us to identify the most frequently occurring similar cases and thus characterize the standard situation, shown in figure 2.

Second question: What are the special cases which demand further study and explanation?
The answer to the first question enables us to remove the standard situations (denoted by the empty boxes in figure 3) from the very complex display in figure 1.

By comparing the remaining situations with figure 2, placed at the top of the page, we undertake a second removal, which permits us to retain only the special situations. The display has become simple and legible (see figure 3). It suggests the objectives of further research.

Cartography is not indispensable in this process. It would be sufficient to linearize the geographic order (figure 4), represent the markets by signs (L: Lisbon; C: Castile; A: Antwerp; G: Genoa), and construct a diagram.
– The two situations, before and after, separated by the wavy line (and completed by the more remote situations) are given on the abscissa *(x)*.
– The cities, repeated for each situation, are also recorded on the abscissa *(x)*.
– The various observations, classed over time, are given on the ordinate *(y)*.
– Largeness (vertical line) and smallness (horizontal line) are represented by a retinal variable (orientation).

One need only add by column to discover the standard tendency (last line at the bottom) and point out the special cases which do not correspond to it (circles).

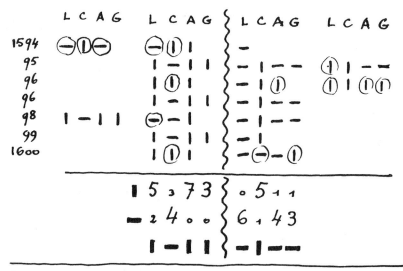

4

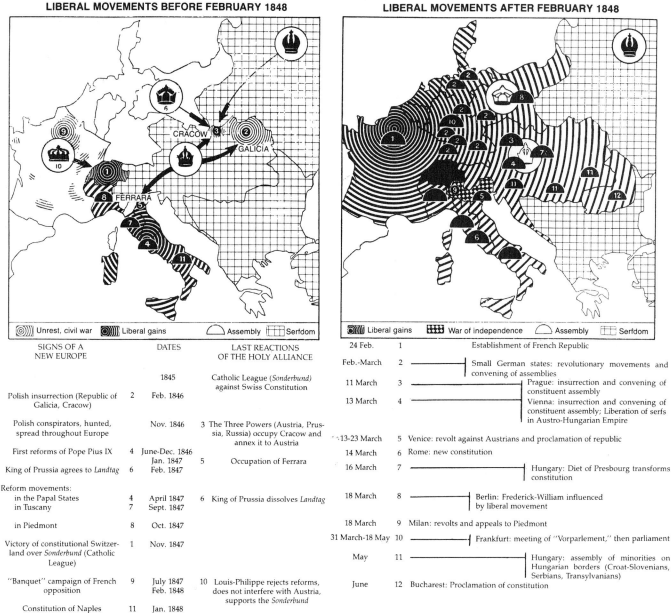

3. Cartographic message (superimposition of simplified images)

Most school maps, pedagogic atlases, blackboard sketches, and synthetic schemas are in this category. The map in figure 2 on page 163 is an example of a cartographic message. It permits us to discover the essential points of comprehensive information in several instants. Remember that to be memorizable, a representation can superimpose only two or three very simple shapes, which must be immediately distinguishable. They must be differentiated:

(1) by differences in visibility (consequently, a dominant shape must be chosen),

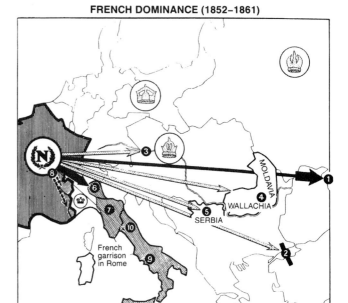

(2) by differences in implantation (gray areas, black lines, points).

A graphic message can be constructed in several separate images, as in the example from A. Siegfried (figure 2, page 405).

Finally, a series of graphic messages can be created to facilitate, for example, the visual retention of a series of historical situations. This is the case with the images above and on pages 410 and 411, which are taken from C. Morazé, P. Wolff, and J. Bertin, *Nouveau cours d'Histoire, classe de philosophie* (Paris: A. Colin, 1950).

For each period a dominant geographic distribution has been suggested by highlighting a particular country or phenomenon: a feeling of encirclement, internal change, expansion of liberal movements, reactionary repression, extension of the sphere of influence by treaty or war....

Attention is focused on the graphic outline, which tends to define each phase and contrast it with neighboring ones. Elementary reading is possible due to the key, but it is controlled by the general shape, which strikes the reader before any consideration of details.

These maps can only be comprehended in relation to one another. They thereby illustrate the reasons for the selection which the authors felt justifiable as a means of understanding the principal traits of contemporary European history.

RISE OF NEW POWERS (1861–1871)

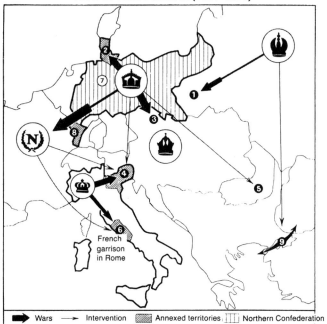

Wars → Intervention ▓ Annexed territories ▓ Northern Confederation

1863	1	Insurrection and crushing of Poland
1864	2	War of Prussia and Austria against Denmark
1866	3	War of Prussia and Italy against Austria (Sadowa)
1866	4	Austria returns Venetia to France, who returns it to Italy
1866	5	A Hohenzollern assumes the throne of Rumania
1867	6	Napoleon III stops Italians, who threaten Rome
1867	7	Formation of the Northern German Confederation
1870–71	8	War between Prussia and France; France loses Alsace-Lorraine
1870	6	Italians take Rome
1871	9	Russians open Straits

GERMAN DOMINANCE (1871–1890)

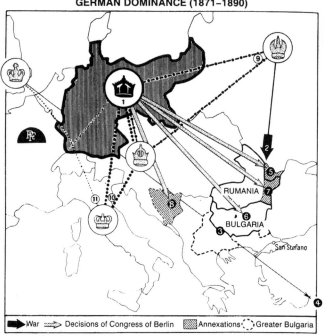

War ⟹ Decisions of Congress of Berlin ▓ Annexations ⸺ Greater Bulgaria

1871	1	William I proclaimed emperor of Germany
1877–1878	2	War between Russia and Turkey
1878	3	Formation of Greater Bulgaria
—	4	England aquires Cyprus

Congress of Berlin

	5	Russia expands on left bank of Danube
	6	Founding of lesser Bulgaria
	7	Expansion of Rumania (Dobroudja)
	8	Austria obtains administration of Bosnia-Herzegovina
1881	9	Alliance of Three Emperors
1882	10	Triple Alliance
1887	11	*Entente* with England

CHECK OF BISMARK SYSTEM (1890–1914)

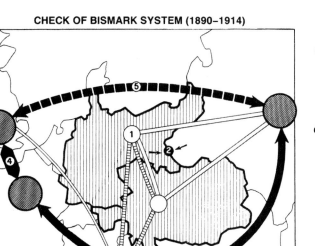

	1	Network of Bismark alliances
	2	German-Russian rivalry in industry and agriculture
1893	3	Franco-Russian alliance
1904	4	*Entente cordiale* between France and England
1907	5	Settlement of litigation between England and Russia

AUSTRO-GERMAN GROWTH (1895–1914)

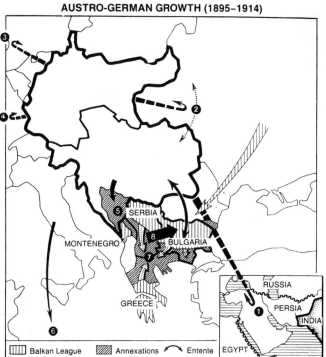

1899–1914	1	Undertaking of Bagdad railroad
1903	2	Failure of Austria-Russia reconciliation (*Mürzsteg*)
1908	3	Failure of British-German talks on sea policy
1902–1911	4	Failure of economic talks with France
1908	5	Austrian annexation of Bosnia-Herzegovina
	6	Italian conquest of Tripoli
1912	7	Balkan League (under Russian influence) against Turkey
		First Balkan War and dividing of Macedonia
1913	8	Second Balkan War. Serbians beat Bulgarians, allies of Austro-Hungarians
1914	5	Serajevo

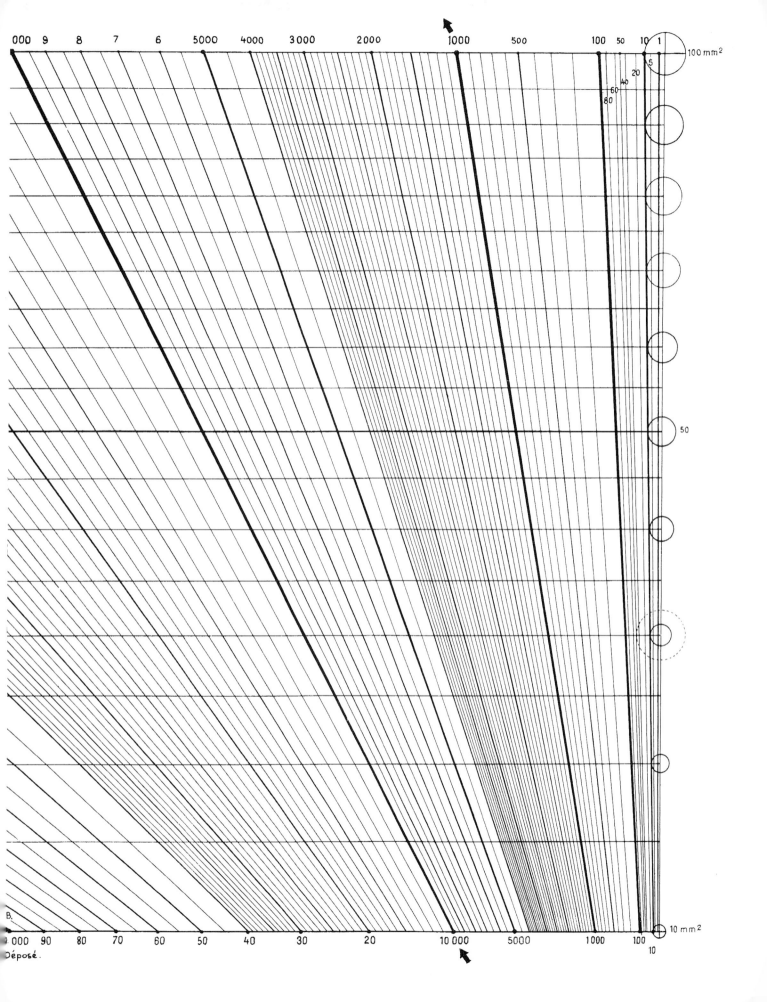

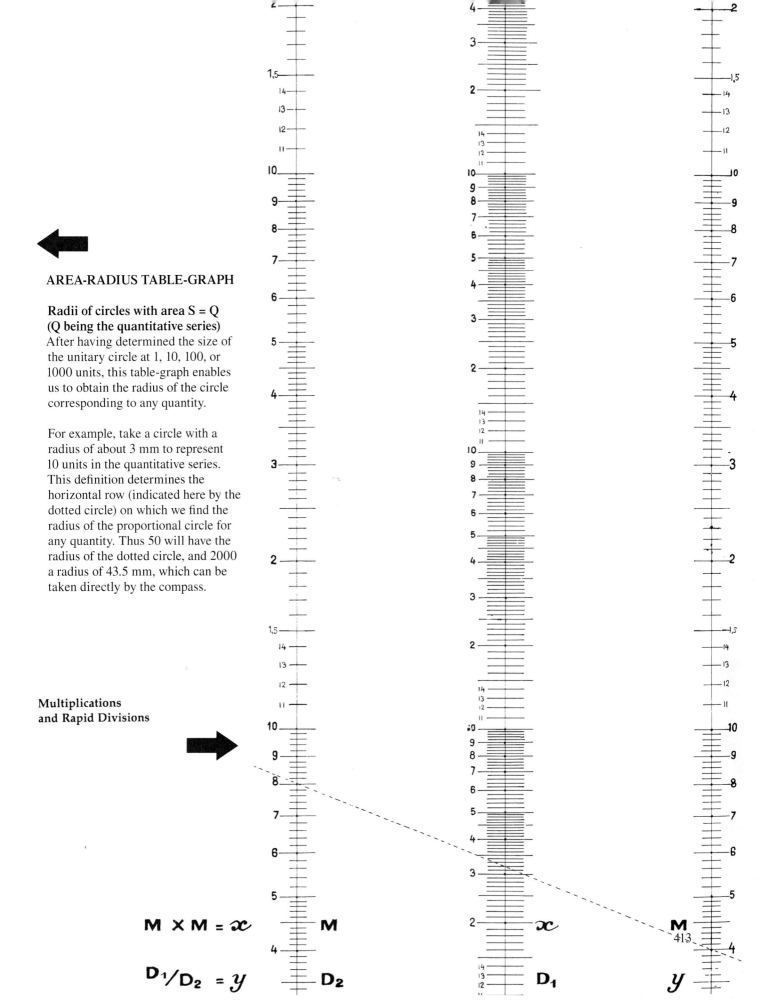

Special thanks to
Karel van der Waarde
for his help redrawing images for the epilogue.

Epilogue

The Origins of *Semiology of Graphics*

A few anecdotes will help to situate *Semiology of Graphics* within its original landscape. In the thirties and forties, when it came to publication, who was responsible for creating the map? The printer! And how was the cost of the map determined? By the number of words appearing in it! After the war, when I asked the administration for military reassignment as a cartographer, to my great astonishment my request was accepted immediately: "Of course, the colonel told me, there's a shortage of printers!"

How should the sea be represented? By horizontal lines. Any other choice was considered an error (note that writing a numerical expression like "5th" is still considered an error by many editors, and yet how many errors does its use avoid!). In a survey done in the seventies, most geographers considered mapping to be a "technique."

In the first edition of *Semiology of Graphics,* the notation "with the support of the CNRS" (Centre National de la Recherche Scientifique) was added on a label. The explanation: the notions of "graphics" and "semiology" were considered by the first reviewers as nonscientific or even unheard of.

Such a landscape amply justified the creation of the School of Cartography at the Sorbonne by Professor E. de Martonne. For a long time drawing was taught by Professor L. Bergelin, and the focus was on topographic, geologic, hydrographic, and geomorphic maps, as well as on wall maps and composite maps. Learning involved how to draw letters and "plot a dot" down to a twentieth of a millimeter, but graphics in the service of mathematical projections, geography teaching, and documentation were practically unknown.

A few years later, at the CNRS and the École des Hautes Études, the problems encountered in graphics were of an entirely different nature and concerned questions of history, ethnography, psychology, archaeology, architecture, economy, linguistics, politics, demographics, geography, medicine, etc., and it was evident that the best graphic solutions were usually the same, regardless of the discipline. There had to be something common to all these graphic problems, something completely independent of the scientific fields involved. This was, I believe, the origin of the discovery of the VISUAL VARIABLES and of the specificity of graphic language.

THE VISUAL VARIABLES

I say "discovery," since the "visual variables" and their properties have their own history.

In *Paris and the Parisian Region* (1952), Paul-Henry Chombart de Lauwe spoke of differences in VALUE, SHAPE, and SCINTILLATION. It was noted that value and "scintillation" (texture) are much more selective than differences in shape, which make it difficult to distinguish one type of sign from another.

In a brochure done for the École des Sciences Politiques, a fourth variable appeared: a variation in DIMENSION with the accompanying notion of visual addition. In 1954 the term ORIENTATION was used.

In 1957 in a publication of the École Estienne, COLOR was added, while "scintillation" became "TEXTURE." The properties of these six variables were termed selection, fusion, hierarchy, and weight; and the first table showing the correspondence between variables and applications was constructed.

In 1967 with the first edition of *Semiology of Graphics,* THE TWO DIMENSIONS OF THE PLANE were added, making a total of eight visual variables. The introduction of the X and Y dimensions of the plane and the location of the six other variables in the "Z" dimension underscored the advantage inherent in graphic language: any component of the data can be represented by the X dimension, the Y dimension, the XY plane (cartography), or the Z dimension of the visual image, the problem being to determine the best application.

The signs (≠) (≡) (O) (Q) serve to designate components that are "different, similar, ordered, proportional." They appeared in the first table classifying the properties of the visual variables: that is, 2DP, Sh, V, T, C, O, Si (the term DIMENSION was replaced by SIZE to avoid confusion with the dimensions of the plane).

In 1977 with *Graphics and Graphic Information-Processing,* a major distinction was made between VARIABLES OF THE IMAGE and DIFFERENTIAL VARIABLES. In 2003 the properties of the variables took the forefront, with the notion of overall perception clearly expressed by the symbols (≠) (≡) for difference and resemblance and (O) (Q) for order and quantified order.

Certain commentators on *Semiology of Graphics* have attempted to add to the list of visual variables. Some have proposed the cinematographic image, forgetting that it's a matter of quite another language with entirely different laws. Blurring (marginal gradient) is the most serious proposal I've seen up to now, along with blinking (limited to the computer), both of which are useful in the representation of variability.

PERMUTATIONS WITHIN THE IMAGE

The basic cartographic and geographic education formerly dispensed in schools of cartography engraved the image of an immutably fixed plane in the mind. That was considered the fundamental property of a map, and no one imagined transforming it. There were, to be sure, a few "graphics" introduced at the end of the school year, which I used to do... before coming to the realization, during a lecture, that what I was saying, what I was showing, was incomplete, incoherent, and illogical!

I had to start over from nothing, assemble collections of "graphics" (a term poorly defined in even the best encyclopedias), multiply experiments, forget most of what I'd been taught, and especially, break the habit of seeing the piece of paper as immutable. In the first edition of the *Semiology* in 1967, a few of the images involved a reordering of rows, even columns. But what audacity! Who had the right to make such changes? At INSEE (Institut National de la Statistique et des Études Économiques), reworking a data table was considered an abomination!

However, the first *Semiology* already considers scalograms, Guttmann scales, the Permutator invented by Pages, and the first "domino" equipment built in the Laboratory of Cartography. The second edition (1973) deals with "the universality of the matrix construction" (260–261), but it's in *Graphics and Graphic Information-Processing* (1977) that the notion of the internal mobility of the image is more fully developed. This allowed for a clear distinction between a **research graphic** (comprehensiveness) and a **communication graphic** (simplicity) and became the basis for a classification of images for the entire field of Graphics.

Meanwhile, the computer fostered the development of mathematical data analysis; the first automated drafting of an "image-file" in France (Golbéry), the first software for matrix permutation using Apple Two (Leduc), and the first automated statistical cartography on IBM.

Competition between graphic processing and mathematical processing is still keen despite the increasing complementarity of the two methods. In the absence of adequate education concerning graphics, the power of visual perception is still underappreciated. But I am pleased to observe today that a color matrix of 365 rows (days) over 162 columns allows the Société Nationale des Chemins de Fer to plan railway traffic **one year ahead of time,** whereas one month ahead of time was the maximum possible with algorithmic processing.

A REGULAR PATTERN OF PROPORTIONAL POINTS

To keep me occupied over the summer break, a colleague gave me an "easy" case to take a glance at before fall. Easy! There were 36 datasets concerning the economic history of the Iberian Peninsula. But also...
1) they were spread out over 430 years, from 1530 to 1960.
2) the monetary units varied—*maravedís,* reals, escudos, pesetas—without any indication of their relationship to each other.
3) the population figures were sometimes by household, sometimes by person.
4) and to cap it off, the geographic regions of the data could differ from one period to another!

Why not add that certain historians considered some of the data to be erroneous!

I'll pass over the different solutions I attempted during this pleasant summer "break" to arrive at my conclusion that proportional points alone offered the subtlety, freedom, and rigor necessary to solve this problem. However, there still remained this variation in geographic regions, which was skewing the data! But why use only one point per area? If I use two or three and divide the data by two or three, I'm still maintaining visual truth. And how else should the points be arranged but in a regular manner, that is, in the form of a regular pattern!

Remaining as well were a few more details such as the sizing of the proportional points from the smallest to the largest, since it's ineffective to determine the radius of each point as a function of the number to be represented: one must choose a scale that is neither too limited (like 5 steps) nor too large. So let's take a simple base, say 20 steps, running between two points that vary from 1 to 10 units in size. We now know that one perceives only relationships (Fechner). The progression is thus the twentieth root of 10; that is 1.122018454. It is on this basis that the plates referred to as "Bertin points" were obtained and applied to the first automated statistical cartography using an IBM printer.

TYPES OF QUESTIONS, LEVELS OF READING

In the mid-sixties most of the preceding elements were in the wind, and it was no longer sufficient to write short pieces about one or another question. There had to be a book, but how should it be organized? At this point, I was asked by the scholarly journal *Études Rurales* to review a German publication dealing with cartographic symbols in agricultural maps. I have always been set against using a lot of symbols, but that wasn't enough; I had to state why!

That is how I came to explore the questions that a "reader" can ask when viewing a map or any other form of graphic. This led to defining types of questions—"questions pertaining to X and questions pertaining to Y"—and, in cartography, "questions concerning geography" *(at a given place, what is there?);* and "questions concerning characteristics" *(a given object, where can it be found?).* I was thus able to demonstrate that the 560 signs proposed in the German work answered the question "at a given place, what is there?" but produced no answer to the question "where can a given object be found?".

These types of questions, soon to be complemented by the notion of levels of reading, led to a theory of the image proposed on page 139 of the *Semiology.* Yet the organization of the book remained confused.

UNITY AND SPECIFICITY OF GRAPHIC LANGUAGE

It is to the linguist Christian Metz, in response to comments he made on the *Semiology* in the journal *Annales* (1971), that I owe the notions of monosemy, polysemy, and pansemy and could thus define with some precision what is meant by the term GRAPHICS compared to the principal languages usually approached by Saussurian analysis.

But the means to bring these concepts to fruition in a book came from the École des Hautes Études en Sciences Sociales. This EHESS support enabled the Laboratory to pursue further experiments in graphics and to publish the French version of *Graphics and Graphic-Information Processing* in 1977. In this book, the following appeared for the first time:
- a graphic information process ranging from data analysis to communication of results (for the explicit difference between data and information see p. 23).
- a general overview of Graphics (synoptic table p. 29).
- a discussion of the problems linked to the elaboration of the data table (chapter D).

But these gains came at the expense of an oversimplification of graphic "grammar" and of cartography.

THE PRESENT STATE

The present state of work in the field of graphics, as of 2004 and summarized in the following pages, brings to bear reflections inspired by communications theory but especially by the development of the MATRIX THEORY OF GRAPHICS. This theory is based on three fundamental questions:
- What are the X, Y, Z **components** of the data table *(what's it about)*?
- What are the **groups** on X and Y formed by the Z data *(what's the overall information)*?
- What are the **exceptions**?

These three questions govern the use of graphic language, and reveal the ineffectiveness of numerous graphics currently being used.

Finally, the function of *precise inventory,* highly developed in geographic information systems (GIS) as well as in other areas (construction, archaeology, mechanics, medicine—wherever a fixed reference plane is involved), assumes a role in the general structure of graphics through the notions of types of questions and levels of reading.

With hindsight, the principal contribution of GRAPHICS seems to me to be in the realm of reasoning, with the precise and constructive visualization of the different stages of a study, *even before* that study is undertaken.

Jacques Bertin (2004)

Brief Presentation of Graphics

diagrams, networks, maps

These few pages present the basic aspects of graphic language as they appeared to me in 2004. To formulate them in as simple and coherent a manner as possible, I relied on one observation: the universal nature of the X, Y, Z visual construction. Any graphic construction can be seen as a variant of the X, Y, Z construction. Any application of Graphics can be analyzed in terms of the visual reordering afforded by this construction. It thus seemed logical to use it as the foundation for a brief presentation of graphic language. This is the "Matrix Theory of Graphics."

CONTENTS

1. DEFINITIONS

2. NATURAL PROPERTIES OF THE GRAPHIC IMAGE
> The three dimensions of the instantaneous image
> The properties of the XY plane
> The properties of Z

3. MATRIX THEORY OF GRAPHICS
> What is the purpose of the graphic? Levels of perception
> Three questions reveal ineffective constructions
> Synopsis of graphic constructions

APPLICATIONS

4. DIAGRAMS
> Double-permutation diagrams
> Single-permutation diagrams
> Non-permutation diagrams: Ordered tables. Collections of images

5. REORDERABLE NETWORKS

6. ORDERED NETWORKS, Cartography

7. REPRESENTATION OF QUANTITIES IN Z

8. SELECTIVITY

9. CONCEPTION OF THE DATA TABLE

10. POWER AND LIMITS OF GRAPHICS

SG refers to *La Sémiologie Graphique* (Paris: École des Hautes Études, 1999 [1967]) Eng: *Semiology of Graphics,* trans. William. J. Berg (Madison: University of Wisconsin Press, 1983)

GR refers to *La Graphique et le traitement graphique de l'information* (Paris: Flammarion, 1977). Eng: *Graphics and Graphic Information-Processing,* trans. William J. Berg and Paul Scott (Berlin/New York: Walter de Gruyter, 1981)

1. Definitions

Graphics utilizes the properties of the visual image to bring out relationships of similarity and order among the data. The graphic language covers the areas of diagrams, networks, and maps.

Graphics applies to a predefined set, represented by the data table, and thus constitutes the rational part of the world of images in a basic classification of fundamental sign-systems:

*		PERCEPTUAL SYSTEMS	
		👂	👁
MEANING attributed to perceptions	The system is open to any meaning; it is PANSEMIC	Music	Nonfigurative image
	The system tends towards the definition of a concept; it is POLYSEMIC	Verbal language	Figurative image
	Transcription of relationships between previously defined concepts; the system is MONOSEMIC	Mathematics	Graphics

It is therefore essential to avoid confusing GRAPHICS—a research tool, which deals only with predefined sets (the data table) and whose laws are natural, indisputable, and learnable—from PICTOGRAPHY, figurative or nonfigurative, which on the contrary, seeks to define a set in the reader's mind. Pictography is an art, which allows for a certain freedom but is always debatable.

Graphics has two objectives:
1. Data processing to understand the data and derive information from them. Take, for example, an engineer who wants to optimize petroleum production: he or she compares numerous figures and numerous characteristics by using tables, maps, and algorithms, in order to determine the conditions for optimal production—that is, *the information*.
2. Communication of this information or an inventory of the data, if applicable. Consider, for example, a reader looking at a curve illustrating the price of butter. The engineer may spend several weeks constructing his or her graphic (or graphics); the reader several seconds reading the curve.

Information processing entails *comprehensivity*. Communication involves *simplification*.

The matrix theory of graphics is the direct application of the fundamental property of human visual perception, which is its ability to understand and memorize the forms within an image constituted by the XY dimensions of the plane with a variation in the Z dimension. The matrix theory takes into account:
– the correspondence between the data table and the image.
– the level of perception (elementary or overall) required by the intended objective
– the mobility or immobility of the image.
  Three questions constitute the bases for this theory.

---

* Note: The written transcriptions of music, verbal language, and mathematics are the visual forms of systems that are fundamentally oral and can't avoid the linear and temporal nature of these systems. The ear can hear an equation over the telephone; it can't hear a map.

2. Natural properties of the graphic image

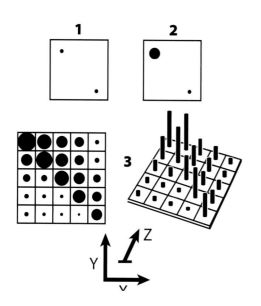

The three dimensions of the instantaneous image

Within the plane (1) a mark can be at the top or the bottom, to the right or the left. The eye perceives two independent dimensions along X and Y, which are distinguished orthogonally. A variation in light energy (2) produces a third dimension in Z, which is independent of X and Y.

The image (3), the meaningful form perceived instantaneously, has three independent dimensions X, Y, and Z. It can thus display the relationships among three independent sets.

The variation in light energy in Z, on a paper medium, is produced by a variation in the Size or Value of the marks. The size and value of marks, along with the XY dimensions of the plane, are thus the visual variables of the image.

The properties of the XY plane

Points or lines: networks or matrices

A "datum" is a relationship between two elements. In the illustrations here, the plane features points and lines. One can thus represent the elements by points and the relationships by lines (4). This is a NETWORK. The X and Y dimensions of the image are not meaningful. We can also represent the elements by lines and the relationships by points (5). This is a MATRIX. The X and Y dimensions are both meaningful.

Whereas the NETWORK is best for representing topographical order, it is highly limited in the representation of reorderable variables: for example, can we perceive the deviant relationship in image (6)? It appears immediately in the matrix (7). The MATRIX, with its three independent variables, provides a natural basis for rational reflection, which is enhanced by the universality of the "double-entry table" and by various means of reordering the data.

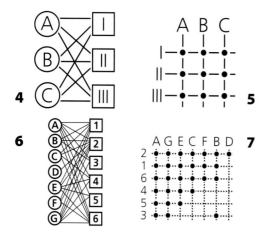

The fixed image and the transformable image

Consider the data table in figure (8), which involves the presence of products A, B, C... in countries 1,2,3. In this form or in its graphic equivalent (9), it resists analysis. Yet, it is just a matter of moving country 2 and product D to uncover groups of similar elements (10) and reduce 25 basic data elements to the 3 groups that characterize this dataset. The internal transformation of the image, by permutation of rows and columns, based on the universal principle of proximity-similarity, defines the REORDERABLE MATRIX, the basis for the matrix theory of graphics. The permutations are represented schematically in figure (11).

Principal planar figures and their standard graphic meanings

Expression of difference (≠) and similarity (≡)
1. pattern   2. proximity, links   3. tree
4. network   5. grid work
6. crossing, cutting, parallelism, axis
7. contact   8. Cartesian grid
9. polarization

Expression of areas of influence
attraction, repulsion, rays

Expression of order—————(O)
horizontal movement, flow, hierarchy

Expression of quantifiable order—(Q)
distance, size, evolution

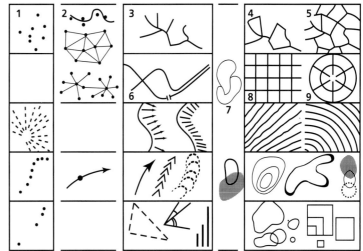

The properties of Z

The eye is sensitive, **along the Z dimension,** to 6 independent visual variables, which can be superimposed on the planar figures: the size of the marks, their value, texture, color, orientation, and shape. They can represent differences (≠), similarities (≡), a quantified order (Q), or a nonquantified order (O), and can express groups, hierarchies, or vertical movements. But, with overall perception, each variable possesses only its specific properties, as described in the table below. Only the plane possesses all of the perceptual properties.

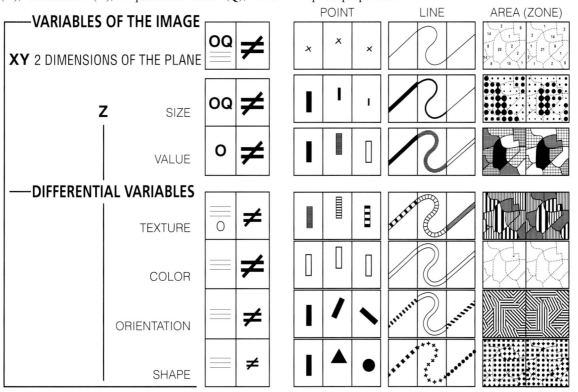

The visual (retinal) Z on the plane is the equivalent of the real Z in space. They come together in trompe-l'oeil drawings, western perspective (a single point of view), exotic perspectives (2 to 4 points of view), and computerized perspective (3 D : n points of view). But only the visual Z allows for the comparison and classing of numerous images.

3. Matrix theory of graphics

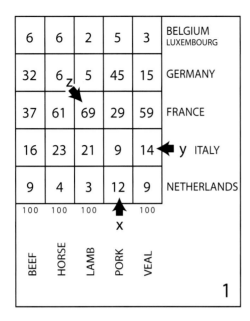

1

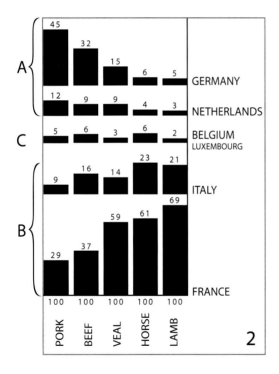

2

What is the purpose of the graphic?

Levels of questions, levels of perception

The main purpose of a graphic is to better understand the data by transforming them. A map or a diagram is a document that is meant to be "questioned." Table (1), for example, which gives meat production in percentages for five countries, can generate questions for all three dimensions:
– In X: a given meat is produced in which country?
– In Y: a given country produces which meat?
– In Z: the highest percentages occur where?
And for each dimension, the questions range from the elementary level to the overall level.

Elementary questions: "How much pork is produced in Italy?" is answered by the number in the cell. That is an elementary datum, the only one we can memorize since we can't absorb the entire set of elementary data, that is, in this case, the 25 numbers in the table. *But understanding involves absorbing the entire dataset.* To achieve that, the data must be reduced to a small number of groups of similar elements. That's the goal of information processing (or data analysis) whether graphical or mathematical.

The overall question: "What groups are formed by the data in X and in Y?" is the essential question. The answer can be found in figure (2), a reorderable matrix, where the rows and columns are permuted to reveal that the 25 numbers in figure (1) can be reduced to 2 groups—A and B—that have opposite structures.
   That is the first piece of information.

   Country C is an exception: it doesn't belong to either group. But in this case the exception is important, since in the political framework of these data, C is equal to its partners and thus has the deciding vote.
   That is the second piece of information.

These essential pieces of information are not visible in figure (1) or in other types of constructions, such as those in figure (3). This is, however, the information that must be brought out. Mathematical or graphical processing must thus precede any written commentary and determine its direction. On the other hand, the publication of documents like those in figure (3) shows that the writer didn't grasp what was important.

Intermediate-level questions correspond to the multitude of subsets that could be defined between the two extremes. And when the overall question is answerable, all of the other questions can be answered.

Three questions reveal ineffective constructions

Bases of the matrix theory:

By definition **any graphic corresponds to a double-entry table,** whose cells can contain binary (yes–no / 0–1) answers, ordinal numbers, cardinal numbers, or question marks (?).

Such a table generates three types of questions: for **X**, for **Y**, and for **Z**. For each type, **the questions can run from the elementary level to the overall level.**

When questions on the latter level can be answered, so can those on all the other levels.

Understanding entails discovering groups; that is, attaining the overall level of perception. Accordingly, the main role of graphics is to produce answers for the following three questions:

1. What are the **XYZ components** of the table? *(What is it about?)*
2. What **groups in X and Y** are formed by the **Z** data? *(What is the overall information?)*
3. What are the **exceptions**?

These three basic questions are applicable to any problem. They enable us to assess the usefulness of any graphic construct and of any graphic innovation. They allow us to avoid ineffective graphics and to detect the weaknesses of constructs like those in figure **(3)**, which provide no answers to the three basic questions! On the other hand, **the reorderable matrix,** seen in figures **(2)** and **(4),** provides answers to all three questions.

With its optimal application of the properties of the image, the reorderable matrix gives concrete form to a chain of logical operations: data-matrix-groups-exceptions-discussion-decision-action (or communication).

The reorderable matrix organizes the thought process, provides a direction for automated procedures, and furnishes the key to a classification of graphics and to the choice of a graphic construction. It is the basic graphic construction.

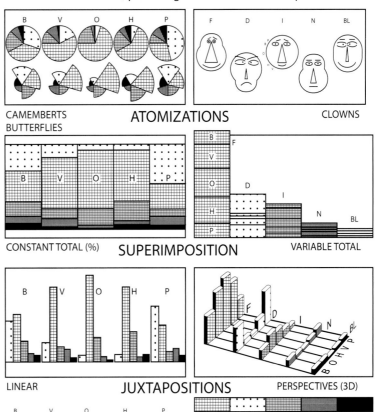

3 Principle ineffective constructions
Mute illustrations, providing no answers to basic questions

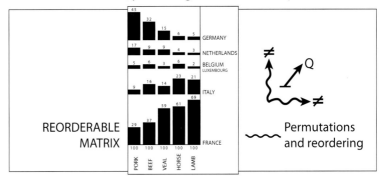

4 Standard construction
Research tool, providing answers to all questions

MATRIX THEORY OF GRAPHICS

Synopsis of graphic constructions

This synoptic table classifies graphic constructions as a function of the elements of the data table. In order to determine the most useful constructions, one has to take into account
- **the number of characteristics all**
- **their nature: ordered —(0) or reorderable ($\neq$)**
- **the presence of a geographic component**

The relationships among characteristics produce diagrams. The relationships among objects produce networks. But networks can also take the form of a matrix (§ 5).

For data tables with three characteristics or fewer, each characteristic occupies one dimension of the image. The groups appear directly with no need for permutation.

Through data analysis, one can construct a type of tree diagram often called a "dendogram," or a "factorial cluster," a form of scatter plot, to bring out the distances between n objects or characteristics, but they obscure the "why": that is, the content of the table.

DIAGRAMS

Reorderable Matrix Standard Construction (Double-permutation diagrams)

Image-file, Array of Curves* (Single-permutation diagrams)

Collection of Images (reorderable)

ORDERED TABLES
(Non-permutation diagrams)

Scatter plot with 3 characteristics

Scatter plot with 2 characteristics

Distribution with 1 characteristic

Dendogram   Factorial cluster

---

* When the slopes are meaningful.

DATA TABLE
n CHARACTERISTICS

	A	B	C	D
1				
2				
3				
4				
5				

DATA TABLE
3,2,1 CHARACTERISTICS

	A	B	C	D
1				
2				
3				

	A	B	C	D
1				
2				

	A	B	C	D
1				

NETWORKS

ORDERED
Topographies
Cartography
Blueprints, anatomical charts

REORDERABLE
Graphs, trees, flow charts

Superimposition
comprehensive or simplified, for maps with 1 characteristic

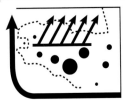

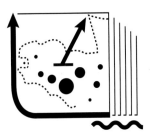

Collection of maps with 1 characteristic

Trichromatic superimposition

Bichromatic superimposition

Map with 1 characteristic

Base map

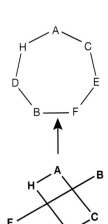

APPLICATIONS
4. Diagrams

Double-permutation diagrams
Reorderable matrix

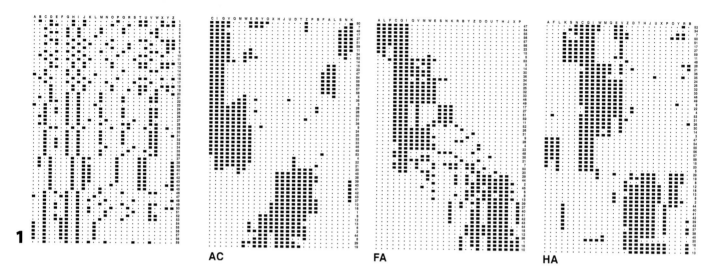

When the two components of the data table are reorderable ($\neq$), the standard construction is the reorderable matrix. Its permutations underscore the analogy and complementarity of algorithmic processing and graphic processing.

Take, for example, a study involving 59 objects (ancient belt buckles) across 27 characteristics (**1**). In applying three algorithmic processes—automated classification (**AC**), factorial analysis (**FA**), and hierarchical analysis (**HA**)—the results are different; they require interpretation:

Beginning with hierarchical analysis (**HA**),
(**2**) inserts separation lines and isolates (a).
(**3**) simplifies (2) by inverting the first three columns of (a) and reordering (b).
(**4**) introduces (a) into (b).
(**5**) simplifies (4) by separating out exceptions for objects and characteristics, which are easily identifiable in comparison to the central structure, which is evolutional and remarkably homogeneous.

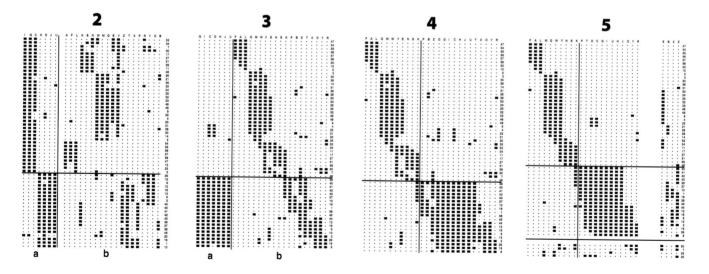

Single-permutation diagrams
image-file, array of curves

When one of the components of the data table is ordered (by time, for example), its transcription onto either of the X or Y dimensions of the graphic eliminates one axis of permutation and thus simplifies the graphic processing. Two constructions are possible:

The Image-file (*fichier*-image)
This construction places an ordered component on X and different reorderable components ($\neq$) on Y, which allows for permuting the image successively, according to each of the ($\neq$) components. Here, the quantities are represented on X.

Take, for example, the question of the homogeneity of a set of 8 insects: To carry out the experiment, three compartments are placed side by side—one lit, one shady, one dark. The 8 insects are placed in a compartment and for each insect we measure, every 5 minutes for an hour, the time spent in that chamber or in a neighboring one. The experiment is carried out 12 times, which produces table **A**. It is a matter of discovering whether 1) the results of the *experiments* are comparable; 2) the *insects* form types; 3) the *times* are homogeneous.

(1) uses an image file,
to place the quantities of time (Q) on X and on Y the 8 insects ($\neq$) multiplied by the 12 experiments ($\neq$).

(2) uses one image per experiment
to discover that the experiments form 2 groups: 5 and 11 differ from the majority and thus need to be studied separately.

(3) uses one image per insect based on the order of the experiments. Three groups appear: slow (A, B, C), indecisive (C, D), rapid (H, E, F).

(4) orders all the times, from the longest to the shortest. Three steps appear: 10, 20, and 45 minutes. Many other observations are possible (see GR, 75).

The Array of Curves (*Éventail de courbes*)
This construction places the ($\neq$) components and the quantities (Q) on Y, which leads to difficulties in permutation, superimposition, and especially scale. In the example in figure **(5)**, entitled "the strongest variations on the Paris stock market," the author implies that a progression from 1900 to 3300 is greater than a progression from 100 to 400! The scale of progressions or logarithmic scale **(6)** allows for correction of this error. The array thus uncovers the best investments; and the classing is nearly opposite from that conveyed by **(5)**!

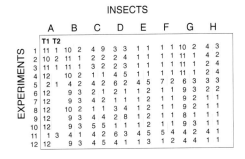

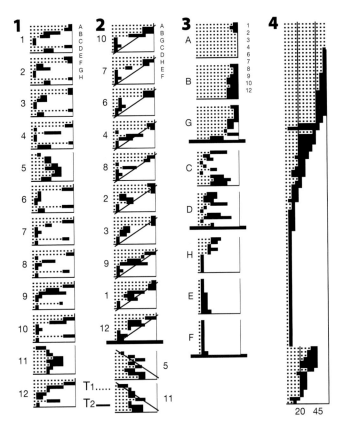

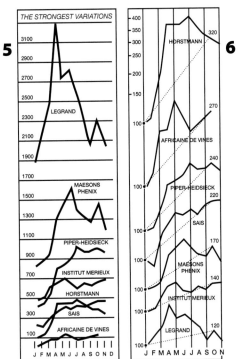

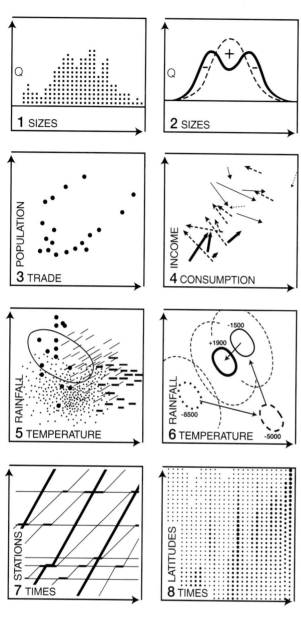

Non-permutation diagrams
Ordered tables, collections of images

A classic area of graphics, ordered tables make groups appear directly, with no need for reordering the rows or columns, as in the following examples:

Data tables with one row produce "distributions":
(1) gives the number of objects by size and displays, for example, errors in manufacturing.
(2) shows the usefulness of a market study (variation between the distribution of sizes manufactured and the distribution of sizes desired by the intended clientele).

Data tables with 2 or 3 rows produce "scatter plots":
(3) reveals two types of cities.
(4) shows the evolution of a correlation between two dates.
(5) superimposes five types of vegetation.
(6) shows the succession of climates revealed by the succession of vegetation types, in a sample.
(7) establishes the succession and schedule of trains on a railway line.
(8) shows the daily evolution of temperature in the stratosphere.

As a collection of images
The ordered table constitutes a remarkable research tool that can be applied to data tables with more than three rows. The identifiers inscribed on each image allow for numerous permutations.
(9) evolution of literacy for men (H) and women (F). The identifiers are place (rural or urban), region, and profession. Three types of evolution appear clearly here.
(10), (11) aging of the population (J=youth; A=adults; V=elderly). One image per village, all periods superimposed, reveals nothing **(10)**; one image per period, all villages superimposed, shows a marked evolution **(11)**.

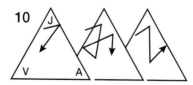 

per village   per period

428

5. Reorderable networks

graphs, trees, and flow charts

These constructions transcribe onto the plane relationships (or connections) between objects (or nodes).

Graphic processing can simplify the image in several ways:

It can simplify the drawing (1).

It can reduce the number of meaningless intersections (2). Graphic theory and the reorderable matrix are useful in finding the best solution.

It can form meaningful groups (3).

It can hierarchize the image by assigning a meaning to the X and Y axes of the plane (4). Order and disorder are powerful visual sensations, often utilized in pedagogical applications.

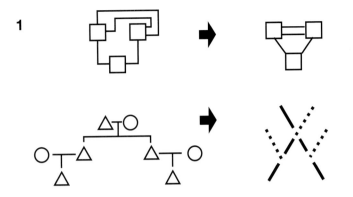

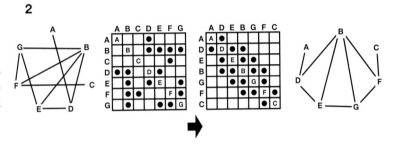

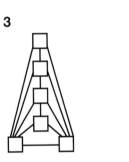  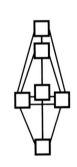

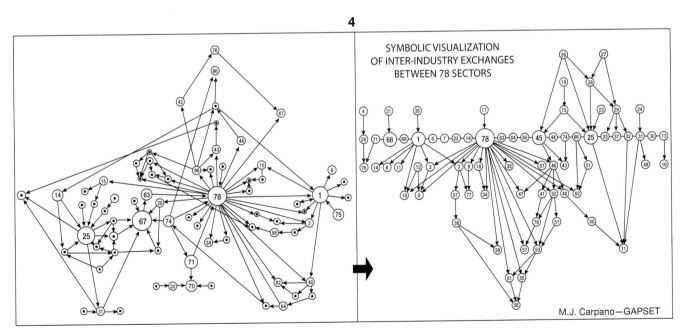

SYMBOLIC VISUALIZATION OF INTER-INDUSTRY EXCHANGES BETWEEN 78 SECTORS

M.J. Carpano—GAPSET

6. Ordered networks
topographies, maps, blueprints, anatomical charts

Table A (center page) represents the matrix form of geography. Here, the geographical elements (states) are on X and the characteristics (products) on Y. To apply matrix theory, the three basic questions can again serve as guidelines.

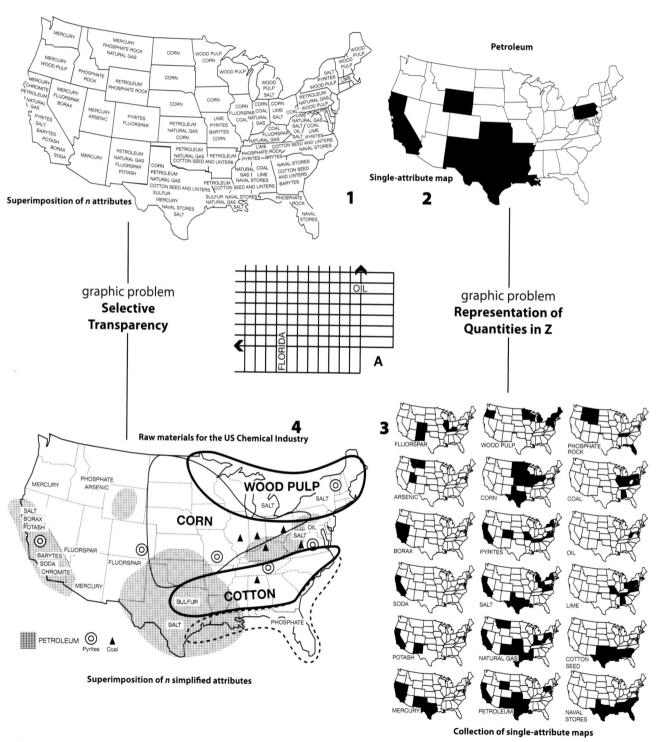

The three basic questions in cartography

The first basic question—*What are the XYZ dimensions?*—determines the title and the legend of the map.

To answer the other two questions—*What are the groups? What are the exceptions?*—one must compare the characteristics, discover similarities, identify regions; that is, answer the questions pertaining to Y: **a given characteristic,** Where is it? (For example, where can petroleum be found?) Figures **(2)** and **(3)** provide instantaneous answers, making the groups appear, along with exceptions and potential explanations. Figure **(1)** does not provide an answer visually.

To fulfill its function as a "precision inventory," based on geographical identification, the map must also provide an answer to the question pertaining to X: **at a given place,** what is there? (For example, What can be found in Florida?) Figure **(1)** provides a complete answer (comprehensivity). Figures **(2)** and **(3)** don't provide any answers visually.
To provide complete answers to all questions, one must therefore use both figures **(1) and (3)**. This problem is solved if one abandons comprehensivity and uses **a simplified map** such as figure **(4)**, which superimposes a few chosen characteristics and represents their distribution schematically.

To choose a graphic formula, therefore, one must define the intended purpose and the relevant question:

If it is **"At a given place, what is there?"** (as with topographical maps, blueprints, anatomical charts, etc.), then geography is the starting point for a detailed reading. The solution is a map superimposing all of the characteristics **(1)**; the problem is one of selectivity.

If it is **"A given characteristic, where is it?"** (as with statistical inventories, determining geographical groupings, correlations, exceptions, etc.), then the characteristic is the starting point for an overall reading. The solution is a collection of maps, one per characteristic **(2)** and **(3)**; the problem is one of representing the quantities in Z.

Are both types of questions relevant?
For cases like a geography course, a pedagogical drawing, or a land plan, for example, both questions are relevant. The solution is *a simplified map,* often referred to as a "synthesis map," which is a superimposition of simplified characteristics **(4)**. The problems involve:
– the choice of characteristics to be retained and the degree of their simplification (how many categories to be retained), choices that are subjective.
– the choice of a data processing method: cartographic **(5)** or matrix, mathematical or graphical **(6)**.
– the choice of a selective graphic formula, a choice that depends on the distribution of the elements, which must thus be studied *before* determining the appropriate graphic formula.
   The simplified map ultimately raises the problem of assessing the proposed reductions when the original data have disappeared **(7)**.

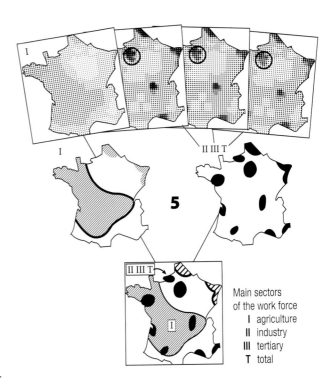

Main sectors of the work force
I agriculture
II industry
III tertiary
T total

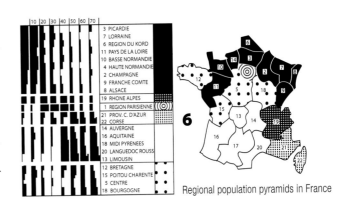

Regional population pyramids in France

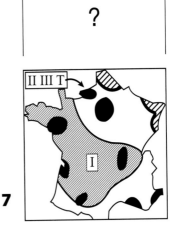

7. Representation of quantities in Z

On the earth's surface, a country's population is a function, among other things, of the extent of its boundaries. Likewise in statistics, the average "height" of a population in an age group depends on the boundaries that constitute each age group (the "bin widths"). Thus the mean height would be different if we chose finely gradated groups (age 1–2) versus wider bounds (ages 0–10). To avoid erroneous representations, in cartography as in statistics, in addition to choosing the finest gradations that are practical, it is necessary:

To place the groups being enumerated on a common scale.
This is an operation that can be mathematical (calculating ratios, densities, percentages, indices) or graphical (grids or contour curves).

To utilize a size variation with points (2) or areas (3).
This variable, coupled with the utilization of *the natural series of graduated sizesv* (GR, 205–209), avoids the irresolvable problem of choosing steps and thus creating misleading images (SG, 77 and 163), a problem compounded by facile solutions (insufficient steps, a priori definitions, screens that are too fine, too unstable, and thus confusing, incomplete analysis of the purpose, etc.).

To determine the width of the categories.
In effect, this means representing quantities in Z amounts after answering two questions:
1) the classic question: **which image** best represents the widths defined by the numbers or the distribution curve?
2) the operational question: **which width** produces the best image, an image that omits islands, resembles another image, covers a given area, underscores a break in the slopes, etc.? The continuous, computerized adjustment of the width, a frequent operation in remote sensing, is an effective solution.

8. Selectivity

The notion of selectivity applies when characteristics are superimposed, and its effectiveness can be measured by its capacity for enabling the reader to disregard everything else. At equal luminosity, selecting squares means disregarding all the other shapes. This is impossible in figure (**1**), where there is no selectivity among the shapes. At equal luminosity, selecting the dark signs means disregarding the light signs, which occurs **instantaneously** in figure (**2**). (Note that selecting the light signs is also instantaneous.) The best selectivity is obtained, in the following order, by:

A difference in intensity of size or value, if order is meaningful.

A difference in implantation, which superimposes point, line, and area symbols, leading to a transparent reading.

Color, but its selectivity is a function of the size of the marks: red and green can't differentiate two pinheads, whereas the eye can probably distinguish nearly a million tints on a wall. The selectivity of color with small points or lines is thusly limited.

Texture, for lines and areas (3 steps).
Orientation (4), with points, lines, or areas, if the texture isn't too fine.

Shape, with points, lines, or areas, has no selectivity for an overall reading, but **for elementary reading,** it is the basis for symbolism and symbolic analyses.

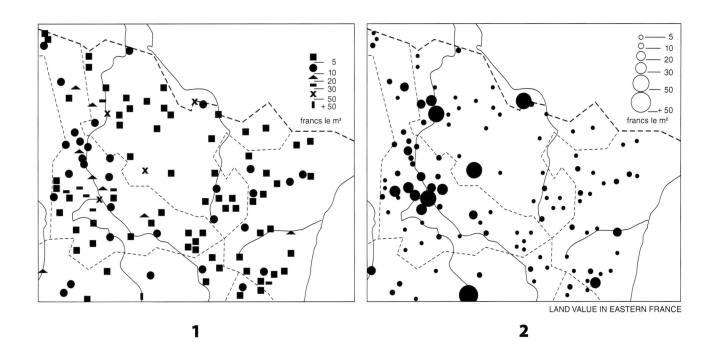

LAND VALUE IN EASTERN FRANCE

1 **2**

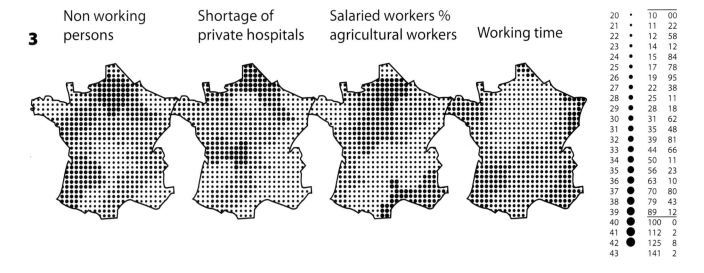

3 Non working persons | Shortage of private hospitals | Salaried workers % agricultural workers | Working time

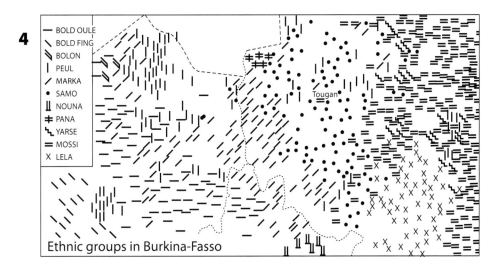

4 Ethnic groups in Burkina-Fasso

- BOLD OULE
- BOLD FING
- BOLON
- PEUL
- MARKA
- SAMO
- NOUNA
- PANA
- YARSE
- MOSSI
- LELA

Tougan

433

9. Conception of the data table

"What type of data table should be constructed?" The matrix analysis of a problem helps in answering this question; it facilitates organization of a study in three successive steps.

Step 1. Formulation of the problem through simple questions: identification of the list of characteristics and objects that it would be useful and desirable to know (without consideration of technical limits); and notation of their relationships. *This is the apportionment table.*

Step 2. Conception of the ideal homogeneous table containing the greatest number of elements from the above list. In other words, "What should be displayed on X so that the greatest number of characteristics can be displayed on Y?" Then assess its usefulness in terms of means, time, and size; and study eventual reductions by aggregation, sampling, and interpolation. *This is the homogeneity table.*

It is at this point that three main options appear: use X for space (maps) or time (curves) or a specific component like individuals, categories, objects, etc. The "length" of each component, the number of modalities (for example, 2 sexes, 3 sectors of the work force, 12 months, 22 regions, 38 000 communes, etc.), plays an essential role here.

Step 3. Verification of the pertinency of the table by noting in the margins the correspondences determined by the initial questions (GR, 245–249). *This is the pertinency table.*

This study obviously precedes the processing itself, but it can be fully undertaken without knowledge of the data analyses, whether mathematical or graphical, and their methods.

Schematization of graphic language

With graphic processing, the "sender" and the "receiver" of the information are either the same person or two "actors" asking the same basic questions. This situation thus lies outside the schema of polysemic communication:
A) sender <——> code <——> receiver

Instead, it's a matter of monosemic communication:
B) actor <——> three types of relationships ($\neq$) O, Q.

These are relationships of similarity and order, which allow for reduction of the data and are not subject to conventions since they are conveyed by the same properties—similarity and order—that characterize the visual variables.

Schema (A) applies only when using verbal communication; that is, to answer the first basic question or to identify a shape in elementary reading.

10. Power and limits of graphics

The three dimensions of the image make visual perception our most powerful system of perception and make graphics an especially effective pedagogical tool, which enables us to concretize, from early education on, problems of information, reasoning, and decision-making. Thanks to its potential for permutation, modern graphics materializes notions that used to be abstract.

Graphics gives a visible form to the stages and methods of a study, which facilitates *organization of the work*. It concretizes the notion of data and highlights problems raised in creating the initial table, problems of pure conception that lie outside of computer usage and are defined by the question, "What should be represented by X?"

Graphics gives a visible form to the notion of "data analysis," which is more accessible in its graphic incarnation than in its mathematical one. Graphics underscores the fact that a study is "scientific" only when its assertions are confirmed by the rigorous processing of an explicit data table. Without such processing, we are stuck on the level of personal opinion.

Finally graphics gives a visible form to the notions of *discussion, reasoning, and understanding,* notions that become precise through the level of relevant questions.

But the image has only three dimensions. The effects of this limit probably exceed our imagination, confined as we are by this natural condition.

Mathematical analysis speaks of n dimensions.

But one notes that computer input listings form a single XYZ table and, when it comes to assessing the results of calculations, we end up with an image that still has only three dimensions, the fourth one being time, which is precisely what should be minimized.

This is why *interdisciplinary studies* will always be difficult, since X will always be reserved for space by the geographer, time by the historian, individuals by the psychologist, social categories by the sociologist, etc. How can there be a "synthetic science" when each department, each institute, each discipline is itself defined by the XYZ components that characterize its field of study? It is the absence of a fourth dimension in the image that in fact prohibits the conception of a synthetic science not divided into various disciplines.

This is also how one can demonstrate the *limits of reasoning*. Useful information processing can exist only within the framework of a finite set: the data table. But there are an infinite number of finite sets. Whatever our efforts at rational thinking, they will always be drowned in the infinite sea of the irrational.

Jacques Bertin 2004

Index

Translator's note
Because the text of *Semiology of Graphics* is extensively cross-referenced, only principal references are given in the index.

Original French terms are given after their English translations.

The Introduction, pages 2–13, also constitutes an annotated table of contents and glossary of terms.

Accuracy, cartographic *(exactitude cartographique),* 298
Angular legibility *(lisibilité angulaire),* 178
Area, representation by *(implantation zonale),* 44
Area-radius table-graph *(abaque),* 413
Arrangement *(semis),* 52, 271
Array of curves *(éventail de courbes),* 263
Arrow *(flèche),* 346
Associative variable *(variable associative),* 48, 65
Automation *(automatisation),* 377

Base map *(fond de carte),* 308
Block diagram *(bloc-diagramme),* 380

Cartogram *(anamorphose géographique),* 120, 285
Cartography, maps *(cartographie, cartes),* 51, 117, 285
Category *(catégorie),* 33
Chart-map *(cartogramme),* 119, 129, 285
Circles, pattern of *(semis de cercles),* 369
Circular construction *(construction circulaire),* 55
Class *(classe),* 33
Classes of representation or implantation *(implantation),* 44
Collection of maps *(collection de cartes),* 268
Collection of profiles *(collection de profiles),* 244
Collection of tables *(collection de tableaux),* 265
Color variation *(variation de couleur),* 85
Combination, meaningful *(combinaison significative),* 189
Combination, redundant *(combinaison redondante),* 187

Combination of variables *(combinaison de variables),* 184
Communication, graphic *(communication graphique),* 162
Comparison of orders *(comparaison d'ordres),* 248
Component *(composante),* 16. *See also* Ordered component; Qualitative component; Quantitative component; Reorderable component
Comprehensivity *(exhaustivité),* 160
Computer *(ordinateur),* 377
Concentration curve *(courbe de concentration),* 110, 205, 247
Construction, circular *(construction circulaire),* 55
Construction, orthogonal *(construction orthogonale),* 54
Construction, polar *(construction polaire),* 55
Construction, rectilinear or linear *(construction rectiligne ou linéaire),* 54
Construction, special *(construction particulière),* 172
Construction, standard *(construction de base),* 172
Construction, triangular *(construction triangulaire),* 232
Contingency table *(tableau croisé),* 223
Contour curve, isarithm *(courbe d'égalité, isarithme),* 385
Criticism *(critique),* 400
Curve *(courbe),* 210

Density, graphic *(densité graphique),* 175
Diagonalization of diagrams *(diagonalisation des diagrammes),* 166, 168
Diagram *(diagramme),* 50, 103, 193
Diagram, triangular *(diagramme triangulaire),* 232
Dissociative variable *(variable dissociative),* 48, 65
Distributions *(distributions),* 110, 205, 247
Domino equipment *(jeu de dominos),* 169

Efficiency *(efficacité),* 139, 146
Elementary level *(niveau élémentaire),* 153
Elevation *(élévation),* 9, 60
Elevation, circular *(élévation circulaire),* 55

Elevation, rectilinear *(élévation rectiligne),* 54
External identification *(identification externe),* 19, 140, 287

Figuration *(figuration),* 144, 151
File, image- *(fichier-image),* 245, 258
File, matrix- *(fichier-matrice),* 263
Files and cards *(fichiers et fiches),* 218
Flow chart *(organigramme),* 269
Frequency *(fréquence),* 208, 357, 400
Fusion, color *(diffusion de la couleur),* 89

Generalization, cartographic *(généralisation cartographique),* 300
Generalization, conceptual *(généralisation conceptuelle),* 300, 302
Generalization, structural *(generalisation structurale),* 300, 302, 305
Graphic information-processing *(traitement graphique de l'information),* 164, 254, 397
Graphic or representation *(graphique ou dessin),* 51
Groups of representation or imposition *(imposition),* 50

Histogram *(histogramme),* 210
Homogeneous series *(séries homogènes),* 26

Identification, external *(identification externe),* 19, 140, 287
Identification, internal *(identification interne),* 19, 24, 140, 298
Image *(image),* 139, 142, 151
Image-file *(fichier-image),* 245, 258
Implantation or classes of representation *(implantation),* 44
Imposition or groups of representation *(imposition),* 50, 148
Index *(indice),* 236
Information *(information),* 5, 16
Information, set of *(ensemble informationnel),* 32
Information processing *(traitement de l'information),* 164, 254, 397
Intermediate level *(niveau moyen),* 153
Internal identification *(identification interne),* 19, 24, 140, 298
Invariant *(invariant),* 5, 16
Inventory *(inventaire),* 160
Inventory maps *(cartes d'inventaire),* 391

437

Inversion *(désaccord)*, 249
Isarithm *(isarithme)*, 385

Legend *(légende)*, 24
Legibility *(lisibilité)*, 175
Length *(longueur)*, 33
Level of organization *(niveau d'organisation)*, 34, 48, 64
Levels of questions *(niveaux des questions)*, 141
Levels of reading *(niveaux de lecture)*, 141, 151
Line, representation by *(implantation linéaire)*, 44
Linear construction *(construction linéaire)*, 54
Logarithmic scale *(échelle logarithmique)*, 240

Maps *(cartes)*, 51, 117, 285
Mathematics *(mathématique)*, 2
Matrix; matrix constructions *(matrice; constructions matricielles)*, 230, 256
Matrix, reorderable *(matrice ordonable)*, 256
Matrix-file *(fichier-matrice)*, 263
Mean *(moyenne)*, 208
Meaningful combination *(combinaison significative)*, 189
Medial *(médiale)*, 208
Median *(médiane)*, 208
Message *(message)*, 162, 408
Migration, regional *(migration régionale)*, 351
Mode *(mode)*, 208
Monosemy *(monosémie)*, 2
Movement, representation of *(représentation du mouvement)*, 342

Natural series of graduated sizes *(gamme naturelle des tailles croissants)*, 369
Networks *(réseaux)*, 50, 269
New or pertinent correspondence *(correspondance originale)*, 5, 16, 140
Number of components *(nombre de composantes)*, 28

Ordered component *(composante ordonnée)*, 37, 67
Ordered variable *(variable ordonnée)*, 48
Ordinal numbers *(nombres ordinaux)*, 37
Orientation variation *(variation d'orientation)*, 93
Orthogonal construction *(construction orthogonale)*, 54
Overall level *(niveau d'ensemble)*, 153

Pattern *(semis)*, 369
Permutation equipment *(permutateur)*, 169

Permutations *(permutations)*, 254
Perspective representation *(représentation stéréographique)*, 132, 283, 378
Pertinent or new correspondence *(correspondance originale)*, 5, 16, 140
Planar dimensions *(dimensions du plan)*, 62
Plane *(plan)*, 44
Point, representation by *(implantation ponctuelle)*, 44
Polar construction *(construction polaire)*, 55
Polygon *(polygone)*, 210
Polysemy *(polysémie)*, 2
Processing *(traitement)*, 164, 254, 397
Profile *(profil)*, 244
Projections *(projections en cartographie)*, 288

Qualitative component *(composante qualitative)*, 36
Quantitative component *(composante quantitative)*, 38, 69, 223
Quantitative variable *(variable quantitative)*, 48

Range *(étendue)*, 33, 182
Ratio, critical *(rapport critique)*, 300
Reading levels *(niveaux de lecture)*, 141
Rectilinear construction *(construction rectiligne)*, 54
Reduction of the information *(réduction de l'information)*, 164
Regular pattern of graduated sizes *(semis régulier des tailles croissantes)*, 127, 137, 369
Relief, topographic *(relief topographique)*, 380
Reorderable component *(composante ordonnable)*, 36
Reorderable matrix *(matrice ordonnable)*, 256
Repartition *(répartition)*, 110, 203, 246
Representation or graphic *(dessin ou graphique)*, 51
Retinal legibility *(lisibilité rétinienne)*, 180
Retinal variables *(variables rétiniennes)*, 9, 60, 71
Rules of the graphic system *(règles du système graphique)*, 99, 190

Saturation, color *(saturation de la couleur)*, 85
Scale, logarithmic *(échelle logarithmique)*, 240
Scale of maps *(échelle des cartes)*, 296
Scatter plot *(diagramme de corrélation)*, 254
Schemas, standard *(schémas de base)*, 56, 172

Schemas of construction *(schémas de construction)*, 52, 56
Screens *(trames)*, 330
Selective variable *(variable sélective)*, 48, 67
Separation, angular *(séparation angulaire)*, 175, 178
Separation, retinal *(séparation rétinienne)*, 175, 180
Shape variation *(variation de forme)*, 95
Size variation *(variation de taille)*, 71
Smoothing *(lissage)*, 170
Special construction *(construction particulière)*, 172
Standard construction *(construction de base)*, 172
Standard series *(gamme normale)*, 357, 362, 370
Steps *(paliers)*, 33
Symbolism *(symbolique)*, 51, 90, 95

Table, contingency *(tableau croisé)*, 223
Table, ordered *(tableau ordonnée)*, 265
Table of levels and impositions *(représentation des niveaux et des impositions)*, 56
Tables of properties of the retinal variables *(tableau des propriétés des variables rétiniennes)*, 97
Tables of the natural series *(tables de la gamme naturelle)*, 370
Texture variation *(variation de grain)*, 11, 79
Time series *(chronique, chronogramme)*, 212, 234
Title *(titre)*, 19
Topographic relief *(relief topographique)*, 380
Transformation of networks *(transformation des réseaux)*, 166, 271
Tree *(arbre)*, 276
Triangular construction *(construction triangulaire)*, 232
Trichromatic analysis *(analyse trichromatique)*, 89
Types of construction *(types de construction)*, 53
Types of imposition *(types d'imposition)*, 52
Types of questions *(types de questions)*, 141

Values, summary *(valeurs typiques)*, 208
Value variation *(variation de valeur)*, 73
Variable, retinal *(variable rétinienne)*, 9, 60, 71
Variable, visual *(variable visuelle)*, 7, 42
Vibratory effect of texture *(effet vibratoire du grain)*, 80
Visibility *(visibilité)*, 65

Related titles from Esri Press

Cartographic Relief Presentation

ISBN: 978-1-58948-026-1

Within the discipline of cartography, few works are considered classics in the sense of retaining their interest, relevance, and inspiration with the passage of time. One such work is Eduard Imhof's masterpiece, *Cartographic Relief Presentation*. Originally published in German in 1965, this book illustrates the need for cartography to combine intellect and graphics in solving map design problems. The range, detail, and scientific artistry of Imhof's solutions are presented in an instructional context that puts this work in a class by itself. Esri Press has reissued Imhof's masterpiece as an affordable volume for mapping professionals, scholars, scientists, students, and anyone interested in cartography.

The Look of Maps: An Examination of Cartographic Design

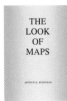

ISBN: 978-1-58948-262-3

Originally published in 1952, *The Look of Maps* documents Arthur H. Robinson's pivotal observation that the discipline of cartography rests at the crossroads of science and art. Based on his doctoral research, this book attempts to resolve the apparent disconnect by covering a range of topics related to the visual characteristics of cartographic technique, including lettering, structure, and color. Robinson offers advice that even the modern cartographer finds relevant: adopt a "healthy questioning attitude" in order to improve and refine the graphic techniques used to present information visually through maps. A classic text, *The Look of Maps* is an essential component to any cartographic library.

Designed Maps: A Sourcebook for GIS Users

ISBN: 978-1-58948-160-2

This companion to Cynthia Brewer's highly successful *Designing Better Maps* offers a graphics-intensive presentation of published maps, providing cartographic examples that GIS users can adapt for their own needs. Each chapter characterizes a common design decision and includes a demonstration map annotated with specific information needed to reproduce the design such as text fonts; sizes and styles; line weights, colors, and patterns; marker symbol fonts, sizes, and colors; and fill colors and patterns. Visual hierarchies and the purpose of each map are considered with the audience in mind, drawing a clear connection between intent and design. With a task index explaining what ArcGIS 9 tools to use for desired cartographic effects, this book is an indispensable resource for all GIS users, from experienced cartographers to those who make GIS maps only occasionally.

Esri Press publishes books about the science, application, and technology of GIS. Ask for these titles at your local bookstore or order by calling 1-800-447-9778. You can also read book descriptions, read reviews, and shop online at www.esri.com/esripress. Outside the United States, contact your local Esri distributor.